Jay Shelton's
SOLID FUELS ENCYCLOPEDIA

by Jay W. Shelton

Illustrations by Susan Gliss

A Garden Way Publishing Book

Storey Communications, Inc.
Pownal, Vermont 05261

Occasionally brand names are mentioned in this book. This is done primarily where the product is either a unique or an especially clear example of a design type, feature, or product. In no case should mention of particular brands in this book be interpreted as a recommendation of the particular product.

This book includes extensive information about the safety of solid fuel heating systems. The reader is cautioned that no practical installed system can be perfectly safe. Safety is a matter of degree, and always involves careful operation and maintenance, as well as sound equipment properly installed. Do not presume that by following any one particular recommendation in this or any other publication that no harm can result.

Extensive material on the recommendations of the National Fire Protection Association (NFPA) and the requirements of many building codes are included in this book. This material is not "official" in the sense of being direct quotations, or of having been reviewed and verified by NFPA or building code authorities. Although the author believes these representations to be correct, in cases where accurate renditions are critical, the primary sources should, of course, be consulted.

In localities with building codes, the only *legal* installations are those in compliance with those local building codes.

All photographs, except where noted, are by the author.

Copyright © 1983 by Jay W. Shelton

For additional information send all inquiries to Storey Communications, Inc., Schoolhouse Road, Pownal, Vermont, 05261

The name Garden Way Publishing is licensed to Storey Communications, Inc. by Garden Way, Inc.

Printed in the United States by Capital City Press
Cover printed by Federated Lithographers
Third Printing, September, 1986

Library of Congress Cataloging in Publication Data

Shelton, Jay, 1942-
 Jay Shelton's Solid fuels encyclopedia.

 Includes index.
 1. Fuel. I. Title. II. Title: Solid fuels
encyclopedia.
TP318.S468 1982   697   82-15648
ISBN 0-88266-307-0

CONTENTS

Acknowledgments

Many people helped in the creation and production of this book. I am grateful for the help of Tony Anthony of Sand Hill, Inc., Charlie Page of Vermont Castings, Bill Moomaw of Williams College, a number of manufacturers (and particularly their engineering staffs), all the chimney sweeps and retailers who tell it like it is in the field, Polly Rose, Andrea Chesman, and all the staff at Shelton Energy Research.

CHAPTER 1

SOLID FUELS FOR HOME HEATING

Wood accounts for over 2 percent of all the energy used in the United States today, and this percentage is increasing rapidly. Two percent may not sound like much, but it is comparable to both the nuclear and hydroelectric contributions to our total energy use (Figure 1-1). It is an oversight that wood is usually left out of pie and bar graphs depicting our current energy mix.

The biggest users of wood energy in the United States are the forest products industries—the lumber, plywood, veneer, reconstituted wood-base panel and board, and paper producers (Table 1-1). In many mills, wood wastes (sawdust, slabs, ends, and so on) are burned to provide heat for buildings and for drying and processing materials. Many plants have been using their wastes

for energy for decades; the cost of other forms of energy and environmental regulations are stimulating renewed interest.

Residential use of wood fuel is smaller than the industrial uses, but it is probably increasing faster. Residential use is difficult to quantify since the fuel never flows through a meter; much of it is never a part of normal commerce, but is gathered by the users. My estimate of the residential use of wood fuel is between one-quarter and one-half of total wood energy use, or between .5 and 1 percent of the total energy budget of the U.S.

Use of wood as energy peaked around 1870 when it constituted about 73 percent of the total, and the actual consumption was about 3 times

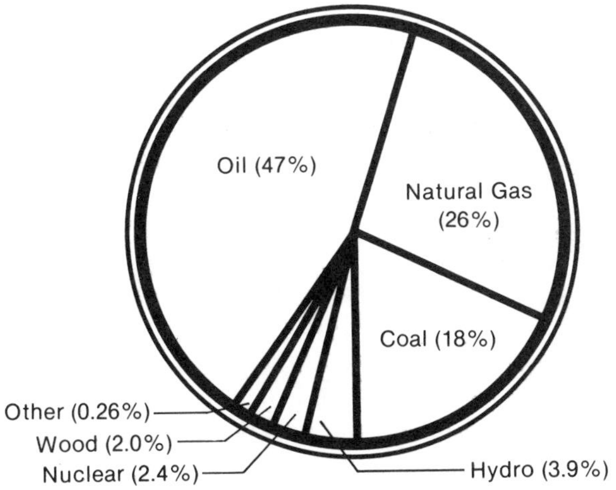

ENERGY SOURCE	PERCENT OF CONTRIBUTION	ENERGY SOURCE	PERCENT OF CONTRIBUTION
Oil	47.0	Urban Waste	0.2†
Natural Gas	26.0	Agricultural	
Coal	18.0	Waste	0.04
Hydro	3.9*	Geothermal	0.01
Nuclear	2.4	Direct Solar	0.01
Wood	2.1		

* This number represents the equivalent amount of fossil fuel energy it would take to produce the hydroelectric energy. This is about 3 times the electric output, since the average efficiency of fossil fuel electric generation is about 32%.

† A significant portion of the energy content of urban waste is wood-based—paper and cardboard. In a sense, the wood contribution in this graph could be increased to include some of the urban waste component.

Figure 1-1. Contributions of energy sources to total U.S. energy use in 1976.

TABLE 1-1
ANNUAL WOOD FUEL UTILIZATION

Wood Fuel Use	1976	2000
	in 10^{12} Btu.	
Pulp and paper industries	980	3,100
Residential firewood	200–400*	300–1,000*
Lumber, plywood, veneer, and other wood-based construction material industries	70	200
Charcoal	15	20
Metallurgical industries	12 ⎫	
Other industries	100 ⎭	300
	1,400	4,200

*These are very rough estimates.

Present and projected U.S. annual wood fuel utilization. Adapted in part from D. A. Tillman, Wood as an Energy Resource *(New York: Academic Press, 1978).*

what it was in 1970 (Figures 1–2 and 1–3). By 1900, wood constituted 20 percent of the nation's energy budget. Since the oil embargo of 1972 and the price increases over the subsequent decade, wood energy use has been increasing dramatically. In 1980, about 100–150 million cords of wood (about 2 × 10^{15} Btu.) were burned, contributing about 2.5 percent to the total energy of the nation.

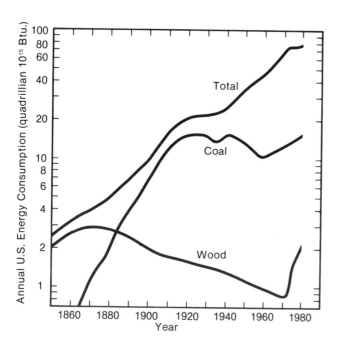

Figure 1-2. Annual U.S. energy consumption.

Use of coal was a tenth that of wood in 1850, but increased so fast that it overtook wood in 1870. Since 1910, coal use has been roughly constant—until the 1970s when its use started increasing.

As we now enter an era when coal is becoming a practical alternative to wood as a home heating fuel, it is interesting to look back a century. Over the period from about 1860 to 1910, coal gradually replaced wood as the dominant fuel for residential heating. Why?

The reasons for the shift were primarily economic. At the same time that wood transportation costs were rising, because forest lands were depleted around population centers, coal became available at reasonable costs to cities served by ship or barge transportation.

SOURCES OF HOME ENERGY

Wood and other "biomass" fuels offer some outstanding advantages compared with other sources of energy. Foremost is their renewability. Wood energy is renewable in the time it takes to grow a tree. The sun is the ultimate source of the energy. As long as the sun shines and the earth is a healthy place for life, wood will be available, if our forests are well managed.

All forms of solar energy, including wood, wind, falling water, and ocean temperature differences, as well as direct radiation, are continuously available or "renewable," while fossil fuels can be used up. But wood is special compared to the other solar options because its energy is so conveniently stored.

House heating with direct solar energy is complicated by the intermittence of sunshine, which creates the need to store heat for use at night and on cloudy days. In trees and other plants, nature has provided both collectors (the leaves with their chlorophyll) and storage. A pound of wood contains more than 50 times the amount of stored energy of a pound of hot water in a solar storage tank, and wood need not be insulated to hold its energy.

The energy efficiency of conversion of direct solar energy to wood fuel via plant photosynthesis is not high; it ranges from 0.05 to 0.2 percent in many forests. Photovoltaic cells have energy efficiencies of 3–12 percent, and flat plate heat collectors have efficiencies of 20–50 percent. But many

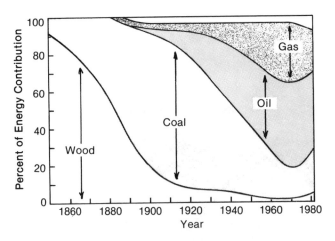

Figure 1–3. Historical trends in U.S. energy sources. The vertical width of each fuel band corresponds to the percentage of all energy coming from that fuel.

forests have no construction and few maintenance costs, and can simultaneously be used for many other things than energy production, such as wildlife habitat, recreation, and timber production. This plus the convenient form in which wood stores solar energy makes the low energy efficiency of tree growth of more academic interest than practical concern.

Coal is increasingly popular as a residential

TABLE 1–2
COAL VERSUS WOOD

Advantages of Coal	Advantages of Wood
Coal takes less space to store. Long-lasting, steady, and high-heat-output fires are easier to obtain. Flue deposits are less with anthracite, and easier to clean with bituminous.	Wood fires are easier to start. Wood has much less ash. Wood ashes can be valuable for garden soils. Wood is cleaner to handle. The odor of wood smoke is more pleasant. Wood flue gases are less corrosive; chimneys last longer. Because wood fires are easier to start, wood is the easier fuel to use in mild climates or in the spring and fall when only a little heat is needed once or twice a day. For many people, wood can be obtained for the effort of cutting and hauling it; one need not buy it. It is easier to learn how to burn wood.

heating fuel. In parts of North America, coal is a better buy than wood. Coal is also a denser fuel than wood; a cubic foot of coal contains more Btu. than a cubic foot of wood. The higher density of coal makes possible less frequent refueling.

CURRENT WOOD ENERGY USE

It is interesting that despite vast increases in our standard of living over the last century, the average amount of energy used *per person* has increased by only a factor of about 4. A century ago, there were no automobiles, trucks, and planes; no electrically powered refrigerators, freezers, electric lights, TVs, or air conditioners; and very few central heating systems. In fact, there were very few of the appliances and machines which are thought of today as significant energy consumers.

But, a century ago, much heating was done with wood burned in open fireplaces, one of the least efficient ways to heat a house. In colonial New England, as much as 20–30 cords of wood per year might be burned in open fireplaces, and even then not much of the house would be very warm. Tighter construction and insulation in contemporary houses reduces the amount of heat needed by 5–10 times. In addition, contemporary closed metal stoves and wood furnaces and boilers are probably 2–10 times more energy efficient than the open colonial fireplace. Today, the same size house in the same climate can be heated more uniformly using only 3–8 cords, which is the sustained annual yield from 5–15 acres of woods in many parts of the country. Chain saws and trucks have eased the human labor involved in acquiring wood, and controlled burning in stoves and furnaces decreases the stoking and reloading effort in the home.

Current wood use in the rest of the world differs from that in America. Many other developed countries do not have the wood resources and forest industries found in the U.S. However, there are some interesting exceptions. Wood supplied about 2 percent of our energy in 1975; it provided 8 percent of Sweden's energy budget; 15 percent of Finland's and 27 percent of Brazil's! But the average for developed nations was about 1.3 percent in 1975.

Developing nations are much more dependent on wood energy, to the average extent of about 9 per-

TABLE 1-3
WORLD FUELWOOD CONSUMPTION

Continent	Fuelwood Consumed in 10^{15} Btu.
North America	2.0
Africa	2.1
Asia	4.4
Latin America	1.2
Europe (including USSR)	0.9
Total	10.6

World consumption of fuelwood in 1975.

cent of their energy budgets (Table 1-3). Even more dramatic is that one-third of the world's people depend almost exclusively on wood for heating and cooking!

LESS CONVENTIONAL SOLID FUELS

Most people think of wood fuel as logs and sticks because this is the dominant form of wood fuel in homes. But wood fuel can have other forms.

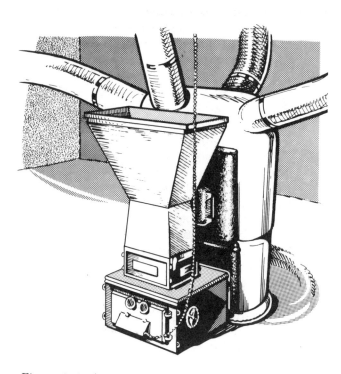

Figure 1-4. A sawdust or wood chip furnace.

Figure 1-5. A hay-burning stove. Twisted hay in the 2 cylinders is forced into the combustion chamber.

Wood-waste fuel in many industrial applications is often "hogged" or chipped, and may also contain sawdust. There is increasing use of pellets or briquettes.

The advantages of processing waste into these forms can be the dryness of the final product (although this can be a disadvantage—see chapter 14), the uniformity of the fuel, the increased density of the fuel, and the greater ease with which this fuel can be fed into burners with stokers. However, the additional processing increases the costs.

The energy cost of chipping and compressing wood into pellets is reasonably low, as is the energy efficiency of all the harvesting and mechanical-processing steps. Energy costs of harvesting, preparing, and transporting wood fuel include the fuel consumed by the equipment and a share of the energy used in manufacturing the equipment. The energy required to transform a tree in a forest into usable fuel made a few hundred miles from the forest is estimated to range from 2 percent to 10 percent of the energy content of the wood fuel. The energy content of the gasoline used by a chain saw to cut a tree into useful stove lengths is roughly only .5 percent of the energy content of the wood fuel. (This percentage is very sensitive to the sharpness of the chain.)

Charcoal is a form of wood energy when it is

manufactured from wood. Making charcoal involves heating wood while denying it enough air to burn freely. Heating wood causes it to transform into wood smoke (vapors, tar, and other ingredients) and charcoal. This process is called *pyrolysis*. To make charcoal briquettes, natural charcoal is compressed.

In England, fuels manufactured from coal are common. Just as charcoal is made by heating wood, coke is made by heating coal. (In practice, the term "charcoal" is sometimes used for coal-based as well as wood-based products.) In England, coke is a common home heating fuel because it is relatively smokeless; the smoke has been extracted in the manufacturing. However, carbon monoxide may still be high depending on how the coal is burned. (Being invisible, carbon monoxide does not contribute to smoke.) Charcoal and coke fuels may become important in North America to allow for better air quality.

PEAT

Peat is a biomass fuel on its way to becoming a fossil fuel. It consists of the tangled roots, stems, and foliage of small plants that grow in bogs and cold swamps. Over thousands of years, this plant matter accumulates into masses as much as 50 feet thick. The lower layers are compressed by the overlying material and are partly decayed; they look like brown coal. In fact, they are on their way to becoming coal.

Peat should be considered as more of a fossil fuel than a renewable resource. The distinction is a matter of degree, but peat bogs grow and accumulate biomass very slowly. Almost any practical harvesting or mining rate will exceed the regrowth rate, and so the resource will be gradually depleted.

Peat is harvested by slicing it into brick-size pieces. Since the moisture content is usually very high—50–90 percent—drying usually is necessary before burning. In Ireland, densified peat is also used.

The peat resource is significant in the world and in the U.S. The largest U.S. peat deposits are in Minnesota, where it could be an important regional source of energy, although peat is unlikely to become a residential solid fuel of national significance. Worldwide, peat is presently used only where coal, oil, and wood are scarce.

WOOD AND COAL GASIFICATION

Wood and coal can be converted into liquid and gaseous fuels. A potential advantage is that these fuels then can be substituted directly for the corresponding fossil fuels with little or no modification of the burners or engines. However, the conversion process is inefficient.

Pyrolysis is the most direct way to get liquid and gaseous fuels from wood. Some of the ingredients of the wood smoke are condensed into a liquid (called creosote when it condenses in a chimney) and the remainder passes through as a gas. Both the liquid and the gaseous components contain combustible ingredients. In some cases, it is necessary to separate the combustible from the noncombustible ingredients (such as water and carbon dioxide) to obtain a useful fuel. Since heat is necessary to cause pyrolysis, and since charcoal (a solid fuel) is a necessary product of pyrolysis, the overall energy efficiency of pyrolysis for producing liquid and gaseous fuels is not high—probably less than 60 percent. The equivalent of 40 percent or more of the original energy content of the wood is consumed or not used in the conversion process.

Modified processing can yield either only gas, or only liquid fuel, and no charcoal residue. Complete gasification involves pyrolysis at a higher temperature, which limits the production of condensible by-products; in addition, water vapor reacts with the charcoal to produce additional gas. The energy density of the gas (usually called synthesis gas) is low compared to natural gas—200 to 350 Btu. per cubic foot as compared to about 1,000 Btu. per cubic foot—but the energy efficiency of the conversion from wood to gaseous fuel is about 80 percent.

If liquid fuel is the desired end product, synthesis gas is usually produced first. The gas may then be converted to alcohol (methanol) in a catalytic process. "Gasohol," a mixture of gasoline and alcohol, can be used in any automobile engine. Also, gas can be converted to a kind of oil, not identical to petroleum, but usable in many of the same ways. With both of these liquidizing processes, the energy efficiencies for the whole conversion from wood are estimated to be roughly 35 percent. This means that almost two-thirds of the original energy of the wood is directly or indirectly consumed in the conversion processes.

Figure 1–6. Harvesting peat in Ireland. (Photo courtesy of Solid Fuel, *a British trade publication)*

Figure 1-7. Carrying turf from the bog after drying it in heavy baskets.

Still other processes involving fermentation and enzymes are being investigated. There is no doubt that wood can be converted to liquid and gaseous fuels, but the practicality of such conversions remains a question.

Whatever process is used, it would seem wise to use liquid fuels primarily to fuel automobiles, buses, trucks, and aircraft. Liquid fuels can be stored compactly and fed automatically with simple, lightweight equipment. For other uses, the bulkier storage and more complex handling characteristic of solid fuels is less of a liability.

NONFUEL USES OF WOOD

Wood is an incredibly versatile resource. It has always been a fundamental building material. The paper this book is printed on was made from wood. Some trees provide us with food—fruits and nuts.

Wood is also a source for industrial chemicals. Chemicals derived from wood were more common in the past than they are today. However, demand for wood for this purpose is rising, since fossil fuels are now the principal source for the chemicals that could be made from wood. Rayon and acetate are now derived from wood, and it is chemically feasible to make plastics, synthetic fibers, and synthetic rubber from wood. The economic feasibility should constantly improve as research continues and fossil fuel prices rise.

Clearly, there will be increased competition for wood. Estimates of the amounts of wood used for various products are indicated in Figure 1–8. Fuelwood is a significant part of the total forest harvest. Most fuelwood used outside of the wood industries comes from logging wastes (tree tops and branches), from forest-improvement thinnings, or from lands not available for logging. But competition between energy and nonenergy uses of wood is inevitable, and is already here for mill wastes. Sawdust, slabs, ends, and chips can be burned for energy or formed into building materials, such as particle and chip boards. The demand for wood as fuel has already driven up the

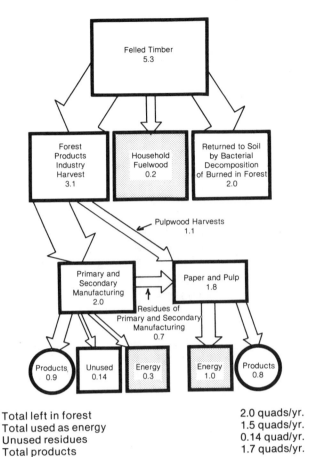

Total left in forest	2.0 quads/yr.
Total used as energy	1.5 quads/yr.
Unused residues	0.14 quad/yr.
Total products	1.7 quads/yr.

Figure 1-8. Material flow for felled timber during the late 1970s (in quads/yr.). (Source: Office of Technology Assessment)

price of pulpwood (for paper production) in some regions.

THE AVAILABILITY OF WOOD AND COAL

The nation's coal resources are extremely large— enough to supply *all* the nation's energy at current use rates for centuries. Availability is only a regional issue, not a national one. In some areas, coal is more expensive because of transportation costs, and in some areas a particular type of coal may not be available. But overall, coal is not a resource whose general availability will limit its use in the next few decades.

However, coal cannot be used at high rates forever because it is a resource that is fixed in quantity. Someday we will effectively run out of coal.

(As the resources dwindle, the price will become prohibitive.)

Wood is a very different sort of resource. Properly managed, it is renewable. The rate at which new trees are growing can balance the rate at which trees are harvested. Thus, unlike fossil fuels, the resource can be available forever.

Unlike fossil fuels, wood is used much more extensively as a building material and for papermaking than as an energy resource. Assessing wood fuel resources must take into account not only the productivity of forests, but also the competitive uses for wood.

Nonetheless, the amount of wood available for fuel is substantial. Fairly straightforward extrapolations of recent activity in the forest industry suggests there will be adequate forest-industry residues to meet increasing demands for both energy and nonenergy uses (Table 1–4). Impressive though the numbers are, they probably understate the availability of wood residues.

Table 1–4 includes the annual production of forest residue—crooked, rotten, diseased, or overmature trees—but not the huge backlog of such noncommercial timber, estimated to be about 16 $\times$ 10^{16} Btu., or 0.8 $\times$ 10^{16} Btu. per year over the next 20 years. Also not included is the wood from converting mixed hardwood stands to more productive species, estimated at about 0.5 $\times$ 10^{16}

TABLE 1–4
WOOD RESIDUE UTILIZATION

Wood Residues	1970	1985 (est.)	2000 (est.)
Residues produced			
Forest residues	1.00	1.10	1.10
Harvest residues	0.38	0.68	0.88
Mill residues	2.20	2.70	3.20
Total residues generated	3.58	4.48	5.18
Residues used for materials			
Pulp and paper	0.39	0.61	0.72
Structural products	0.12	0.23	0.31
Chemicals	0.06	0.11	0.12
Total residues used	0.57	0.95	1.15
Total residues available for fuel	3.01	3.53	4.03
Residues utilized for fuel	1.27	2.07	3.62
Surplus residues	1.77	1.46	0.41

(column header: Wood Utilization in 10^{15} Btu. or Equivalents)

Annual wood residue production and utilization to the year 2000. Data adapted from David A. Tillman, Wood as an Energy Resource (New York: Academic Press, 1978).

Btu. per year over the next 20 years. Finally, the harvesting residue only includes stemwood—portions of the main trunk left behind in the woods. If branches are included, this residue approximately doubles. And if leaves or needles and roots are included also, it can be 4 times as large. (Leaves and needles could be used directly in some industrial burners, or incorporated into densified fuel pellets.) This, however, may not be wise environmentally.

The potential supply of wood cut in the forest explicitly for residential fuelwood consumption (as opposed to forest industry waste) is much more difficult to estimate. Estimates can be made of the sustained yield of fuelwood from national and state forests, but such estimates must be made in the context of land accessibility and competing uses (timber, pulp, recreation, wildlife, and so on). The estimates vary substantially.

The potential fuelwood supply from small private lands is even more elusive to quantify, the biggest variable being the desires and feelings of the owners. Many small landowners would rather leave their forests in the "natural" state. Others would not mind some managing and harvesting of their forests but are afraid, legitimately, of the mess sometimes left behind. Concerns for liability also can preclude tapping some fuelwood resources. All of this is by way of saying that there may be a huge difference between the amount of wood that is out there in the woods and that portion which is available in practice.

In some regions (Colorado and Maine, for example), there are huge amounts of fuelwood available due to insect infestations in the forests. In some towns, landscaping residues are significant sources of wood fuel. Whereas nonfuel uses of wood require certain species and minimum quality, *all* species of wood in virtually *any* condition will burn and provide heat.

What does all this add up to? The potential supply of fuelwood is large, but quantitative predictions are not reliable. Fuelwood is an unusually "soft" resource—the "supply" can expand or contract substantially depending on the demand for fuelwood and depending on competing uses for both harvested wood and natural forests. A serious national shortage seems unlikely, although local shortages will be felt.

On the other hand, if everyone tried to heat with wood, the resource certainly would be inadequate. But not everyone will. Heating with chunk wood will always take much more effort and commitment than heating with oil, gas, or electricity. In many parts of the nation, fuelwood is either unavailable or prohibitively expensive for serious heating. In urban areas, air pollution regulations are already limiting the amount of wood heating.

Forest productivity could be increased substantially, thus increasing the availability of wood fuel. Current growth rates in most natural forests range from ¼-2 cords per acre per year. (However, in some regions, such as the mountains of the Southwest, natural growth rates may be only 1/50 cord per acre per year!) Sustained yields in natural forests probably could be doubled by forest management—thinning, cutting of less productive individual trees, selecting for (or planting of) more productive species and strains, and harvesting trees at the optimum time. Some species can send out new sprouts from their cut stumps; the already established root system provides such abundant supplies of nutrients that the sprouts grow much more quickly than new seedlings would. The result can be a higher yield of wood per acre.

Tree Farming

In the South, tree *farming* is common (Figure 1-9). Particularly high-yield strains are planted in rows and nurtured as carefully as many food crops. These trees are used mostly for wood pulp, although much of the residue material is burned for heat.

There has been much talk about energy plantations—large acreage enterprises growing trees or other plants exclusively for their energy value. The fuel could be used for any purpose, but one purpose often considered is electricity generation. It is estimated that a 100-megawatt electric plant (about 1/10 the size of typical new conventional plants) could be run forever from the growth from about 35 × 35 miles of natural forest, or about 12 × 12 miles of an energy plantation. An energy plantation can have more than 10 times the productivity of a natural forest. However, it is not clear whether energy plantations are economically competitive. Land, fertilizer, water, and financing costs are considerable. Energy plantations may someday compete with food production for land. Although some theoretical studies suggest economic viability, no such plantations are in existence yet.

Figure 1–9. A tree farm. (Photo courtesy the Arkansas Forestry Commission)

Although energy plantations producing *wood* fuel are not yet a reality, plantations producing *alcohol* are, most notably in Brazil. Alcohol can be blended with gasoline to yield gasohol. In some cases, alcohol can be used by itself as a liquid fuel. Alcohol is a renewable and clean-burning fuel, but its production competes directly with food production when it is made from crops such as corn, sorghum, sugar cane, and wheat. The potential exists for an economic-moral conflict as demands for fuel and food continue to rise.[1]

It is interesting that at this time the desired balance between combustion and new growth of plant matter in the world is absent. Because of vast clearing and burning operations in the southern hemisphere for creating new agricultural and development land, wood is being burned faster than it is regrowing. However, in America growth exceeds consumption as fuel.

1. See L. R. Brown, *Food or Fuel: New Competition for the World's Cropland,* Worldwatch Paper 35 (Washington, D.C.: Worldwatch Institute, 1980).

ENVIRONMENTAL STRESSES OF WOOD FUEL PRODUCTION

Two particular environmental problems with wood fuel production are soil erosion and nutrient loss. Soil erosion is increased by almost any harvesting method; clearcutting can be devastating with certain soil types and slopes, and access roads frequently cause significant erosion. If tree roots are harvested also, as has been proposed, the soil is further disturbed. Each site needs individual consideration to determine if and how erosion can be controlled.

Harvesting trees for decades from the same land can result in the disappearance of important nutrients from the soil. The result would be decreasing land productivity. Three remedies can be useful. Where the wood is used as a fuel, returning the ashes to the forest returns most of the nutrients; the major exception is nitrogen. A significant part of the nitrogen content (and many other nutrients of a tree) is in the leaves and twigs; leav-

Figure 1-10. Erosion resulting from forest clearing in Brazil. (Photo by Douglas R. Shane)

ing these parts of the tree in the forest helps complete the nutrient cycle. Finally, applying conventional fertilizers occasionally may be necessary.

In many countries, there is a very serious firewood crisis far exceeding just a short supply.[2] The Indian subcontinent and central Africa have been hardest hit, but parts of Latin America are also suffering.

Many towns in these areas are dependent on wood for heating and cooking. A few decades ago, the task of collecting a family's daily wood took only a few hours. But as populations have grown, the cutting has been so intense that a treeless landscape extends out as far as 30 miles from the towns, making a fuel-gathering trip all but impossible to complete in a day. Reforestation has not been attempted generally, and even where it has, the need for fuel is so great that the new seedlings sometimes have been pulled up and burned.

Serious side effects accompany such intense pressure on wood resources. The exposed soil erodes, loses nutrients, and can hold less water, making it more difficult to reforest. The larger and quicker runoff of precipitation intensifies flooding downstream. Perhaps most serious, as wood becomes too expensive and hard to obtain, people turn to dung as fuel—dung which previously was spread on the land as fertilizer for

2. See E. Eckholm, *The Other Energy Crisis: Firewood,* Worldwatch Paper 1 (Washington, D.C.: Worldwatch Institute, 1975).

crops. In areas where chemical fertilizer replacement is not feasible, food production declines and people starve.

These serious problems are unlikely to arise in America for a number of reasons. Because of our economic strength and natural resource base, we have many sources of energy to choose from—oil, natural gas, coal, uranium, geothermal, tidal, and the many forms of solar energy. We will never be dependent totally on only 1 energy source. This flexibility permits shifting our energy mix in response to changing supplies, environmental problems, and economics. The day will never come when everyone depends on wood, or any other single source, for heating. Wood heating will always be a regional or local phenomenon. In rural forested areas, wood heating is likely to become more common than it already is. In New York City, wood is never likely to be the most reasonable source of heating energy. Wood heating may not be viable in parts of the Southwest due to the scarcity of trees (which is related to the scarcity of water).

Also, some of this country's forested areas are very stable ecologically. For instance, in New England, soil and climate conditions are such that trees are weeds. If the land is not managed, it becomes a forest. Pastures revert to woodlands in a few decades. Finally, environmental concerns are so strong in this country that destructive forest practices will be curbed.

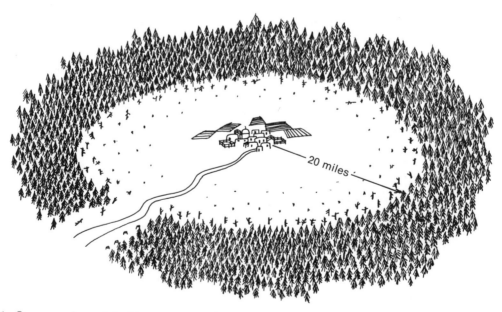

Figure 1-11. Overuse of wood fuel in some areas has led to situations such as this.

Figure 1-12. Wood-smoke air pollution.

WOOD VERSUS FOSSIL FUELS[3]

Although there are a few dissenting opinions, there is general agreement that air pollution from residential use of coal and wood fuel needs to be taken seriously (Figure 1-12). Many communities are already experiencing substantial decreases in winter visibility due to wood smoke – Vail, Colorado; Waterbury, Vermont; Missoula, Montana; Yosemite Valley in California; and Taos, New Mexico, are examples.

The evidence for detrimental health effects is not as visible. However, many ingredients of wood and coal smoke are carcinogenic to test animals, and may well be carcinogenic to humans. This should not be surprising since there are many similarities between wood smoke and cigarette smoke; both originate from smoldering plant material. And, unfortunately, many of the harmful chemicals in smoke are incorporated into

particles which are too small to be ejected easily by the natural mechanisms in human lungs. Smoke is very complex chemically. In addition to carcinogenic compounds, it contains toxic gases such as carbon monoxide, lung and eye irritants, and compounds which tend to destroy the ability of the lungs to cleanse themselves of larger smoke particles ("cilia toxic" agents).

However, the fact that smoke contains unhealthful ingredients does not imply there is a *significant* health effect. Most of the quantitative aspects of the health effects of wood and coal smoke are not well known. For example, what is the amount of each component in smoke? (Of course it will be different with different appliances, fuels, and so on.) Once emitted out a chimney into the atmosphere, how does the material disperse and change form? What actual concentrations are people exposed to? How much exposure or dosage is actually dangerous? Appropriately, the U.S. Environmental Protection Agency and the U.S. Department of Energy are sponsoring research in these areas.

Clearly, the effects of wood and coal smoke on health are not catastrophic or our ancestors would have noticed it. Yet if many people were to heat with wood in a densely populated community where the air can stagnate, there could be some long-term health effects.

There is not yet sufficient information to be able to *compare* wood and coal on the basis of emissions. An *overall* comparison may not ever be meaningful since the effects of appliance design and operation on emissions is so important. However, wood definitely emits less sulfur than coal. Wood contains only about 0.01-0.05 percent sulfur, whereas coal (and also oil) typically contains 0.5-3.5 percent (Table 1-5). Sulfur emissions from coal and oil combustion produce "acid rain," which, over a period of years, decreases the productivity of forests. This is a serious, long-

TABLE 1-5
SULFUR AND ASH CONTENTS

Fuel	Sulfur	Ash
	(in grams of material per million Btu. of fuel)	
Natural gas	less than 1	0
Wood	1-12	20-150
Oil	15-100	less than 5
Coal	50-200	300-600

The sulfur and ash content of wood compared to common fossil fuels.

---

3. Some of the material in this section is adapted from J. Shelton, *Air Pollution from Residential Woodburning Appliances, An Annotated Bibliography* (Portland, Ore.: Western Solar Utilization Network, 1980).

term, and unsolved environmental problem. The weak, nonsulfur acids in wood smoke are probably not significant contributors to acid rain. Coal emissions also contain a small amount of radioactive material. Coal-burning power plants emit more radioactivity than *normally* operating nuclear power plants.

Nitrogen oxide emissions from residential solid fuel heaters do not seem to be a problem because of the relatively low temperatures achieved in combustion.

When and where air pollution from wood stoves *is* a problem, there are a number of possible solutions.

Education. The way a solid fuel heater is operated has a very large effect on emissions. Smoldering, large fuel loads with long duration burns, not only pollute the air, but tend to cause substantial creosote buildups. Smaller fuel loads and adequate combustion air result in cleaner-burning fires. If more people were aware of this, many could reduce emissions and creosote buildup by changing how they operate their equipment when practical.

Cleaner-Burning Appliances. A number of design innovations are being developed and introduced that could help reduce emissions (and creosote) and may increase overall energy efficiency as well.

Fuel Type. London's air quality was improved dramatically when coal was banned as a residential fuel. Coke replaced coal in many homes. In this country, replacing wood and coal with charcoal or coke would have a similar effect. The only potentially significant pollutant from these fuels is carbon monoxide.

Limit Use of Existing Appliances. Albuquerque, New Mexico, already does this. When air pollution levels are high, residents are requested not to use their fireplaces and solid fuel heaters.

Limit the Number of Appliances. Some communities in Sweden and Vail, Colorado, restrict the number of particular types of wood-burning appliances that may be installed. In some Swedish communities, one cannot install a stove if there is another one within 25 meters (about 80 feet).

In industrial-size wood burners, air pollution is not as much of a problem. Combustion is usually much more complete—most potential pollutants are burned up before they enter the stack. In ad-

dition, conventional pollution control technology can handle the particles—fly ash and soot.

The most interesting and perhaps significant comparison between wood and fossil fuel emissions concerns carbon dioxide—a compound not normally considered a pollutant. If the amount of carbon dioxide in the atmosphere changes significantly, the climates of the earth may change. Carbon dioxide absorbs infrared ("heat") radiation coming from the surface of the earth. The loss of energy from the earth in the form of infrared radiation prevents the planet from overheating. Increasing the amount of carbon dioxide in the atmosphere inhibits the loss of infrared radiation from the earth, which could lead to global warming. This is often called the "greenhouse effect," since glass also absorbs infrared radiation.

The consequences for climate changes around the world are impossible to predict because of the incredible complexities of the atmosphere and its interactions with land and oceans. A warming trend is the most common guess, but this could be accompanied by cooling over parts of the earth, due to shifts in global circulation patterns. Precipitation would be likely to increase in some regions and decrease in others. The consequences of an increase in carbon dioxide in the atmosphere are unpredictable even if carbon dioxide were the only atmospheric pollutant; but other pollutants, such as particles, may also affect climates.

Although the changes could be either detrimental or benign in the long run, they are likely to be disruptive, particularly for agriculture. As food supplies tighten with increasing population and rising material living standards, more and more marginal agricultural areas will be brought into production. In such areas, a slight change in climate could result in large decreases in yields. It is estimated that a decrease in the mean annual temperature of 1° C. in Iceland would decrease the yield of forage by 27 percent. The observed 60 percent decrease in carrying capacity for cattle (and formerly bison) in the U.S. western high plains is apparently attributable to a 20 percent decrease in annual precipitation.[4] Of course, in some areas, climate changes would undoubtedly increase production. However, a general warming

4. R. A. Bryson, "A Perspective on Climatic Change," *Science* 184 (1974), p. 753. See also W. W. Kellogg and S. H. Schneider, "Climate Stabilization: For Better or Worse?" *Science* 186 (1974), p. 1163, and J. Hansen et al, "Climate Impact of Increasing Atmospheric Carbon Dioxide," *Science* 213 (1981), p. 957.

trend ultimately could result in the flooding of much prime agricultural land (as well as many cities) because of a rising sea level caused by the melting of the Antarctic and other ice caps.

Adding carbon dioxide to the atmosphere may or may not contribute to mass starvation. But it is certain that we will not know for many decades what the effects will be, by which time it may be too late to do anything about it. Like so many environmental problems, the carbon dioxide "problem" involves the *possibility* of a grave situation in the future. Almost no reliable predictions now exist about its probability or seriousness. Yet, if preventative action is to be taken, it must be started now.

Shifting from fossil fuels to wood where feasible would help but not because wood emissions are free of carbon dioxide. Both fossil fuels and wood contain carbon, and when burned, most of the carbon is incorporated in carbon dioxide molecules. The more complete the combustion, the more carbon dioxide is produced. Separating and storing or chemically altering the carbon dioxide is unfeasible.

The critical difference between wood and the fossil fuels is whether the carbon dioxide would have been generated and released even without our burning the fuel. In the case of wood, it would. If plant matter is left on the ground and allowed to decay, the same amount of carbon dioxide is released as would have been if the plant had been burned. Burning wood releases the carbon dioxide a few years sooner, but there is no long-term effect on the amount of carbon dioxide in the atmosphere. There is a slight one-time increase in atmospheric carbon dioxide whenever wood combustion rates are increased, because of the earlier release of carbon dioxide. However, there is no constantly increasing carbon dioxide burden when the rate of use of wood fuel is in balance with regrowth.

All of the carbon in plants is obtained from carbon dioxide taken out of the atmosphere by growing plants. In using wood on a sustained-yield basis (harvesting and burning it at no more than the rate at which new growth is occurring), as much carbon dioxide is taken out of the air by growing plants as is released in burning them.

When any fuel is burned, oxygen from the air is consumed. Significant oxygen depletion is not a serious possibility. However, burning fossil fuels results in a net disappearance of oxygen from the atmosphere, using wood as a fuel does not. A growing tree releases the same amount of oxygen as it will consume when burned. If left to decay in the forest, a tree consumes the same amount of oxygen in the process of rotting.

Heat is released when any fuel burns. The climates of large urban areas (for example, parts of the Northeast) are a few degrees warmer because of this heat. Wood, again, is unique compared to fossil fuels. Whether wood is burned or left on the ground to rot has no net effect on the amount of heat released. As wood decays, heat is very slowly released, but the total amount is the same as if the wood were burned. The combustion of fossil fuels releases heat which otherwise would have been locked up forever.

FINAL THOUGHTS

Nature has been growing and decaying trees for hundreds of millions of years. The solar energy locked into chemical form in the plants through photosynthesis always has been released mostly as heat during decay. By taking the wood and burning it in a stove, the heat is released in a home rather than on the forest floor. The stored solar energy ends up in the same form in either case; burning the wood merely reroutes the energy through a house on its way back into the atmosphere and eventually back into space. The amounts of carbon dioxide and oxygen in the atmosphere are not significantly affected, whether wood is burned or left to decay. The net effects are the changes in the forests and their watersheds due to harvesting, and the emissions of chemicals and particles from incomplete combustion of wood—problems which need to be taken seriously, but which have solutions.

Wood should not be thought of as an energy source for all purposes and for all people. For that matter, neither should any other energy resource. There is strength in diversity. Wood energy makes sense in some regions and not others. The same is true of coal. Wood has always contributed, and will always contribute, to energy needs; but we should not ask it, or coal, to do too much.

CHAPTER 2

FUNDAMENTALS: ENERGY, TEMPERATURE, AND HEAT

Heating successfully with solid fuels is not complex or difficult. But operating a solid fuel heating system efficiently is more involved than flicking a switch on a furnace. If you start with a firm grasp of the principles of heating, you will have a clearer understanding of why stoves and other heat systems work the way they do, and you will be better equipped to make wise decisions.

ENERGY

Most of the energy used to toast bread in American homes came from the sun. It originated there as nuclear energy of hydrogen. When hydrogen nuclei combine deep in the sun (in nuclear reactions similar to those in hydrogen bombs), nuclear energy is released mostly as heat. This makes the sun hot, which in turn causes it to radiate energy (sunlight) out in all directions in the solar system. The earth intercepts a minute fraction of this radiant energy. Some of the radiant energy is absorbed in green plants and converted by them into the chemical energy contained in their own molecules. Long ago this organic matter began to accumulate on the ground or the bottom of lakes and ultimately was buried under mud and sand, which became rock in geological time.

This process continues today while we also mine the accumulated organic material, since transformed into oil or coal. Burned at a power plant, the chemical energy is released as heat

again, the same form it once had on the sun. The heat is transferred into water to make steam. When passed through a turbine generator, some of the energy in the steam is converted into electricity.

The small wires in an electric toaster convert the electric energy back into heat for a third time. This makes the wires glow red hot, and some of their heat is then radiated to the bread to toast it. Ultimately, all the heat generated by the toaster finds its way out of the house and into the outdoor air. Its final fate is to travel as infrared radiation back into space. Since space is relatively empty, the energy is unlikely to encounter any matter or undergo any more transformations.

Forms of Energy

Of the many forms of energy, the most fundamental is motional or *kinetic* energy. Kinetic energy is energy by virtue of motion. There are three other fundamental forms of energy which are different kinds of *potential* energy: gravitational, electrical, and nuclear.

Kinetic Energy. Any moving object has kinetic energy—a traveling car, Mars (due to its orbital motion about the sun), in the air, and even individual molecules in this book. The higher the speed of an object and the heavier it is, the more kinetic energy it has.

Individual atoms and molecules are in constant, random, jostling motion. In solids, each atom vibrates around a fixed location, although its speed can be very high. In liquids, the molecules can migrate, but most of the motion is essentially vibrational, as in solids. In a gas, such as air, molecules are far enough apart that each molecule can travel a distance of many molecular diameters before colliding with another and bouncing off in a new direction. All these molecular motions are essentially perpetual; on an atomic scale, there is *no* friction in the usual sense.

Potential Energy. Potential energy can be converted into kinetic energy. A rock held above the ground has gravitational potential energy. If the rock is released, that potential energy is converted into kinetic energy.

Solid fuels represent a form of electrical potential energy called *chemical energy*. When the fuel is burned, atoms in the fuel and oxygen from the air rearrange themselves into new molecules. This process is a chemical reaction. During the rearrangement, strong electrical forces act, releasing stored potential energy. The forces accelerate the atoms, which increases their kinetic energy.

Energy is never lost or created. The total amount of energy (of all kinds) is always the same before and after a reaction or transformation. When a pound of wood is burned, all of its energy can be accounted for in various forms, including light, infrared radiation, chemical energy (combustion is never perfectly complete), latent heat, and ordinary (or "sensible") heat.

HEAT AND TEMPERATURE

Heat is a form of energy. In fact, it is the kinetic energy of molecules due to their random jostling motions.[1] The amount of heat in any object is the sum of the kinetic energies of all its molecules.

Temperature is the intensity of heat on a molecular scale. The temperature of a substance is directly proportional to the average kinetic energy per molecule. The faster the average speed of molecules in their random motions, the higher is their temperature.

---

1. Technically, heat is the form of energy which is *transferred* by virtue of a temperature difference and molecular collisions. This more technical definition is not used in this book, because such use would be confusing.

Heat and temperature are different concepts. A group of molecules with a high average speed (kinetic energy, technically) is hot—they have a high temperature; but the quantity of heat they represent depends on how many of them there are. If there are only a few molecules, there is very little heat (their total kinetic energy is small). A group of molecules whose average speed is low has a lower temperature, but if there are many of them, the amount of heat they represent can be large. For example, the same amount of heat can be stored either in a small tank of hot water or in a large tank of warm water. Aluminum foil in a hot oven is as hot as the oven; but because the foil is so thin, it doesn't contain much heat. This is why it can be touched without causing a burn.

When fuel is burned, the chemical forces causing the atomic rearrangements accelerate the atoms, which increases their speed. Since the average speed of each atom is increased during the reaction, the temperature of the reactants increases (flames are hot!). Since the total kinetic energy of all the molecules has increased, there is more heat (combustion generates heat).

Once released, energy in the form of heat does not change easily into other forms of energy. As the heat in a candle flame spreads through the air, the temperature of the heat decreases. On a molecular level, the high kinetic energy per molecule in the flame is shared with the vastly larger number of molecules in the environment, increasing their energies very slightly. While the total amount of kinetic energy (heat) is not changed, because it is shared among a much larger number of molecules, each has less. The temperature of the heat is less as it spreads out.

The heat used to warm houses is constantly moving out of the house and into the outdoor air. Its total quantity is unchanged in the process, but its temperature decreases. Trying to heat a poorly insulated house is frustrating and expensive because it is almost literally an attempt to heat the outdoors.

All objects contain heat because their molecules have some kinetic energy. The colder matter becomes, the less kinetic energy it has per molecule, and hence the less heat it has. (There is a special temperature, absolute zero, or $-460°$F., at which molecules have virtually zero kinetic energy.) It may seem strange to speak of a cold object containing heat, but an object, say at $0°$ F., could be warmed by touching a $32°$ F. object.

THE MOVEMENT OF HEAT

Heat moves from one place to another by conduction, convection, and radiation. These principles help you understand the relationships between stove design and heat transfer efficiency.

Conduction

Heat is conducted up a silver spoon from a hot cup of tea to your fingers. Most of the heat transferred through solid airtight walls in a house is also via conduction.

Heat is always conducted from hotter to cooler regions by molecular and electron collisions and interactions. Whenever an energetic (hot) molecule collides with, or vibrates against, a less energetic one, their energies tend to be equalized in the collision.

In the case of the silver spoon, the lower end of the spoon is the hottest. Throughout the length of the spoon, energy tends to be transferred upwards in all the little collisions and interactions.

Materials differ in their inherent ability to conduct heat—in their *conductivity*. Thermal conductivity is a measure of the rate at which heat is conducted through a slab of material whose 2 sides are maintained at a constant temperature difference. Conductivities of some materials are given in the middle column of Table 2-1. The range of values is enormous, from 245 for silver to .02 for glass-wool insulation. Materials, such as glass-wool, which have low conductivities, are good insulators. Cast iron and steel have virtually identical conductivities. Wood is a moderately good insulator, and charcoal and ashes are even better.

Convection

In contrast to conduction, convection *does* involve transport of matter along with heat. With convection, heat is transported by moving hot matter.

Hot air heating systems distribute their heat via convection—by circulating the hot air. The air molecules and heat move together. If a blower or fan (or pump in the case of liquids) is involved, the convection is *forced*. If the hot air moves only under the influence of its own natural buoyancy, the heat transport is called *natural* (or gravity) convection. Natural convection tends to move the warmest air to the ceiling of a room. Forced convection from a fan in the room can distribute the heat evenly.

TABLE 2-1
PROPERTIES OF MATERIALS

Material	Specific Heat (Btu./lb. °F.)	Thermal Conductivity (Btu./hr. sq. ft. °F./ft.)	Density (lb./cu. ft.)
Wood	.5-.7	.06-.14	25-70.0
Charcoal (wood)	.20	.03	15.0
Ashes (wood)	.20	.04	40.0
Glass (soda-lime)	.18	.59	154.0
Glass-wool insulation	.16	.02	3.3
Cast iron	.12	28.0	450.0
Steel (mild)	.12	26.0	490.0
Silver	.056	245.0	654.0
Building brick	.2	.4	120.0
Firebrick	.2	.6	110.0
Insulating firebrick	.2	.08-2	20-50.0
Soapstone	.2	3.0	175.0
Water	1.0	.35*	62.3
Air	.24	.014*	.074†

*This is conductivity only, without any convection.

†At 70° F, standard atmospheric pressure, and 50% *relative humidity.*

Properties of materials. Specific heats are related to the amount of heat that can be stored in a given mass of material. Thermal conductivities are related to the rate at which heat can be conducted through materials.

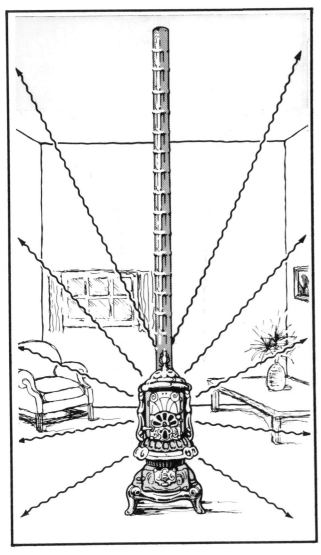

Figure 2-1. Relatively hot air rises via natural convection; radiation travels in straight lines.

Radiation

Energy can also be transported via radiation. Technically, there is really no such thing as radiant *heat*. Radiant *energy* is infrared electromagnetic waves. Light, radar, X-rays, microwaves, and radio broadcast waves are all forms of electromagnetic waves.

Infrared radiation cannot be seen, for it is beyond the deepest red of the visible spectrum. It is not heat, since it has nothing to do with the motion of molecules. In fact, infrared radiation travels most easily in a vacuum. However, if some of it is absorbed on a surface, such as your skin, its energy becomes heat, which you can then feel. Radiant energy is thus a form of *latent* or potential heat.

Infrared radiation travels at the same speed as light and radio waves—so fast that in practical circumstances it arrives at essentially the same instant it is emitted. (Conduction and convection of heat usually take perceptible amounts of time.)

Infrared radiation cannot penetrate most solids; it is usually mostly absorbed on their surfaces, where it becomes heat. Usually only a small fraction of infrared radiation is reflected, unless the surface is a smooth bare metal, in which case most of the radiation is reflected. A few materials allow some of the radiation to pass through them—certain plastics and minerals are examples. Most types of glass do not let much infrared radiation through, but absorb most of it.

All matter is constantly both emitting and absorbing infrared radiation. The intensity of the emitted radiation depends on the surface temperature of the object and its emissivity. The hotter a surface is, the more energy it radiates. Increasing the surface temperature of a stove from 200° F. to 400° F. more than quadruples the net radiation from it. Infrared photography works by the same principle: warmer objects emit more radiation and hence come out brighter in infrared photographs.

At any given temperature, a surface with an emittance of 1.0 is the very best possible emitter of infrared radiation. An emittance of 0.1 means only a tenth the maximum amount of radiation will be emitted. Polished silver, with an emittance of about 0.02, emits very little radiation when hot (Table 2-2).

The amount of infrared energy emitted by an object is affected only by the surface layers of molecules. A thin coat of paint on polished silver raises its emittance to 0.85-0.95. Tarnishing or a little dirt on a metal surface can have a big impact on its emittance.

Our environment is always full of high levels of infrared radiation because most surfaces at room temperature emit radiation at a rate of about 130 Btu. per hour, per square foot, which is half the intensity of direct sunlight. Why are you not more aware of this radiation?

Since you cannot see the radiation, you can sense its presence only if it results in a change in the temperature of your skin. Normally, there is a near balance between the radiation absorbed on your skin, which was emitted by the surroundings, and the radiation emitted by your skin. If

TABLE 2-2
EMITTANCES OF VARIOUS SURFACES

Material	Approximate Emittance
Iron	
polished	0.08
steel plate, rough	0.97
cast, oxidized	0.66
galvanized, new	0.30
galvanized, dirty	0.45
Nickel, aluminum, chromium, highly polished	0.05-0.15
Copper, polished	0.05
oxidized	0.83
Silver, polished	0.02
Stainless steel, type 18-8	
polished	0.18
weathered	0.85
Building materials and finishes: wood, stone, brick, plaster, tile, paper, all types and colors of common paints except bright metallic paints	0.80-0.95
Metallic paints—aluminum, bronze, or gilt	0.40-0.65
Glass	0.85
Water	0.96

Approximate emittances of various surfaces. Emittances are highly dependent on surface condition, moderately dependent on temperature, and vary somewhat with definitions (normal versus hemispherical). Except for water, the emittances are given at typical stove surface temperatures (roughly 500° F.). Data adapted from F. Kreith, Principles of Heat Transfer *(New York: Intext Educational Publishers, 1973); and* ASHRAE Handbook and Product Directory 1977 Fundamentals *(New York: American Society of Heating, Refrigerating and Air-Conditioning Engineers, 1977), p. 2.9.*

the same amount of energy leaves as enters, there is no net change in energy; your skin temperature is unaffected and neither the arriving nor leaving radiation is noticed. But in the vicinity of hot objects, such as an operating wood stove, your skin is warmed because it absorbs more radiation than it emits, resulting in a net energy gain.

If you were in a house with a large window area on a cold night, your face would feel cooler when facing the windows than when facing an interior wall. Because of the relatively cool temperature of the glass and its low emissivity, your face receives less radiation from the glass than it radiates away towards the glass. The radiation deficit cools your skin.

Test this effect for yourself. Stand 5 or 10 feet back from the window so that the effect will not be confused by the cooler air next to the window.

The radiation exchange between 2 surfaces is always a net energy flow from the warmer surface to the cooler, just as in the case of conduction.

When surfaces are sufficiently hot, the emitted radiation starts to have a visible component. At about 900° F. most objects appear deep red. As the temperature increases, the brightness of the glow or emission increases, and the color shifts to orange, to yellow, and finally to white (at about 2,800° F.). This phenomenon can be used to make a rough estimate of temperatures between 900° F. and 2,900° F. (Table 2-3). Most of the radiation, however, remains in the (invisible) infrared part of the spectrum.

Radiation is emitted outward in all directions from every point on most surfaces. Anywhere in front of a flat radiating surface, some radiation is received, even towards the sides. The intensity of the received radiation at any point is approximately proportional to the apparent size of the radiating surface as seen from that point (technically, the solid angle subtended, or angular area). Radiation is most intense directly in front of, and up close to, a hot surface, for then the surface practically fills the whole field of view. The intensity is less as you move away from or to the side of the hot surface, and is in proportion to its smaller angular size. The radiation from a red hot coal easily will ignite a piece of paper held half an inch away, but will have virtually no effect on paper a few feet away. These geometrical considerations are especially important for safety of installations; radiant intensities on nearby combustible walls and floors must be kept sufficiently low to avoid ignition.

TABLE 2-3
RELATIONSHIP OF COLOR TO TEMPERATURE OF IRON OR STEEL*

Color	Temperature
Dark blood red, black red	990° F.
Dark red, blood red, low red	1,050
Dark cherry red	1,175
Medium cherry red	1,250
Cherry, full red	1,375
Light cherry, light red	1,550
Orange	1,650
Light orange	1,725
Yellow	1,825
Light yellow	1,975
White	2,200

* Adapted from T. Baumeister, ed., *Marks' Standard Handbook for Mechanical Engineers*, 7th ed. (New York: McGraw-Hill, 1967), pp. 4-7.

ENERGY UNITS

Energy can be measured in many different units. The basic unit usually used in the heating and air-conditioning field is the *British thermal unit*, or Btu. A Btu. is defined as the amount of energy it takes to increase the temperature of 1 pound (1 pint) of water by 1 degree Fahrenheit. Thus, to heat 1 pound of water from 60° F. up to the boiling point, 212° F., requires 212 − 60 = 152 Btu. Even though the basic definition of a Btu. is in terms of heat, Btu. can also be used for any other form of energy, since all forms can be converted into heat.

A Btu. is a relatively small amount of energy on a human scale. It takes about 75 Btu. to heat a cup of water to boiling (starting at about 60° F.). About 370,000 Btu. are needed to boil down 30–40 gallons of sap to make 1 gallon of maple syrup. The 3,000 calories of food energy that a typical adult eats each day contain 12,000 Btu. of chemical energy. A 2-minute shave with an electric shaver uses about 1 Btu. of electrical energy. Portable electric space heaters generate a maximum of about 4,000 Btu. of heat for each hour they are on.

POWER

A piece of wood contains a given amount of chemical energy. If it is burned completely, a given amount of heat will be released. If burned quickly, the heat will be produced at a high rate for a short time; if burned slowly, the same amount of heat will be produced at a slower rate for a longer time. The rate at which energy is converted from one form to another, or transported from one place to another, is called *power*.

Power has units of energy divided by time. Common units are Btu. per hour and watts. (A watt is 1 Joule per second, where a Joule is a unit of energy equal to 1/1,054 Btu.)

Conventional furnaces are usually rated in Btu. per hour. A 100-watt light bulb converts electrical energy into light and heat at a rate of 341 Btu. per hour. (1 watt = 3.41 Btu. per hour.) If oven-dry wood (with an energy content of 8,600 Btu. per pound) is burned at a steady rate of 1 pound per hour, the rate of burning is about 8,600 Btu. per hour. If 50 percent of the energy in the wood becomes useful heat in the house (that is, the energy efficiency of the stove is 50 percent), then the heating power of the stove in this circumstance is half of 8,600 Btu. per hour, or 4,300 Btu. per hour.

The human body is constantly converting chemical energy from food into other forms, mostly heat. For a sitting person, this conversion rate is about 500 Btu. per hour. If 500 Btu. of food energy are "burned" each hour for a day, the total energy used is 12,000 Btu. (500 Btu./hr. × 24 hr.), which is the same as 3,000 calories, a typical daily food-energy consumption. The heat output of the human body can be a significant contribution to house heating needs, particularly in a crowded or extremely well-insulated house.

HEAT STORAGE

The capacity of a stove to store heat affects its performance. The more massive a stove is, the more heat it can store, and the more heat it takes to warm it up. The heat capacity of an object is the amount of heat it takes to raise its temperature by 1 degree. The ability of different materials to store heat is usually described as *specific heat*, which is the heat capacity per pound. The specific heat of iron (and steel) is about 0.12 Btu. per pound per degree Fahrenheit (Table 2–1); it takes 0.12 Btu. to warm 1 pound of iron 1° F. The heat stored in a 100-pound iron stove with an average temperature of 470° F. (referenced to room temperature, 70° F.) is 4,800 Btu. (0.12 Btu./hr. °F. × 100 lb. × (470−70° F.) = 4,800 Btu.).

The term "heat" is often called *sensible heat*, to distinguish it from latent heat and radiant energy. Only sensible heat is related directly to the motional energy of molecules and atoms. To feel sensible heat requires physically touching an object or substance; it can only be sensed by direct contact (on a molecular level). Radiant energy (or sometimes "radiant heat") becomes sensible heat when absorbed.

LATENT HEAT

There is another form of energy that, like infrared radiation, is often associated with heat because of the ease with which it is converted into heat, and vice versa. Latent heat is a form of potential energy which is contained in water vapor. When water is at the boiling point, energy in the form of heat must be added to make it boil away. Its temperature does not change in the process, but its physical state changes from liquid to vapor. The water vapor contains the energy as latent heat; if it condenses back to liquid water, the energy it took to boil it away is released again as heat.

Evaporating water requires roughly the same amount of heat energy as boiling it away. With evaporation, the necessary sensible heat is absorbed from the surroundings, mostly the remaining liquid water. This is why evaporation is a cooling process. In either case, boiling or evaporation, the amount of energy required to change a liquid to a vapor is about 1,000 Btu. per pound of water (1,050 Btu. at room temperature, 970 Btu. at 212° F.). Whenever water condenses, the same amount of heat is released. Water vapor is said to contain latent (concealed) heat because sensible heat is released when it condenses.

TABLE 2-4
ENERGY TRANSFER

Temperature of Surface (° F.)	Total Heat Transfer (Btu. per hr. per sq. ft.)	Radiation* (percent of total)	Natural Convection† (percent of total)
80	13	70	30
100	47	63	37
150	156	59	41
200	294	58	42
400	1,150	63	37
600	2,620	70	30
800	4,980	75	25
1,000	8,560	80	20
1,200	13,700	84	16
1,400	21,000	87	13

*Computed in Btu. per hour per square foot as $0.9 \times 1.71 \times 10^{-9} \times (T_s^4 - 530^4)$ where Ts is the temperature of the surface in degrees Rankine. This differs slightly from Table 2-3 in *The Woodburners Encyclopedia* (Waitsfield, Vt.: Vermont Crossroads Press, 1976) in that here an emittance of 0.9 instead of 1.0 is used.

†Computed in Btu. per hour per square foot as $0.19 (T_s - 70)^{4/3}$ where Ts is the temperature of the surface in degrees Fahrenheit.

The amount and type of energy transferred from an exposed hot surface, such as a radiant stove, to its surroundings.

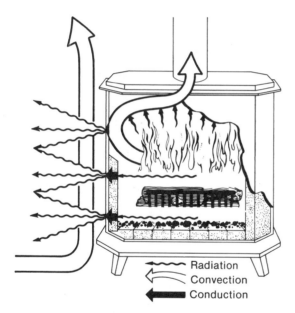

Figure 2-2. Heat from radiant stoves. Radiation travels outward in straight lines in all directions from a hot stove, until it is absorbed on a hot surface. Air is heated by coming in contact with the stove and stovepipe. Then, the heated air rises and spreads out along the ceiling. As the heated air loses its heat, it flows down along the walls and returns to the stove at floor level. (See chapter 9.)

HEAT FROM STOVES

In a fire, chemical energy is converted into radiation and heat. In a closed stove, the radiation is absorbed by the inner surface of the stove walls. Some of the sensible heat is convected and conducted to the stove wall; the rest convects up the chimney. The heat at the inner surface of the stove walls generally conducts to the outer surfaces of the stove. There some of it radiates away, and some is convected away in the air (Figure 2-2).

With most stoves (all nonconvective types), more energy is transferred by radiation than convection. The percentage depends on the temperatures of the outer surfaces of the stove, and ranges from 60 percent to 70 percent (Table 2-4) for surface temperatures up to about 800° F., the approximate maximum surface temperature of most stoves.

CHAPTER 3

FUELWOOD

Much has been written on the attributes and liabilities of various kinds of wood for fuel, and on the available heat from wood. Unfortunately, much of this information has been misleading or even wrong. But once you understand the fundamental properties of wood as a fuel, you will be better equipped to evaluate different types of wood as fuel sources.

ENERGY CONTENT

Is the fuel energy of white oak higher than that of white pine? The answer can be either yes or no depending on the basis for comparing the fuels.

The *energy content* (also called "higher heating value" or "caloric value") is the absolute maximum amount of heat that can be obtained by completely burning a given quantity of the fuel.

When comparing the energy contents of different types of wood, you can compare equal weights or equal volumes of different species. The results are quite different.

To within a few percents, equal weights of oven-dry (zero-moisture content) wood of almost all species have the same energy content—about 8,600 Btu. per oven-dry pound.[1] (See p. 23.) This

is because the chemical compositions of different species of wood are more nearly the same than they are different. They are all basically wood!

The differing resin (pitch) content of various species accounts for most of the small differences in energy contents. Resins have an energy content of roughly 15,000 Btu. per pound. There are species, such as pitch pine, which have so much pitch that their energy contents exceed 11,000 Btu. per oven-dry pound. Individual samples of more normal conifers, such as Douglas fir, may also have energy contents in excess of 11,000 Btu. per oven-dry pound, if they are unusually high in pitch due, for example, to the tree having been injured near where the sample came from. Some conifers also have very high pitch content in the heartwood of the bowl of the tree (the part of the tree near ground level). Knots also can have unusually high energy content.

A very high pitch content is readily identifiable—the wood appears saturated with pitch and tends to burn with a black smoke. But, aside from these extreme cases, the normal range of energy contents both among species and among samples from different locations is usually 8,400–9,000 Btu. per oven-dry pound.

So why are woods such as oak and hickory usually preferred over woods such as pine and poplar? Because a *cord* of white oak has almost twice as much energy as a *cord* of white pine. On a *volume* basis, woods differ significantly from each other. A cord is a volume measure, and for equal volumes, white oak weighs almost twice as much

1. E. T. Howard, "Heat of Combustion of Various Southern Pine Materials," *Wood Science* 5 (1973), pp. 194–197; S. W. Parr and C. N. Davidson, "The Calorific Value of American Woods," *J. of Indus. and Eng. Chem.* 14 (1922), p. 935; and Gottlieb, *J. Prakt. Chem.* 28 (1883), p. 414.

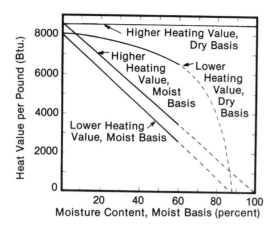

Figure 3-1. Higher and lower heating values per pound *for most kinds of wood. Values for moisture contents above 60 percent are shown dashed since very few woods are ever this moist.*

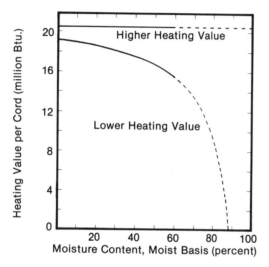

Figure 3-2. Higher and lower heating values per cord *for Douglas fir. Other woods would have similarly shaped curves but at various heights on the graph. Values for moisture contents above 60 percent are shown dashed since few woods are ever this moist.*

as white pine (with both woods at zero moisture content). Since each pound of oven-dry wood of any species has essentially the same amount of energy, the cord of white oak contains about twice the energy.

How are energy contents measured? Actual measurements of energy contents of woods are made by completely burning very small pulverized samples in what is called a bomb calorimeter. The combustion occurs in a closed container (the "bomb"), and essentially all the heat generated is measured by the resulting rise in temperature of the bomb and its surroundings (the calorimeter).

The bomb is pressurized with oxygen before the wood sample is ignited to help ensure complete combustion.

In fact, since good measurements on a large variety of firewoods are lacking, most lists of energy content per cord (including Table 3-1) are computed from the assumption of uniform energy content per unit weight of wood (of the same moisture content), and use the difference in weight per cord to estimate the difference in energy per cord. In this approximation, a table of relative densities is essentially equivalent to a table of relative energy contents per cord—the denser the wood, the more energy a given volume of it contains.

BURNING QUALITIES

Softwoods often are said to burn faster and hotter than hardwoods and are usually preferred for kindling. Hardwoods are generally said to last longer in a fire and generate more coals.

Careful experimental measurements support these notions. But the critical variable is the density of the wood, not its classification as a softwood or a hardwood.

Denser woods last longer in a fire. But as the wood burns, the rate of decrease in weight is very nearly the same for all woods (Figure 3-3). Since all woods have nearly the same energy for the same weight of wood, equal weight loss rates imply equal energy release rates. So, the heat output

Softwoods Versus Hardwoods

Native American trees can be divided into 2 groups: hardwoods, which have broad leaves, and softwoods, which have leaves in the form of rounded needles (for example, pines) or flat needles (for example, cedars). Softwoods are also called conifers since all native species have cones. Most softwoods are evergreen; the exceptions are tamarack, larch, and cypress. Most hardwoods are deciduous; they drop their leaves and grow new ones each year. Softwoods are not necessarily softer than hardwoods. For instance, Douglas fir, a softwood, is as hard as yellow poplar, a hardwood, and aspen, a hardwood, is in fact softer than white pine, a softwood.

TABLE 3-1
PROPERTIES OF WOOD SPECIES

Species	Energy Content* (million Btu./cord)	Moisture Content of Green Wood (percent)† Heartwood	Moisture Content of Green Wood (percent)† Sapwood	Relative Density or Specific Gravity‡
Hardwoods				
Alder, red	17.6	..	49	.41
Apple	..	45	43	..
Ash:				
Black	21.0	49	..	.49
Blue	24.9	..	37	.58
Green	24.0	..	..	.56
Oregon	23.6	..	..	.55
White	25.7	32	31	.60
Aspen:				
Bigtooth	16.7	..	..	.39
Quaking	16.3	..	..	.38
Basswood, American	15.9	45	57	.37
Beech, American	27.4	35	42	.64
Birch:				
Paper	23.6	47	42	.55
Sweet	27.9	43	41	.65
Yellow	26.6	43	42	.62
Butternut	16.3	..	..	.38
Cherry, black	21.4	37	..	.50
Chestnut, American	18.4	55	..	.43
Cottonwood:				
Balsam poplar	14.6	..	..	.34
Black	15.0	62	59	.35
Eastern	17.1	..	..	.40
Elm:				
American	21.4	49	48	.50
Cedar	...	40	38	..
Rock	27.0	31	36	.63
Slippery	22.7	..	..	.53
Hackberry	22.7	38	39	.53
Hickory, pecan:				
Bitternut	28.3	44	35	.66
Nutmeg	25.7	..	..	.60
Pecan	28.3	..	..	.66
Water	26.6	49	38	.62
Hickory, true:				
Mockernut	30.9	41	34	.72
Pignut	32.1	42	33	.75
Red	..	41	34	..
Sand	..	40	33	..
Shagbark	30.9	..	..	.72
Shellbark	29.6	..	..	.69
Honeylocust	28.3	..	..	.66§
Locust, black	29.6	..	..	.69
Magnolia:				
Cucumbertree	20.6	..	..	.48
Southern	21.4	..	..	.50
Maple:				
Bigleaf	20.6	..	..	.48

* Higher heating value, assuming 80 cubic feet of solid wood per cord and 8,600 Btu. per pound of oven-dry (zero moisture content) wood. In practice, values may vary at least 20 percent due to varying packing densities (the volume of solid wood per cord) and due to varying density of the wood. For very resinous woods, higher heating values are a few percents higher. Higher heating values per cord do not depend on moisture content except through the shrinkage of the wood as it dries. The values in this table are for 12 percent moisture content. A green cord would have a smaller heating value by a few percents.

† Moisture contents are on the moist wood basis.

‡ Densities are relative to the density of water, or equivalently, grams per cubic centimeter. To convert to pounds per cubic foot, multiply by 62.3. Densities are averages; 10 percent variations among different samples of the same species are common. The densities are based on oven-dry weight and volume at 12 percent moisture content.

§ Estimates.

Species	Energy Content* (million Btu./cord)	Moisture Content of Green Wood (percent)†		Relative Density or Specific Gravity‡
		Heartwood	Sapwood	
Maple, *continued*				
Black	24.4	..	..	.57
Red	23.1	..	..	.54
Silver	20.1	38	49	.47
Sugar	27.0	39	42	.63
Oak, California black	..	43	43	..
Oak, red:				
Black	26.1	..	..	.61
Cherrybark	29.1	..	..	.68
Laurel	27.0	..	..	.63
Northern red	27.0	44	41	.63
Pin	27.0	..	..	.63
Scarlet	28.7	..	..	.67
Southern red	25.3	45	43	.59
Water	27.0	45	45	.63
Willow	29.6	45	43	.69
Oak, white:				
Bur	27.4	..	..	.64
Chestnut	28.3	..	..	.66
Live	37.7	..	..	.88
Overcup	27.0	..	..	.63
Post	28.7	..	..	.67
Swamp chestnut	28.7	..	..	.67
Swampy white	30.9	..	..	.72
White	29.1	39	44	.68
Sassafras	19.7	..	..	.46
Sweetgum	22.3	44	58	.52
Sycamore, American	21.0	53	54	.49
Tanoak	27.4	..	..	.64§
Tupelo:				
Black	21.4	47	53	.50
Swamp	..	50	52	..
Water	21.4	60	54	.50
Walnut, black	23.6	47	42	.55
Willow, black	16.7	..	..	.39
Yellow-poplar	18.0	45	51	.42
Softwoods				
Baldcypress	19.7	55	63	.46
Cedar:				
Alaska-	18.9	24	62	.44
Atlantic white-	13.7	..	..	.32
Eastern red-	20.1	25	..	.47
Incense-	15.9	29	68	.37
Northern white-	13.3	..	..	.31
Port-Orford	18.4	33	49	.43
Western red-	13.7	37	71	.32
Douglas fir:				
Coast	20.6	27	53	.48
Interior West	21.4	..	..	.50
Interior North	20.6	..	..	.48
Interior South	19.7	..	..	.46
Fir:				
Balsam	15.4	..	..	.36
California red	16.3	..	..	.38
Grand	15.9	48	58	.37
Noble	16.7	25	53	.39
Pacific silver	18.4	35	62	.43
Subalpine	13.7	..	..	.32
White	16.7	49	62	.39
Hemlock:				
Eastern	17.1	49	54	.40
Mountain	19.3	..	..	.45
Western	19.3	46	63	.45
Larch, western	22.3	35	52	.52
Pine:				
Eastern white	15.0	..	..	.35

| Species | Energy Content* (million Btu./cord) | Moisture Content of Green Wood (percent)† | | Relative Density or Specific Gravity‡ |
		Heartwood	Sapwood	
Pine, *continued*				
Jack	18.4	..	..	.43
Loblolly	21.9	25	52	.51
Lodgepole	17.6	29	55	.41
Longleaf	25.3	24	51	.59
Pitch	22.3	..	..	.52
Pond	24.0	..	..	.56
Ponderosa	17.1	29	60	.40
Red	19.7	24	57	.46
Sand	20.6	..	..	.48
Shortleaf	21.9	24	55	.51
Slash	25.3	..	..	.59
Spruce	18.9	..	..	.44
Sugar	15.4	49	69	.36
Virginia	20.6	..	..	.48
Western white	16.3	38	60	.38
Redwood:				
Old-growth	17.1	46	68	.40
Young-growth	15.0	..	..	.35
Spruce:				
Black	17.1	25	56	.40
Engleman	15.0	34	63	.35
Red	17.6	25	56	.41
Sitka	17.1	29	59	.40
White	17.1	25	56	.40
Tamarack	22.7	33	..	.53

Properties of wood species. Different common names are often used for the same species. The scientific names corresponding to the common names used in this table are in the reference below for most species, as well as some alternative common names. Most fruit trees are not included in the above list. Most fruit wood is relatively dense, and is considered to be excellent fuel. Data from Forest Products Laboratory, *Wood Handbook*, Agricultural Handbook No. 72 *(Washington, D.C.: U.S. Department of Agriculture, 1974), Tables 3–3 and 4–2.*

of a stove, under steady burning conditions, does not depend critically on the kind of wood used, assuming equal moisture contents and equal piece sizes. (Temperatures were also measured during some of the experiments, and the maximum temperatures achieved were all about the same.)

TABLE 3-2
COMPOSITION OF WOOD FUELS

Ingredient	Percent
Carbon	50–53
Oxygen	38–43
Hydrogen	6–7
Nitrogen	0.1–0.4
Sulfur	0.01–0.04
Ash	0.3–2

Approximate composition of typical residential wood fuel. Composition varies with species and with tree anatomy. Very pitchy woods, such as pitch pine, fall outside these ranges. The composition of leaves, twigs, and bark also may fall outside these ranges.

Low-density woods tend to ignite more quickly —they can sustain their own flames sooner (Figure 3–5). This makes light woods preferable as kindling, and it means that a stove will heat up more quickly with low-density wood as the fuel. This quickness of ignition may be responsible partly for the impression of a hotter fire. Figure 3–6 indicates that when the flames die out, the mass of wood (mostly charcoal) still remaining is highest for dense woods; dense woods produce relatively more coals. In the case of light woods, more of the charcoal is burned while the gases are still burning.

Although density is the single most important property of wood species that differentiates burning characteristics, there are other factors. Wood containing relatively large amounts of pitch have slightly higher energy contents and probably burn with larger flames. Use of pitchy wood also may result in more creosote. Some wood (such as oak) have relatively concentrated regions in their structure containing pores or open channels

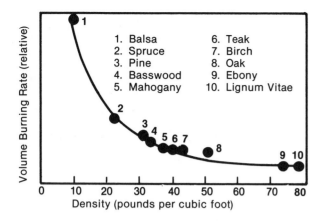

Figure 3-3. *Volume burning rate as a function of density.*[2]

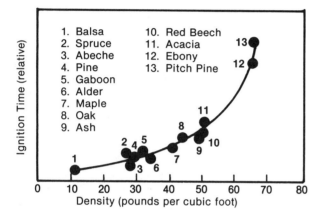

Figure 3-4. *Mass burning rate as a function of density.*[3]

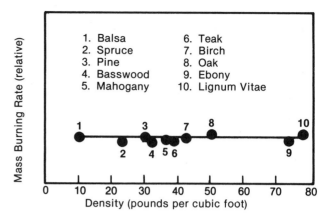

Figure 3-5. *Ignition time as a function of wood density.*[4]

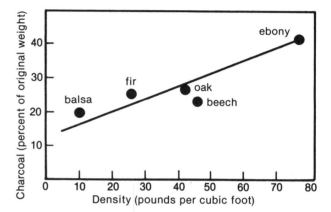

Figure 3-6. *Percentage of original weight left as charcoal after large flames have died out.*[5]

which may affect the way they burn (such wood are said to be ring-porous). The mineral content of wood also may affect burning characteristics.

Some wood burn noisily. Common sounds are hissing and crackling, but wood with structures that are relatively impermeable to gases burn explosively. When heated, gases inside the wood are generated and try to expand. In porous wood, the gases are gradually released as they are generated, sometimes with hissing sounds, but nothing more violent. In impermeable wood, particularly those with structures containing pockets of liq-

uid, the pressure of the gases builds up until the wood ruptures, relieving the pressure, letting out the gases suddenly, and making sharp sounds. Resinous wood tend to burn noisily.

Some people are very choosy about the type of wood they prefer to burn. If you can afford to be choosy, that is fine; there are some differences. But if you have not heated with wood before, or if you do not have a choice, do not worry. All wood types are satisfactory as fuel. Some types burn more quickly, requiring more frequent refueling, some generate longer-lasting coals, some burn more noisily, some require less time to season, and some ignite more easily and are especially suitable as kindling. But no common wood type is truly unsuitable as fuelwood. In practice, people burn whatever is most readily available.

2. L. Metz, "Fire Protection of Wood," *Z. Ver. deut. Ing.* 80 (1963), pp. 660–667; L. Metz, *Holzschutz gegen Feuer* (Berlin: VDI-Verlag, 2nd ed., 1942). Apparent inconsistencies in densities are probably due to variability among the particular samples used. A given tropical wood type may vary in density by as much as 50 percent depending on growing conditions.

3. Ibid.

4. Ibid.

5. Ibid.

Figure 3-7. Equal weights of almost all kinds of wood at the same moisture content have about the same energy content. Zero moisture content is assumed in the figure, but the principle is valid at any moisture content.

DEFINITION OF MOISTURE CONTENT

The standard method for determining the moisture content of wood is to measure the weight loss of the wood when it is dried. Essentially all moisture can be driven off if the wood is placed in an oven at about 212° F. for sufficient time (hours to weeks depending on the size of the piece). When the wood stops losing weight perceptibly, it is called oven-dry and, by definition, has a moisture content of zero percent.

If a piece of wood weighs 10 pounds before oven drying, and 8 pounds after, 2 pounds of water were driven off, and the moisture was 20 percent of its original weight. Its moisture content is thus 20 percent. The moisture content of green wood is typically 30–60 percent. Seasoned wood typically has a moisture content of 15–30 percent.

Unfortunately, there are 2 commonly used definitions of moisture content. The more intuitive and, therefore, the preferred definition was just given. A second definition compares the weight of the moisture to the oven-dry weight of the wood, rather than the original weight of the wood. This leads to the possibility of moisture contents exceeding 100 percent, which does not make much

intuitive sense. The more natural definition – the moist basis moisture content – is used throughout this book. The other definition is called the dry basis moisture content.[6] (For converting from one to the other, see Appendix 2.)

SOURCES OF FUELWOOD

You can obtain fuelwood from a great variety of sources at a great range of prices. The most expensive and most convenient way to get wood is to buy it cut, split, and "seasoned," or air-dried. Prices typically fall between $50 and $250 per standard cord. Green wood is sometimes cheaper. A year is ample for drying (or seasoning). If you have storage space and can buy a year in advance, you can take advantage of the lower prices some sellers offer on green wood.

Sawmill slabs are the wastes from making round, tapered trees into square, straight lumber. Some mills, particularly smaller ones, sell (and in some cases deliver) slabs at a very reasonable cost. Some people prefer these mill wastes because they do not need splitting, and they are inexpensive. However, they usually need cutting because they tend to come in long lengths (typically 8–16 feet). Hardwood wastes, which some people prefer, are less common than softwood wastes. Prices, of course, may rise and availability decline as demand grows, especially the demand for chips. Increasingly, mills sell or use all their wood wastes. Many mills burn their wastes to heat their buildings and dry their lumber. Chips are also used to make particle board. The older practice of burning wastes just to make disposal easier is declining due to air pollution laws and the rising value of mill residues.

If you are equipped and willing to haul wood,

---

6. In *The Woodburners Encyclopedia*, dry basis moisture contents were used. I now feel it is easier (more intuitive) to use the moist basis.

Seasoned Wood   Green Wood

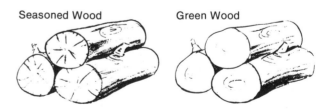

Figure 3-7. Seasoned wood is usually checked. It has cracks in the ends.

Determining the Moisture Content
Of Your Wood

1. Select a typical piece of wood. This is important. Small pieces are usually drier than large ones. Pieces on the top are usually drier than pieces at the bottom of the pile. Select a typical size piece located 1–2 feet below the top of the pile. If the pile has different species, or wood cut at different times, select a typical piece of each.

2. Cut the piece into slabs about 2 inches thick, or split the piece into small sticks about 1 inch in diameter. These dimensions are not critical. You are trying to decrease the drying time by increasing the surface area.

3. Clean off any sawdust or loose bark. Then weigh the pieces.

4. Put the pieces on a tray in the oven at about 212° F. The exact temperature is not critical.

5. Leave in the hot oven for 6 hours.

6. Reweigh the wood. Subtract the dry weight from the original weight to obtain the amount of moisture the wood had in it. Divide this weight of moisture by the original weight and then multiply by 100 to obtain the percent of moisture content.

The time necessary to completely dry the wood varies considerably; piece size, original moisture content, and species can be important factors. To verify that the wood is thoroughly dry, you can put it back in the oven for another hour or 2 to see if any more moisture is driven off.

It is important to weigh the wood as soon as it is taken out of the oven; oven-dry wood *absorbs* moisture out of the air fairly quickly.

Using only 1 or 2 slabs or sticks from the piece of wood may give misleading results since wood pieces are drier on the outside than the inside. For the same reason, electronic moisture gauges can give misleading results.

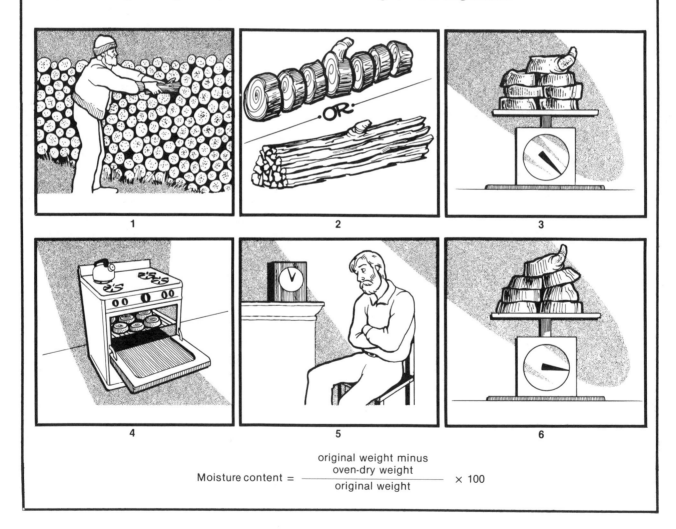

$$\text{Moisture content} = \frac{\text{original weight minus oven-dry weight}}{\text{original weight}} \times 100$$

and to cut and split it, local dumps, landfills, and construction sites can be sources of already felled and trimmed wood. Landscapers and arborists may be happy to dump their wastes at your home. Neighbors who do their own landscaping may give you felled trees. Both electric and telephone companies clear trees and large branches from their overhead lines, and often leave usable fuelwood under the lines for the public. Demolition sites can be a source of very dry, and sometimes dirty, lumber. Road construction and improvement projects sometimes have felled and limbed trees which need disposal, and many state and national forests have programs permitting cutting of selected trees by private citizens.

Some loggers, particularly the smaller ones, will deliver a logging truckload of green full-length logs to residences. This wood has little or no value as timber or pulpwood; the species are often mixed, the logs crooked and often contain knots and some rot. But all wood burns, and sometimes this is one of the cheapest ways, if not *the* cheapest way, to get delivered fuel. The hardest job has been done—getting the wood out of the woods. Cutting and splitting can be done as time allows at home. However, be warned that the driver's estimate of how many cords are delivered may not coincide with reality, and you will not know you have been misled until the wood is cut, split, and stacked. Overestimates by a factor of 2 are common.

Figure 3-8. A logging truck delivery.

Good and free advice on how to manage your woodlot is usually available from state extension services.

MEASURING FUELWOOD

Fuelwood is usually measured by the *cord* (or standard cord), which is a 128-cubic-foot pile of neatly stacked (parallel) wood (Figure 3-9). The 128-cubic-foot volume includes some air between the pieces of wood. The actual volume of solid wood can range from 60 cubic feet to 100 cubic feet depending on both the shape of pieces and how they are stacked. If the pieces of wood are crooked and bumpy, the air spaces will be relatively large, and the cord will contain less solid wood. If straight pieces of a variety of diameters and shapes are used, much more wood can be piled into a cord, if care is taken to use the smaller pieces to fill in the spaces between the larger, as in a carefully constructed stone wall (Figure 3-10).

There is another older definition of a cord which essentially is being legislated away. This definition specifies 4-foot lengths for the logs—not that they must be 4 feet long when sold, but that when they were 4 feet long, they occupied 128 cubic feet. This is not splitting hairs. When a 128-cubic-foot pile of 4-foot lengths is cut into shorter lengths and restacked, a 25 percent volume reduction is possible. Some wood is lost as sawdust. More important, shorter pieces generally fit closer together, by an amount dependent on how crooked and irregular the 4-foot pieces were.

This is a legitimate reason why some fuelwood sellers claim that a cord is "really" 96 cubic feet or some other number less than 128. This leads to confusion since the actual volume of what was originally 128 cubic feet depends on many variables. It generates ill feelings since the purchaser has no way to verify that the original pile of 4-foot lengths was a full cord.

A number of states have passed laws redefining a cord to be 128 cubic feet of neatly parallel-stacked wood *as delivered*, whatever length that might be. This is the best definition. Buyers can verify the amount of wood themselves. Sellers then have 2 legitimate reasons for charging more for a cord of short pieces: a cord of shorter pieces contains more actual solid wood; and more labor is required to cut, split, load, unload, and stack shorter pieces.

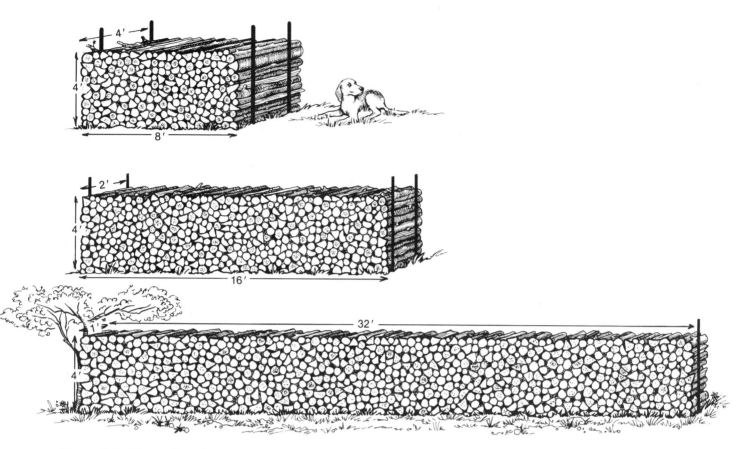

Figure 3-9. Three piles of wood, each of which is standard cord.

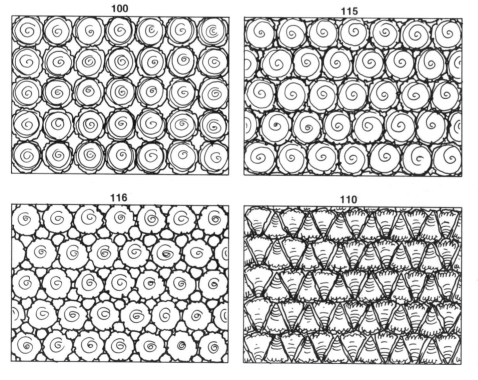

Figure 3-10. How wood is stacked affects the amount of actual solid wood in a cord. The numbers above each sketch are the relative amounts of wood in each pile and very nearly the number of cubic feet of solid wood in a cord stacked in that manner. With real wood, stacking is always less dense than in these idealized cases.

Some states have defined the cord to be less than 128 cubic feet for pieces less than 4 feet long. It does not really matter as long as both buyer and seller are using the same definition.

A *face cord* (Figure 3–11) is the amount of wood in a pile 4 feet high and 8 feet long, period. The width of the pile (or length of the pieces) can be anything. A face cord of 2-foot-long pieces (a 2-foot face cord) would be a pile 4 feet by 8 feet by 2 feet and would be half a cord.

A *thrown cord* is a 128-cubic-foot pile of randomly oriented wood. The Maine Energy Office estimates that it takes 160–190 cubic feet of such loosely piled wood to be equivalent to a standard cord. It is not a recommended measure. A normal pickup truckload of randomly piled fuelwood usually holds roughly half of a standard cord.

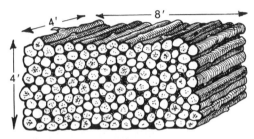

A Face Cord of 4-Foot Logs
A Standard Cord

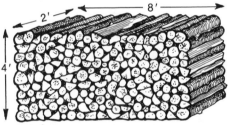

A Face Cord of 2-Foot Logs
Half a Standard Cord

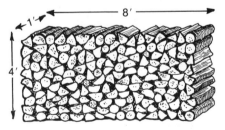

A Face Cord of 1-Foot Logs
A Quarter of a Standard Cord

Figure 3–11. Three piles of wood, each of which is a face cord, but each contains a different amount of wood.

Measuring Your Own Woodpile

If you would like to know how much wood you burn, or if you want to check whether the cord of wood you just bought really is a full cord, here is how to measure your woodpile.

First, neatly stack your wood, with the pieces all parallel to each other and the ends of the pieces aligned with each other as well as possible. It is convenient, though not necessary, to have the ends of the pile be vertical rather than sloped.

Use a tape measure to determine the height and length of the pile and the average length of the pieces. Multiply these 3 numbers together to get the volume of the pile, and compare this volume to 128 cubic feet to obtain the number of cords.

It is most convenient to work in feet. If your tape measure is not marked in feet and decimals of feet, divide the number of inches by 12 to obtain the dimension in feet. Then multiply the 3 dimensions together to get the volume in cubic feet. Finally, divide by 128 for the number of cords.

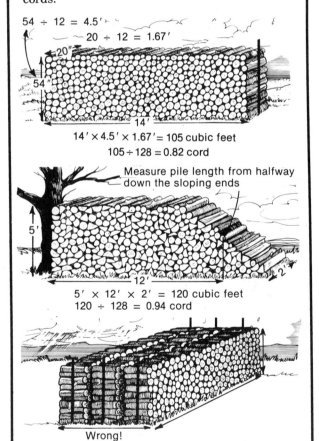

$54 \div 12 = 4.5'$
$20 \div 12 = 1.67'$

$14' \times 4.5' \times 1.67' = 105$ cubic feet
$105 \div 128 = 0.82$ cord

Measure pile length from halfway down the sloping ends

$5' \times 12' \times 2' = 120$ cubic feet
$120 \div 128 = 0.94$ cord

Wrong!
Close parallel rows inevitably have space between them. Use 3 times the average log length rather than the full width of the three rows.

STORING WOOD

Freshly cut wood from a living tree is said to be *green*. After it is cut, green wood dries (or seasons) quickly at first, and then at a slower and slower rate. Because wood is hygroscopic (it attracts water), it never dries to zero moisture content in any normal storage situation. Rather, its moisture content gradually approaches an equilibrium value which depends on the storage conditions. When wood is covered, the relative humidity of the surrounding air is the main determining factor (see Table 3-3). In the Southwest in the summer, the moisture content of wood stored outdoors may be as low as 6 percent; in parts of the Northwest in winter, the moisture content of (covered) seasoned wood cannot be much less than 18 percent.

The term *seasoned* has no precise meaning when applied to fuelwood, although it means some degree of dryness compared to green wood. If the term had the same meaning as it does when applied to paper, seasoned wood would be wood which is as dry as it can get, given its surroundings (Table 3-3).

The *time* it takes for wood to season depends principally on the size of the pieces, on how freely air can circulate around the pieces, and on the temperature. Splitting accelerates drying; the smaller the pieces, the faster they will dry. Air circulation is better in free-standing woodpiles than in piles next to buildings. Also, if rows are placed immediately next to each other, air will circulate less freely than if they are spaced apart a few feet. Piling wood with alternating directions (criss-crossed) lets more air through. Direct exposure to sunlight helps speed drying by its warming effect, as does indoor storage in a heated room.

Generally, firewood will be very close to its equilibrium moisture content (as dry as it can get) after 2 years of seasoning. Drying is fastest at the beginning; after 6 months of drying, wood is often called seasoned, and usually has an acceptably low moisture content of 25 percent or less. If you cut and split your firewood in the spring, it will usually be ready to use the next winter.

Covering wood can decrease both the final moisture content and the drying time, particularly in climates with relatively heavy precipitation. A shed without tight walls, or with no walls, is excellent, since it protects against precipitation but does not inhibit air flow. However, in very dry climates, a shed may not help because its shading of the wood may offset its protection against the small amounts of precipitation. A plastic sheet that completely covers a woodpile down to the ground is usually worse than no cover since it inhibits circulation and tends to hold moisture in, including moisture coming out of the ground (which can be considerable).

One simple way to achieve some protection against precipitation is with a cover made of 2-3-foot-wide strips of plastic, fastened along their long edges to poles or small boards. The weight of the poles hanging over the sides keeps the plastic strips in place on top of the pile.

Ideally, woodpiles should be kept a foot or more above the ground; concrete blocks can be used for this purpose. Not only is the ground itself often damp in most climates, but the humidity of the air close to the ground is also usually higher than the humidity in the rest of the air.

TABLE 3-3
EQUILIBRIUM MOISTURE CONTENT OF WOOD

Relative Humidity (percent)	Equilibrium Moisture Content (percent)
5	1.2–1.4
10	2.3–2.6
20	4.3–4.6
30	5.9–6.3
40	7.4–7.9
50	8.9–9.5
60	10.5–11.3
70	12.6–13.5
80	15.4–16.5
90	19.8–20.0
98	26.0–26.9

Equilibrium moisture content of wood as a function of the relative humidity of the surrounding air. The actual moisture content of very thin pieces of wood varies daily and even hourly as the relative humidity of the air changes. The moisture content of wood logs varies much more slowly, following monthly and seasonal changes in average relative humidity. There is a slight temperature dependence in the range of typical ambient temperatures; equilibrium moisture contents are 5–10 percent (not percentage points) lower at 90° F. than at 30° F. for the same relative humidity. Data from Forest Products Laboratory, Wood Handbook, *Agricultural Handbook No. 72 (Washington, D. C.: U.S. Department of Agriculture, 1974), pp. 3–8.*

Storing wood in a heated space, such as a heated basement or garage, will speed up the drying process. The equilibrium moisture content will be lowered, particularly if the air is not humidified. Wood stored indoors in a heated space in New England will have a moisture content of less than 10 percent after a year or 2 because the air is very dry in a heated but unhumidified structure in cold weather. However, indoor storage is not always advisable. Insects that are detrimental to wood buildings, such as termites, may be living in the fuelwood and may spread to the house. State extension services or the U.S. Department of Agriculture should be able to give advice on the advisability of indoor storage for particular regions of the country.

Wood can get *too* dry. Wood drier than 15 percent can burn with a lower energy efficiency and produce more creosote than wood with a 20–25 percent moisture content (see chapter 14).

If stored for too long a time, wood may rot. Rotting is caused by the growth of fungi; the fungi convert the wood into water, carbon dioxide, and heat, just as a fire does. Thus, rotting decreases the energy of wood. The fungi are most productive when the temperature is between 60° F. and 90° F., the moisture content of the wood is above 30 percent, and ample oxygen is available.

Virtually no rotting takes place in cold winters because the temperature is too low for the fungi. Wood in the interior of a house does not rot because it is too dry. Wood kept continuously underwater seldom rots, partly because of a lack of oxygen, and partly due to the usually cool temperature. On the other hand, wood that is lying on the ground, especially in summer, usually provides a productive environment for fungi because the ground keeps the wood sufficiently moist. Rot will appear first at the bottom of a woodpile placed directly on the ground. The most practical way to retard rotting is to dry fuelwood quickly and to keep it dry, especially during the warm summer months when most decay occurs.

Woods differ in their natural resistance to decay. The sapwood of virtually all species is highly susceptible to decay, but there is a considerable difference among heartwoods, as indicated in Table 3–4. Given the choice, the more susceptible species should be burned first so that less of their energy content will be dissipated by the fungi. The most susceptible wood can decay significantly in 2 years of outdoor, uncovered exposure in climates with warm, humid summers.

TABLE 3-4
DECAY RESISTANCE OF HEARTWOOD

Most Resistant	Highly Resistant	Moderately Resistant	Least Resistant
Locust, black	Baldcypress	Douglas fir	Alder
Mulberry, red	Catalpa	Honeylocust	Ashes
Osage-orange	Cedars	Larch, western	Aspens
Yew, Pacific	Cherry, black	Oak, swamp chestnut	Basswood
	Chestnut	Pine, eastern white	Beech
	Cypress, Arizona	Southern Pine:	Birches
	Junipers	Longleaf	Buckeye
	Mesquite	Slash	Butternut
	Oak:	Tamarack	Cottonwood
	Bur		Elms
	Chestnut		Hackberry
	Gambel		Hemlocks
	Oregon white		Hickories
	Post		Magnolia
	White		Maples
	Redwood		Oaks (red and black species)
	Sassafras		Pines (other than longleaf, slash, and eastern white)
	Walnut, black		Poplars
			Spruces
			Sweetgum
			True firs (western and eastern)
			Willows
			Yellow-poplar

Relative resistance of the heartwood of various species to decay. The sapwood of all species has little resistance to decay. From U.S. Forest Products Laboratory, Wood Handbook, *Agriculture Handbook No. 72 (Washington, D.C.: U.S. Department of Agriculture, 1974), pp. 3–17.*

PROCESSED WOOD FUEL

Sawdust, wood chips, densified wood, and newspaper logs are all fuels based on wood. Sawdust and wood chips are rare as residential fuels, although there used to be thousands of sawdust-burning heaters years ago in the Northwest. These fuels are not very practical or safe to use except in appliances designed for them.

Densified or pelletized fuel is highly compressed sawdust, chips, and sometimes other organic waste material. The compression results in a product roughly 2–3 times denser than natural wood. It is produced as briquettes or as solid cylinders with diameters ranging from a fraction of an inch up to about 4 inches (Figure 3–12). The

Figure 3-13. A densified-wood log.

Figure 3-14. A sawdust wax log.

larger diameter size often is used as a fireplace fuel.

Wood pellets emit less sulfur when burned compared to coal. Pelletized wood also has a lower and more uniform moisture content compared to most other forms of wood, which is particularly advantageous in industrial applications. Coal burning systems often can be switched to pellets with very little modification since the pellets can be made to have the same size as coal.

The residential use of pelletized wood is not as easy. The equipment must be designed for this fuel, or at least for coal. Because the fuel is dense, it can burn with a very hot fuel bed, as coal does. This can damage a stove not designed for it. The low moisture content can result in poor combustion efficiency when the pellets are burned in typical airtight equipment. In a well-designed automatic stoker appliance, the dryness and density of the fuel can become an advantage rather than a possible liability. Then the potential advantages are automatic operation of the whole system, low emissions, low creosote, and high efficiency. However, such a system clearly will be relatively expensive, will not operate without electricity, and requires regular fuel deliveries. Thus, some of the simplicity and reliability of a wood stove is lost.

Normally no binder is necessary to hold the wood together in pellet form. But if the pellets get wet, or even if the air gets very humid, the pellets may fall apart and expand substantially. Binders can be added, but this increases the cost.

The interest in making wood pellets is high, but the process itself consumes energy and increases the cost of the fuel significantly. It remains to be seen under what circumstances the advantages are substantial enough to be worth it. Waste paper, garbage, and agricultural wastes also can be made into fuel pellets.

Wax-sawdust logs (Figure 3-13) consist of sawdust and/or small wood chips bound together with paraffin. The energy content is high because of the wax, but so is the cost. The logs generally are recommended only for use in open fireplaces and only 1 at a time. The wax content can generate considerable soot. But the logs are clean to store and handle and need no paper or kindling for ignition.

Homemade newspaper logs are more talked about than used. In some cases, paper is just rolled into log form and bound with wire or string. Sometimes soaking the paper (or rolled log) in a detergent solution or other materials is part of the process, followed by drying or "seasoning."

Very few users find they can burn newspaper logs alone—a 50/50 mix with ordinary wood logs is common. The main issue is whether or not making the logs and trying to keep them burning is worth the effort, and this must be a personal decision.[7]

Glossy paper magazines are rarely practical as fuel. The gloss comes from a thin coating of very fine clay, which makes it harder for the paper to burn and results in a large ash residue.

Charcoal and Coke

Charcoal (derived from wood) and coke (derived from coal) are relatively clean-burning manufactured fuels. The wood or coal is pyrolyzed (heated but not allowed to burn) in the manufacturing process. The smoke is driven off and used for other purposes, such as chemicals or gaseous and liquid fuels. The remaining char is mostly carbon and can only become carbon dioxide or carbon monoxide when burned. Hence, creosote cannot form, and neither can organic emissions. The fuels are said to be "smokeless." Since natural charcoal and coke can be physically fragile and not very dense, they are often compressed and sometimes combined with other ingredients to make a denser, easier-to-handle fuel.

These smokeless fuels could be very important should strict emissions regulations come into being. However, without such regulations their high cost will restrict their use.

Charcoal-based and coke-based fuels in their typical forms have some of the same properties as coal and thus should be used only in appliances designed for coal or densified biomass pellets. The fuel bed can get very hot; a grate may be necessary to get enough air into the fuel bed, and, in the case of coke, considerable ash may be produced, which could clog some kinds of grates.

---

7. One reason newspaper logs are hard to keep burning is their layered structure. Sustaining vigorous flaming combustion requires that heat from the flames and glowing outer surface penetrate the log; this heat then pyrolyzes more of the fuel, thus generating more gases to sustain the flames. The small air spaces between sheets of newspaper inhibit heat flow. Thus, combustion does not easily penetrate to the center of a newspaper log. The glowing combustion of the newspaper char is similarly inhibited. The soaking can help by "gluing" the sheets together, either with an added binder or just by making the paper mushy so that the sheets lose their definition and the dry log becomes more nearly a homogeneous mass of wood pulp.

ASH

Wood ash is all the solid residue from the complete combustion of wood. Both the total amount of ash and its composition depend not only on wood species, but also on the amount of bark included, on the ratio of heartwood to sapwood, and whether the wood came from the trunk or branches of the tree. Ash content typically is between 0.3 percent and 2 percent. Ash content is rarely an important consideration in choosing firewood types.

The composition of wood ash is indicated in Table 3-5. Wood ash is useful as a fertilizer, particularly because of its potassium content. Potassium carbonate can be extracted from wood ash and used in making soap. Wet ash is caustic, as is lye, and can be used as a degreasing cleanser. Ashes will decrease the acidity of garden soil. Areas with high rainfall tend to have acid soils. But in dry regions, soils tend to be alkaline, so adding ashes can exacerbate an existing problem. Plants vary in their needs, and the composition of the soil is also important. If in doubt, consult an extension agent about the advisability of using ashes on your soils.

TABLE 3-5
COMPOSITION OF WOOD ASH

Element	Assumed Form	Typical Quantity (percent of total ash)
Calcium	CaO	30–60
Potassium	K_2O	10–30
Sodium	Na_2O	2–15
Magnesium	MgO	5–10
Iron	Fe_2O_3	1– 2
Silicon	SiO_2	2– 7
Phosphorus	P_2O_5	5–15
Sulfur	SO_3	1– 5

Composition of wood ash. These figures are determined under conditions which ensure complete combustion and retention of all ash generated. The percentage composition is calculated assuming the elements are in the stated chemical form; the actual chemical forms may be different. Data interpreted from L. E. Wise, ed., Wood Chemistry *(New York: Reinhold Publishing Corporation, 1944), p. 435.*

CHAPTER 4

COAL AS A FUEL

Heating with coal is more complicated than heating with wood. Virtually any kind of wood can be burned easily in any ordinary wood heating appliance. Coal is not as versatile. It may not be practical to use a coal appliance at all with the wrong coal. Unlike with wood, most coal variables are not controlled by the person fueling the stove.

ORIGIN OF COAL

Coal started as plant matter in dense forests and lush swamps and marshes. Plant debris—wood, bark, leaves, ferns, mosses, seeds, and pollen—accumulated in the ground or under water. New plants grew in the decaying remains of the old. Gradually, substantial layers of organic material built up. The process is slow on a human time scale. It is estimated that the organic material builds up at a rate of roughly 1 foot per century.

Peat is the first step in the transformation of plants to coal. Peat is slightly decayed plant debris, with recognizable undecayed plant parts still remaining. Peat usually has a very high moisture content: 40–50 percent on the moist basis, wetter than many types of green wood. It must be dried before burning.

Over geological periods of millions of years, peat deposits become buried under layers of mud, sand, and more organic material. The resulting pressure and heat transforms the peat into coal. Roughly 3 feet of peat are necessary to make 1 foot of coal. The material generally becomes denser, drier, and higher in carbon content (with an accompanying loss of hydrogen and oxygen).

The type of coal formed depends mostly on the pressure, temperature, and time involved. This is the progression of transformations:

vegetation → peat → lignite → sub-bituminous → bituminous → anthracite.

Lignitic coals are the least transformed (least metamorphosed) from peat, and anthracite is the most changed. Anthracite is generally found deeper in the earth, which is consistent with its requiring higher temperatures and/or pressures for its formation. The more metamorphosed coals are said to be of *high rank*; those less transformed from peat (lignite and sub-bituminous) are called *low rank coals*.

Some of the properties of the various types of coals are indicated in Figures 4-1 and 4-2, and Table 4-1.

PHYSICAL PROPERTIES OF COAL

If you were to buy cordwood, split and cut to length, you would probably ask the dealer 3 questions. What species of wood are you selling? Is the wood green or seasoned? What lengths are the pieces?

Coal is another story. There are many physical properties of coal that influence its efficiency as a home heating fuel. If you already own a coal stove, buy the type and size coal recommended by your stove manufacturer. Beyond that, your

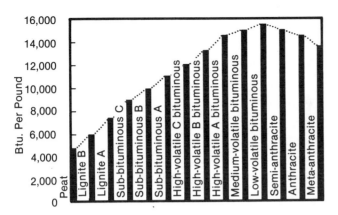

Figure 4-1. Higher heating values of coals and peat on a moist, mineral-matter-free basis.

choice of coals will be limited by what is available in your area. Still, you can shop around to determine where you can purchase coals with the highest heating values, the appropriate volatile content, the least ash content, and the highest ash-fusion temperature.

Some dealers will provide analyses of the physical properties of the coals they sell.

Heating Value

The higher heating value of coal ranges from less than 6,000 Btu. to more than 14,000 Btu. per pound as delivered. The coals most often used in home heating range from about 11,000 Btu. to 14,000 Btu. per pound. On the dry basis, where the heating value of wood is about 8,600 Btu. per pound, coal produces about 12,000–15,000 Btu. per pound. Combined with the fact that coal has a higher weight density, this means that coal has much more energy in a given volume than wood. Coal can be expected to last longer at the same heat output rate compared to wood loaded into the same size fuel chamber. Enough coal to last a heating season needs less storage space than a season's worth of wood.

Volatile Matter and Fixed Carbon Contents

When coal is heated, some of the coal is transformed into gases and vapors which leave the coal. This process is called pyrolysis or thermal

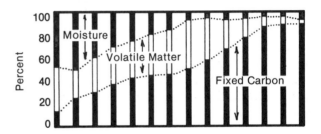

Figure 4-2. Proximate analysis of coals and peat on a moist, mineral-matter-free basis.

TABLE 4-1
CLASSIFICATION OF COALS BY RANK

Class	Group	Fixed Carbon Limits. Percent (dry, mineral-matter-free basis)	Volatile Matter Limits. Percent (dry, mineral-matter-free basis)	Calorific Value Limits. Btu. per Pound (moist, mineral-matter-free-basis)	Agglomerating Character
I. Anthracitic	1. Meta-anthracite	98–100	0– 2	... ...	nonagglomerating
	2. Anthracite	92– 98	2– 8	... ...	
	3. Semianthracite	86– 92	8–14	... ...	
II. Bituminous	1. Low volatile bituminous coal	78– 86	14–22	... ...	
	2. Medium volatile bituminous coal	69– 78	22–31	... ...	commonly agglomerating
	3. High volatile A bituminous coal	Less than 69	31 or higher	14,000 or higher	
	4. High volatile B bituminous coal	... ...	... ...	13,000–14,000	
	5. High volatile C bituminous coal	... ...	... ...	11,500–13,000	
				10,500–11,500	agglomerating
III. Sub-bituminous	1. Sub-bituminous A coal	... ...	... ...	10,500–11,500	
	2. Sub-bituminous B coal	... ...	... ...	9,500–10,500	
	3. Sub-bituminous C coal	... ...	... ...	8,300– 9,500	nonagglomerating
	1. Lignite A	... ...	... ...	6,300– 8,300	
	2. Lignite B	... ...	... ...	Less than 6,300	

Adapted from ASTM Standard D388-77, American Society for Testing and Materials, Philadelphia.

Bases for Evaluating Coal

As is true for wood, there are different bases used when discussing certain properties of coal. One dimension is identical to that for wood – the dry versus the moist basis. For example, higher heating values reported on the moist basis are the higher heat content of a sample divided by the weight of the sample, *including* the weight of its moisture content. On the dry basis, one divides by the oven-dry weight of the sample. Both bases are in common use within the coal and wood industries. A conversion table for getting from one to the other basis is in Appendix 3. In this book the moist basis is used throughout unless otherwise indicated.

For coal, there is another dimension involved in choosing a basis. The ash content of coal is much larger than that of wood – roughly 5-15 percent of coal, as compared to usually less than 1 percent of wood. The ash content is also highly variable, even within coal of a particular type. Since ash is considered less critical than some other properties (such as rank), and since its amount is so variable, it is often clearer and simpler to use an ash-free basis when discussing these other properties. Most discussions of coal properties are on the mineral-matter-free basis (approximately the same as the ash-free basis), such as the data in Figures 4-1 and 4-2.

This is not the most logical basis for the home-owner, whose coal is delivered in its natural state, including the mineral content. Making the crude assumption that your coal has an ash content of 10 percent, and if its higher heating value is 12,000 Btu. per pound on the ash-free basis, then its heating value is 10 percent less, or 10,800 Btu. per pound on the ash-included or as-delivered basis. The heating value decreases since part of each pound is ash, which has zero heating value. (The as-delivered basis includes both the ash and moisture contents in the denominator.)

(Mineral matter content and ash content are not precisely the same concept since the mineral matter is altered somewhat as it becomes ash during combustion. However, they are sufficiently close that they are treated as identical in this book.)

decomposition and happens whenever coal or wood burns.

The *volatile matter content* of coal is defined as the percentage of weight loss (other than that attributable to the moisture content) that occurs when a small sample is heated quickly to a temperature of about 1,740° F. without oxygen being present. It represents, approximately, the portion of the fuel that will burn in the gaseous state, as flames. What remains after this heating is mostly carbon, and its percent of the original weight of the sample is called the *fixed carbon content*. It represents, approximately, that portion of the fuel which will be burned in the solid state, as glowing coke.

Anthracite has only up to 10 percent volatile matter, while most bituminous coals have 30-40 percent volatile matter. Wood has a volatile matter content of about 80 percent when measured the same way (very quick heating), but the slower heating of logs in an actual fire results in substantially less volatile matter being generated (see chapter 3).

The volatile content of fuels affects appliance efficiency, creosote accumulation, and ease of ignition. Anthracite burns with only short flames, more like charcoal, because of its low volatile content. Wood and bituminous coals burn with much longer flames. To achieve maximum combustion efficiency, fuels with long flame lengths must be burned in appliances designed to provide adequate oxygen, good mixing of the oxygen with gases, and a hot environment over the entire volume where the gases will be burning.

Coal with high volatile matter content has greater potential for generating large flue deposits. Flue deposits from coal tend to be soft and sooty or cobweblike, rather than tarry and hard as wood deposits are. Anthracite flue deposits are often red brown or gray, indicating the presence of ash and relatively little combustible material.

High volatile content makes ignition easier, with wood as well as coal. As soon as the fuel is hot enough to release significant amounts of volatiles, the volatiles can burn by themselves. The flaming volatiles then heat the original piece of fuel as well as others, generating more gases, and a self-sustaining fire is achieved.

With low volatile fuels, such as anthracite, charcoal, and coke, flaming is so weak that the necessary mutual heating of fuel pieces must be supplied by glowing combustion of a number of nearby pieces of fuel. To heat up this larger mass of fuel requires more heat. Thus, ignition is more difficult.

There is no single optimum volatile content. But, some coal stoves are designed to burn efficiently with high volatile coals, others with low volatile fuels. Follow your stove manufacturer's suggestions concerning the type of coal that will burn best in your stove.

Moisture Content

The moisture content of all natural (in-the-ground) coals is much less than that of green wood, and usually less than that of seasoned wood. Only sub-bituminous and lignite coals have moisture contents greater than about 20 percent on the moist basis. (Some coals have moisture contents of less than 2 percent.)

Drying, or "seasoning," coal usually is not necessary. What is more, the moisture content of anthracite and bituminous coals cannot be changed very much or very quickly by drying; coal is relatively impervious to moisture exchange, and also is not stored easily in a well-aerated situation (and it shouldn't be, as discussed later).

If coal is rained on, it can get wet on the surface, but not much water penetrates. Wood, on the other hand, behaves more like a sponge; considerable moisture can be absorbed into wood.

Despite all this, it is still best to store coal in a dry, covered place. The coal will burn somewhat more easily, and other potentially more serious problems will be avoided.

Kinds of Moisture in Coal

The moisture in coal is often divided into 2 categories. The *inherent,* or "bed," moisture is the moisture content when the coal is in moisture equilibrium at 96-97 percent relative humidity and at 86° F. This is approximately the moisture content of the coal in its seam underground if the seam is not wet.

Extraneous moisture is wet, surface moisture, due to ground water or to the use of water in mining, processing, or transporting the coal.

Actual moisture content can be less than the inherent moisture content if the coal has no extraneous moisture, and if storage conditions have permitted some of the inherent moisture to evaporate.

Ash Content and Composition

Ash is the noncombustible solid residue remaining after complete combustion of a fuel. The ash content of coal is very high, and very variable, even within the same type of coal. Ash content of coals typically used in home heating varies from less than 5 percent to over 15 percent, which is far greater than the ash content of wood.

The inherent mineral (noncombustible) content of pure plant matter, such as wood, is small, usually less than 1 percent. Although coal is derived from plant matter, most of its mineral content is a combination of clay, sand, and dust deposited with the vegetative matter in the original swamp or forest (carried there by the wind or by water), plus animal remains and minerals, which moved into the coal seam from neighboring seams of rock.

Ash is generally a liability, and it accumulates much more quickly in a coal stove than in a wood burning stove, making frequent ash removal necessary. Unfortunately, coal ash is not generally beneficial to add to garden soil. Unlike wood ash, its potassium and phosphorus content is very low. (Potassium, phosphorus, and nitrogen are the 3 most important plant nutrients.) Coal ash also contains small amounts of elements such as beryllium, boron, cobalt, germanium, arsenic, and even some radioactive elements, such as uranium — some of which are toxic to plants and people.

Also, ash can trap unburned carbon; the amount of unburned fuel thrown away with the ashes tends to be higher for coals with higher ash content.

Agglomerating or Caking Character

When heating up in a fire, some coals have a tendency to fuse together. The coal softens and swells, and individual pieces become joined together as one. This "caking" of the coal can inhibit air flow and combustion, and it can cause "bridging," preventing the coal from gravity feeding into the primary combustion zone.

Of the 4 major classes of coal, only bituminous tends to be agglomerating (Table 4–1), although some bituminous coals are not. Anthracite does not tend to agglomerate.

Some coal heaters are designed for agglomerating coals, but others are not. Agglomerating coal can cause great difficulty in an appliance not designed for it.

TABLE 4-2
BITUMINOUS ASH COMPOSITION

Component	Percent
Silica, SiO_2	20–60
Aluminum oxide, Al_2O_3	10–35
Ferric oxide, Fe_2O_3	5–35
Calcium oxide, CaO	1–20
Magnesium oxide, MgO	0.3–4
Sodium oxide, Na_2O	0.2–3
Potassium oxide, K_2O	0.2–4
Titanium oxide, TiO_2	0.5–2.5
Phosphorus pentoxide, P_2O_5	0.0–3
Sulfur trioxide, SO_3	0.1–12

Typical limits of ash composition for bituminous coal. As is traditional, ash components are all reported as oxides. In practice, oxides predominate, but other chemical forms are also present. Adapted from W. A. Selvig and F. H. Gibson, Analyses of Ash from United States Coals, *Bureau of Mines Bulletin 567 (1956).*

The agglomerating character of coals is given by their "free swelling index," a number from 1 to 9, where a value of 1 indicates no agglomerating tendency and 9 indicates maximum agglomeration can be expected.

Coals which do not tend to agglomerate (or cake) are said to be *free-burning*.

Ash Fusion Temperature

As coal ash is heated, it becomes soft and sticky. At sufficiently high temperatures, it becomes fluid. (The fluid stage is rarely reached in residential coal heaters.)

This softening and fusing together of the ash is not desirable. It results in the formation of *clinkers*, strong agglomerations of ash that are too large to fall between the grates. As the clinkers build up, they and smaller ash particles will impede air flow, and the fire may go out.

The tendency of coal types to form clinkers is usually indicated by the *ash softening temperature*. As coal ash is heated, it passes through various stages on the way to becoming fully fluid; these stages are called the *initial deformation temperature*, the *softening* (or spherical softening) *temperature*, the *hemispherical softening temperature*, and finally the *fluid temperature*. Ash does not have just 1 well-defined melting temperature, as do most pure substances, partly because it is a mixture of different compounds.

These temperatures depend primarily on the composition of the ash. And since both the amount and composition of ash varies significantly from seam to seam even within the same type of coal, the ash fusion characteristic of coals is an independent parameter.

In comparing coals to each other, be careful to use the same fusibility concept. The ash softening temperature is the most commonly used measure of fusibility. A softening temperature of 2,000° F. is low, 2,300° F. is about average, and 2,700° F. is high.

Wood ash can also fuse, but it rarely occurs in wood heaters, because wood ash has a higher softening temperature and wood fuel beds tend not to be as hot.

Sulfur Content

Some coals contain very large amounts of sulfur—more than 5 percent. For most coals, the sulfur content ranges from 0.5 percent to 4 percent. (The sulfur content of wood is typically less than 0.1 percent.)

Sulfur is a serious air pollutant (in the forms SO_2, SO_3, and H_2SO_4). These sulfur compounds aggravate respiratory ailments, cause damage to agricultural crops, cause corrosion of almost everything—metal, stone, concrete, and clothing—and contribute to acid rain. Also, sulfur imparts an unpleasant odor to coal fumes; SO_2 is part of what makes rotten eggs smell rotten. Finally, sulfur can be very corrosive to the appliance and its flue. Sulfuric acid (H_2SO_4) as a vapor is not corrosive; but if it condenses on the flue wall, the corrosion potential is substantial. Coal also contains chlorine, which becomes hydrochloric acid, which is also extremely corrosive.

Sulfur occurs in coal principally in 2 forms. "Organic" sulfur usually accounts for 30–70 percent of the total sulfur. It is part of the organic material of the coal itself and when burned generally becomes SO_2 or SO_3. Most of the remaining sulfur is in the form of iron pyrite (FeS_2), which is fool's gold, a mineral. When burned, the sulfur also becomes SO_2 and SO_3, and the iron becomes iron oxide in the ash. Since much of the iron pyrite is in the form of relatively large pieces of pure mineral, which is distinct from and much denser than the coal itself, it is possible for some of it to be removed from the coal by mechanical means before combustion. Some coal is partly desulfured in this manner before being burned in some power plants.

Piece Size

Most coal burning equipment is quite sensitive to the piece size of the fuel, and names have been given to various piece size ranges (Table 4-3). These names are well established nationally for anthracite, although local variations in nomenclature exist. The terminology for bituminous coal piece sizes is more variable, but sometimes is the same as for anthracite.

In addition to the standard terms defined in Table 4-3, there are other piece size terms. *Slack* coal is all the coal that passes through a screen of a given size. It is usually designated by the screen size. Two-inch slack coal consists of piece sizes from two inches on down—a mixture of stove, nut, pea, and buckwheat. *Stoker* coal sometimes means pea coal, and in other cases may mean slack coal, usually with pea as the largest piece size in the mix.

TABLE 4-3
ANTHRACITE PIECE SIZES

Name	*Size (in inches)*
Run of the mine	Whatever mix of sizes comes straight from the mining operation before separation into specific sizes
Lump	over $4\frac{3}{8}$
Broken	$3\frac{1}{4} - 4\frac{3}{8}$
Egg	$2\frac{7}{16} - 3\frac{1}{4}$
Stove	$1\frac{5}{8} - 2\frac{7}{16}$
Nut	$\frac{13}{16} - 1\frac{5}{8}$
Pea	$\frac{9}{16} - \frac{13}{16}$
Buckwheat No. 1	$\frac{5}{16} - \frac{9}{16}$
2 (Rice)	$\frac{3}{16} - \frac{5}{16}$
3 (Barley)	$\frac{3}{32} - \frac{3}{16}$
4	$\frac{3}{64} - \frac{3}{32}$
5	less than $\frac{3}{64}$

Piece size nomenclature for anthracite.

Anthracite stoves typically use nut or pea coal. Stove coal, despite its name, is appropriate only for some of the largest coal stoves. Buckwheat and rice coal are often appropriate for stokers. However, follow the manufacturer's instructions if they are available.

Friability

The tendency of coal to break and crumble when handled is called its *friability*. Most coal appliances work best on a uniform size of coal. If small broken pieces are mixed in, as is inevitable with relatively friable coal, they can impede air flow through the bed or drop through the grate unburned. Friable coal will accumulate in a pile of unusably small pieces of coal at the bottom of your coal bin. Also, you may order one size of coal and end up with much smaller pieces by the time it is transported to you.

Generally, lower volatile bituminous coals are the most friable, while anthracite coal is the least.

SUMMARY OF COAL TYPES

Anthracite coals are divided into 3 types.

Meta-anthracite. This coal approaches graphite in structure and composition. It is very hard and dense. It has a very low volatile content and is generally difficult to ignite. Meta-anthracite is not a common form of coal, and so it is not usually available.

Anthracite. A hard, dense coal, anthracite creates little dust and is relatively clean to handle. The low volatile content of anthracite means it is relatively difficult to ignite; however, it also means anthracite tends to be relatively smoke-free. Because of this, anthracite tends not to create large creosote or soot deposits as other coals do. Low volatile content fuels also burn with short flames. Anthracite is also noncaking and tends to have a relatively low sulfur content. Overall, anthracite is one of the best coals for home heating; its major liabilities are somewhat difficult ignition and generally higher ash content than most other coals.

Semi-anthracite. With properties intermediate between those of anthracite and bituminous coals, semi-anthracite coal, compared to anthracite, is not as hard, and it has a slightly higher volatile content, so it is a little easier to ignite. However, it still burns with a short flame and is noncaking. Overall, its uses and attributes are roughly those of anthracite.

Bituminous. Also called soft coal, bituminous coals include a large range of coals. They tend to be dirtier to handle. Their higher volatile content makes them easier to ignite; they burn with longer flames; and they are more prone to smoky emissions and soot deposits in chimneys. Many bituminous coals are susceptible to caking.

The 5 types of coal under the general class

"bituminous" span a wide range of properties. The 5 types are:

- Low volatile bituminous.
- Medium volatile bituminous.
- High volatile A bituminous.
- High volatile B bituminous.
- High volatile C bituminous.

The lower volatile bituminous coals are sometimes also called *semi-bituminous*. They are also sometimes known as *smokeless coals* because of their low volatile content compared to other bituminous coals. However, smokeless is a relative term, and these low volatile bituminous coals are not as smokeless as anthracite.

Low-volatile bituminous coal tends to be friable. Most of the higher volatile bituminous coals are stronger and less friable.

Sub-bituminous. This is a coal in a distinct class from bituminous. Sub-bituminous coal deposits are mostly in the West. Sub-bituminous coals tend to be high in moisture content and tend to break up as they dry. They are noncaking and relatively easy to ignite. In fact, they are especially susceptible to spontaneous ignition in storage. They tend to be cleaner burning than bituminous coals.

Lignite. The lowest rank coal, lignite has the highest moisture content, the lowest heating value, the greatest tendency to disintegrate as it dries or burns, and the greatest tendency to ignite spontaneously in storage. Lignite often is brown in color, so it is sometimes called *brown coal*.

OTHER COAL CLASSIFICATIONS

Coal may be further classified according to certain visual aspects.

Banded Coal. This common type of bituminous and sub-bituminous coal has alternating layers of irregular thickness of colors from gray to black and differing in reflectivity or texture from dull to shiny. These bands are thought to be due to different conditions during the decay of the plant matter that became the coal.

Nonbanded Coal. Cannel and *boghead* coal are 2 related varieties of nonbanded coal. These coals tend to break along curved shell-like surfaces.

They are noncaking, ignite relatively easily, and have a very high volatile content, burning with long yellow and potentially smoky flames. It is thought that nonbanded coals are formed from particular types of organic matter, such as algae and spore coats.

Cannel coal burns so intensely that it can destroy grates quickly and overheat appliances. The very high volatile content makes cannel coal prone to backflashing (see chapter 14) and very large soot deposits. Cannel coal should be used only in open fireplaces on appropriate grates. (Cannel and boghead coals do not fall within the rank classification system.)

Packaged blocks of coal mixed with other ingredients are now available. They feature the same clean and convenient storing, handling, and lighting of densified wood logs and wax-sawdust logs. However, unlike the wood-based logs, they are intended for use in stoves as well as fireplaces.

AVAILABILITY OF COAL

Coal fields are not uniformly distributed across the U.S. (Figure 4-3). By far the largest anthracite deposits are found in Pennsylvania, but small deposits occur in Arkansas, Colorado, New Mexico, Virginia, and Washington. Bituminous coal is found in 29 states, distributed from the East Coast to the West Coast and Alaska.

Coal is a very abundant energy resource. U.S. identified coal reserves are estimated at about 1,500 billion tons. At the current consumption rate, this would last more than 1,000 years. Anthracite reserves are about 100 times smaller than bituminous reserves, but so is the demand for anthracite. Anthracite reserves could also last many hundreds of years at current anthracite consumption rates.

Because transportation costs for coal can be significant, and because demand for certain types of coal may be low, the type of coals available in a region are usually those that are mined in the region. Anthracite is not available in many parts of the country, at least not at competitive prices.

This regional variation in available coals creates challenges for manufacturers and stove users. Very few, if any, coal appliances can handle all types of coal equally well, although this is the objective of some manufacturers. Before purchasing a coal appliance, be sure the appliance will perform well on the coal available.

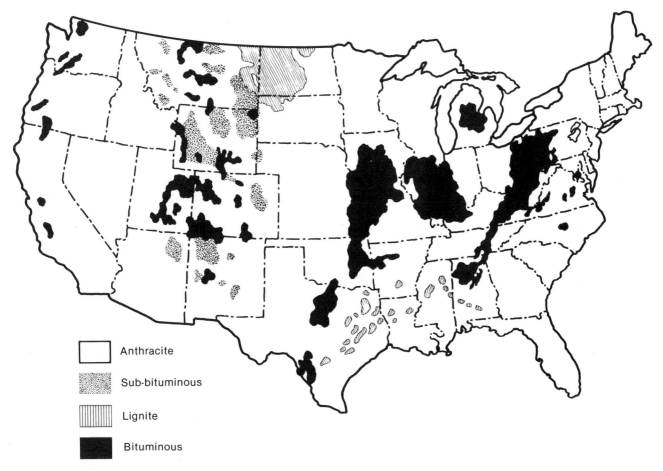

Figure 4–3. Major coal deposits in the continental U.S.

Legend:
- Anthracite
- Sub-bituminous
- Lignite
- Bituminous

PURCHASING AND STORING COAL

Coal is increasingly available in small packages in grocery, hardware, stove, feed, and fuel stores. However, it is rarely economic to heat with coal purchased this way. Typically, people who seriously heat with coal have the fuel delivered by the ton to their homes. If you have a truck, you can sometimes save a little by picking it up yourself from the supplier, but you will need a strong back and some time to transfer it to your storage area.

A reasonable amount of coal to purchase at a time is a few tons. Coal consumption varies considerably, but 50–100 pounds a day is typical. A 2-month supply would then be 1½–3 tons.

How much space is required? Much less than for an equivalent amount of wood because coal is denser. Since the density of loosely packed coal

averages about 52 pounds per cubic foot (44–59 is the typical range), each ton occupies about 38 cubic feet and will fit in a bin measuring 3 feet by 3 feet by 4 feet. About 3–4 tons can be stored in a 4-foot by 4-foot by 8-foot bin. A 55-gallon drum will hold roughly 250 pounds of coal.

Because coal is dusty and dirty, store it outside of living spaces. But for convenience, keep it close to the coal appliance: in a basement, garage, or contiguous shed.

Your storage area should be accessible to the delivery truck. Unlike oil, coal cannot be piped 75 feet from the truck through a hose. The coal truck should be able to get very close to your storage area. In addition, many trucks do not have the means to lift coal up (such as a conveyor belt), so the storage area should be low enough to allow the coal to flow down into it from the truck.

Coal does not need to be dried or seasoned. Unlike wood, coal does not need air circulation around it. This is 1 reason coal is stored in bins, large piles, 55-gallon drums, and so on.

Coal should be protected from rain and snow; added moisture generally does not enhance combustion. Also, coal pieces that are frozen together are not easy to shovel. Finally, when coal is alternately wetted and dried, the chances for spontaneous ignition are enhanced.

Spontaneous Ignition of Stored Coal

Spontaneous ignition is more likely with lower rank coals. Anthracite is not prone to it. Although the detailed mechanisms for spontaneous ignition are not well understood in all cases, experience has revealed what is necessary to decrease the chances of the problem.

When coal is exposed to air, especially fresh coal or coal with freshly broken faces, some oxidation occurs. The oxidation is very, very slow compared to burning, but the chemical reactions have similarities, and heat is released. If the heat is not dissipated, it builds up and the temperature rises. This, in turn, increases the oxidation rate. The process can create temperatures high enough to ignite the coal.

To minimize spontaneous heating, there must be no heat sources very near the coal. Steam and hot water pipes should not run through or under the coal. The ambient temperature should not exceed 75° F. No oily rags, hay, or any other materials susceptible to spontaneous ignition should be stored in contact with the coal.

Limit the amount of air that can get into the coal pile. Encouraging heat dissipation is useful, but not by letting air pass through the pile, as this will stimulate heat generation. (Most residential coal piles are small enough that heat dissipation is adequate.)

Finally, it is useful to keep the fuel either dry or very wet (underwater storage is used in some very large-scale operations—*not* home heating). Small amounts of moisture can enhance spontaneous heating, so storing coal directly on damp ground is not advisable. One safe way to store coal is in covered 55-gallon drums, where air and moisture are easily excluded.

CHAPTER 5

COMBUSTION

An understanding of the basic processes of combustion is helpful for understanding how to start a fire, keep it going, control the rate of burning, minimize smoke and creosote, and successfully burn green wood.

Combustion is also an inherently fascinating subject. Blowing on a smoldering wood fire can ignite the gases, while blowing on an already burning candle will snuff out the flame. Why? All fuels need oxygen to burn, so it is reasonable that blowing on a fire increases the combustion rate because more oxygen is provided. But why does blowing on a candle flame have the opposite effect?

Open wood fires normally burn with a yellow-orange flame, but sometimes a little blue color is also present. If the same gases (wood smoke) are burned in a properly adjusted laboratory bunsen burner or a cooking range burner, the flame color is almost all blue. What is the difference? What determines flame color, and what does the color indicate about the efficiency of combustion?

Wood fires often have no flames. At the end of a fire the charcoal can burn flamelessly, and even wood can smolder and smoke and generate heat without flames.

When wood is smoldering, why doesn't the very high temperature of the glowing portions of the wood always ignite the combustible gases? Many stoves have "secondary air" inlets intended to add enough oxygen to gases coming off the main fire to permit complete combustion. Many of them fail to help the gases burn. Why?

FUNDAMENTALS

Combustion involves energy conversion—stored chemical energy in the fuel is converted into heat, light, infrared radiation, and other forms of energy. Oxygen is required, and it is consumed (incorporated into other molecules) in the process. Many common fuels, including wood, are made mostly of carbon, oxygen, and hydrogen. When burned completely, these elements are transformed into carbon dioxide and water vapor.

Combustion is different from some other chemical reactions because it is *self-propagating*—if one corner of a piece of paper is ignited, adjacent portions will ignite, and ultimately the whole piece is burned. Explosions are like combustion in this respect, but are sometimes distinguished from combustion by the velocity of the reaction; in an explosion of gas or vapors, the velocity exceeds that of sound, so that a shock wave is formed. In combustion or burning, the velocity is less than the speed of sound.

For all ordinary kinds of combustion, there must be fuel, oxygen, and high temperatures. Wood in air does not burn unless first ignited—that is, unless it is raised to a high temperature. Similarly wood, no matter how hot, will not burn if no oxygen is present. In commercial charcoal production, wood chars, and gases are released from the wood, but there are no flames and little or no heat is produced. Wood only burns if it is both sufficiently hot and free (gaseous) oxygen is present.

A Simple Example

A burning candle provides a simple example of the combustion of a solid fuel. The candle's fuel is wax, or paraffin. The surrounding air contains oxygen, about 21 percent by volume. The ignition temperature of paraffin is about 1,000°F. But as the temperature of paraffin is increased, the paraffin first melts and then vaporizes before it reaches this temperature. Only the wax vapor burns.

The energy necessary to melt and vaporize the wax comes from the candle flame. Radiant energy from the flame melts the wax. The liquid wax then climbs up the wick, just as water is soaked up by a sponge. Radiant energy also evaporates wax from the wick. The wick is in the middle of the flame, but the flame (the glowing gases) does not actually touch the wick (except near its tip). The wax molecules constantly evaporating from the wick create a wax-vapor wind which, in a sense, prevents the flame from touching the wick. What actually happens is that not until the wax vapor is a little distance away from the wick will sufficient oxygen mix with it to make combustion possible. The wax-vapor flow keeps oxygen (and hence flaming) away from the wick except near the tip.

For a wax molecule to start burning, it must have a sufficiently energetic collision with an oxygen molecule. The bonds within each molecule must be weakened enough in the collision that new bonds between oxygen and the fuel form.

This is why heat must be supplied to light a fire or ignite a flame. High temperature means high

Figure 5-1. Burning process in a candle flame.
1. Heat radiation from the flame melts the wax.
2. Melted wax climbs up the wick by capillary action.
3. Heat radiation from the flame vaporizes the wax.
4. Wax vapor and air diffuse into each other.
5. Flaming combustion occurs wherever there is sufficient vaporized fuel, sufficient air, and high enough temperatures.
6. Heat output is in 2 forms: rising hot gases and radiation.

TABLE 5-1
COMBUSTION PROPERTIES OF WOOD SMOKE COMPOUNDS

Compounds in Wood Smoke		Ignition Temperature (° F.)	Flammability Limits (percent gas in mixture with air)	Minimum Air Needed For Complete Combustion (air volume) / (gas volume)	Flame Velocity (feet per second)
Hydrogen	H_2	1,000	4–75	2.4	0.2–10
Carbon monoxide	CO	1,100	12–74	2.4	0.3–1.5
Methane	CH_4	1,200	5–15	9.5	0.1–1.3
Acetic acid	CH_3COOH	1,000			
Formaldehyde	HCHO	800			
Pine tar		670			

Combustion properties of some compounds in wood smoke. Data from F. L. Brown, Theories of the Combustion of Wood and Its Control, *Forest Products Laboratory Report No. 2136 (Washington, D.C.: U.S. Department of Agriculture, 1963); C. G. Segeler, ed.,* Gas Engineers Handbook *(Industrial Press, 1969); and R. M. Fristrom and A. A. Westenberg,* Flame Structure *(New York: McGraw-Hill, 1965).*

speeds of the molecules in their random motions; when a collision occurs, it is more violent. At sufficiently high temperatures, the collisions are energetic enough to loosen the old bonds and allow new bonds to form.[1] The chemical reactions of combustion then can take place, releasing chemical energy as heat. This heats the gas further, which helps more reactions to occur, and a flame results.

It is this release of large amounts of energy that makes combustion self-sustaining. The heat released prepares more fuel so that it too can react and release more heat. Heat from the candle flame melts and then evaporates the wax, and also raises the temperature of the wax and oxygen molecules so their collisions will be energetic enough to stimulate the chemical reactions.

In order for a fuel gas to be flammable, it must have some oxygen mixed with it, but not too much. In a candle, the bulk of the wick and the ad-

jacent wax vapor do not burn because of insufficient oxygen; the mixture is too fuel-rich. Just beyond the luminous regions there is no combustion despite the presence of some fuel (incompletely burned wax) because the fuel is too dilute. In between, the mixture of fuel and air is flammable and combustion occurs. The flammability limits of a gaseous fuel are the concentrations of the gas in air which are the leanest and richest mixtures that can burn.

Table 5-1 gives the limits of flammability of some gases found in wood smoke. The ranges of flammability are quite large; most of the gases are flammable with concentrations (in air) anywhere from about 10 percent to 60 percent.

When burning any fuel for heat, it is desirable to burn it completely. Complete combustion means the conversion of all the carbon to carbon dioxide (CO_2) and all the hydrogen to water (H_2O). In practice, a considerable amount of extra air must be supplied to achieve relatively complete combustion, mostly because there is never perfect mixing of fuel and air; a fuel-rich region will run out of oxygen and cannot use the excess oxygen in a distant lean region. It is common in oil and natural gas burners in furnaces to supply 20–40 percent excess air to assure moderately complete combustion. However, supplying excess air is no guarantee that combustion will be complete. For instance, a candle flame has virtually no limit to its air supply and yet sometimes smokes prodigiously, indicating unburned carbon particles. The same is true of a smoldering piece of wood in an open fireplace. The 2 additional necessary con-

---

1. It is in this process of new bond formation that energy is released. For instance, a hydrogen atom and an oxygen atom are strongly attracted when they are near each other. The force pulls them together violently, so when they arrive at their normal bonded separation, they have a large amount of excess speed and, hence, energy. If this energy remains in the form of motional (kinetic) energy, it will be shared eventually with the other molecules in the gas, and that means that the temperature of the gas will be increased (flames are hot). Some of the energy may be emitted as radiation (visible, ultraviolet, or infrared). These are the major types of energy released in flames—sensible heat (which is the motional energy of molecules) and radiation. A little energy is also in the form of latent heat since water vapor is 1 of the products of combustion.

ditions for complete combustion are good mixing of the air with gases, and high temperatures for the mixture. (The high temperatures must also be maintained for sufficient time so that the reactions can be completed.)

As a flammable substance (solid or gas) is gradually heated in air, it will burst into flame at a temperature that is called its *spontaneous ignition temperature*. Spontaneous ignition temperatures of solids are often determined by heating the surface of the solid with radiation (such as in igniting leaves with a magnifying lens in sunlight). Approximate values for some gases are given in Table 5-1. These are approximately the temperatures at which the collisions between molecules are energetic enough for new bonds with oxygen to form at a sufficient rate for the energy release to exceed the heat losses away from the reaction region. The reaction region warms up still more, enabling more reactions to take place: a flame has started.

If a flame laps against a solid surface, combustion can be hindered or not, depending on the temperature of the surface (Figure 5-2). A cold surface in contact with burning gases quenches the flames. Gases within the "quenching distance" of the surface are below their ignition temperature despite the hot burning gases next to them, be-cause of the loss of heat to the cool surface on their other side. The quenching distance, over which no flames exist, depends on the temperature of the surface. It is largest for cold surfaces (a few millimeters) and decreases with increasing temperature, becoming zero for temperatures around the ignition temperature.

Combustion of flames near large surfaces, at any temperature, can be suppressed due to inadequate oxygen. Wood-fire flames are usually diffusion flames—oxygen gets to the fuel by diffusing into the gaseous fuel from the air around the flame. When a flame laps against a surface instead of standing free in air, the normal oxygen supply can be restricted. By contrast, the flames in gas ranges (and most other gas appliances) are "premixed flames"—the fuel and air are thoroughly mixed together before they start to burn. In these cases, proximity of a surface does not deprive the flame of oxygen.

As a consequence of quenching and/or inadequate oxygen, combustion in flames next to a surface can be incomplete. Soot may deposit on the wall or leave the fire as visible smoke. Other products of incomplete combustion also may be formed. But if the surface is hot and there is adequate oxygen, combustion can be enhanced.

Flammable mixtures of gases have character-

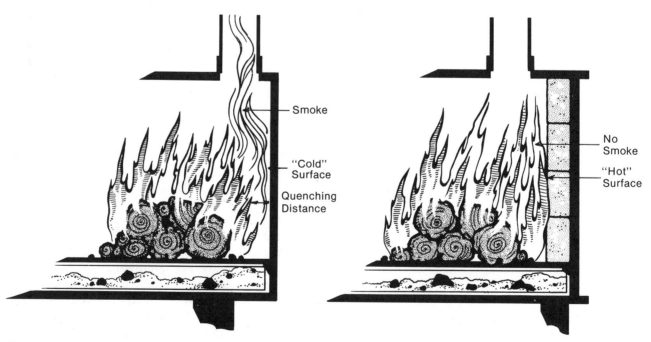

Figure 5-2. The effect of surface temperature on combustion. The quenching distance is exaggerated for clarity.

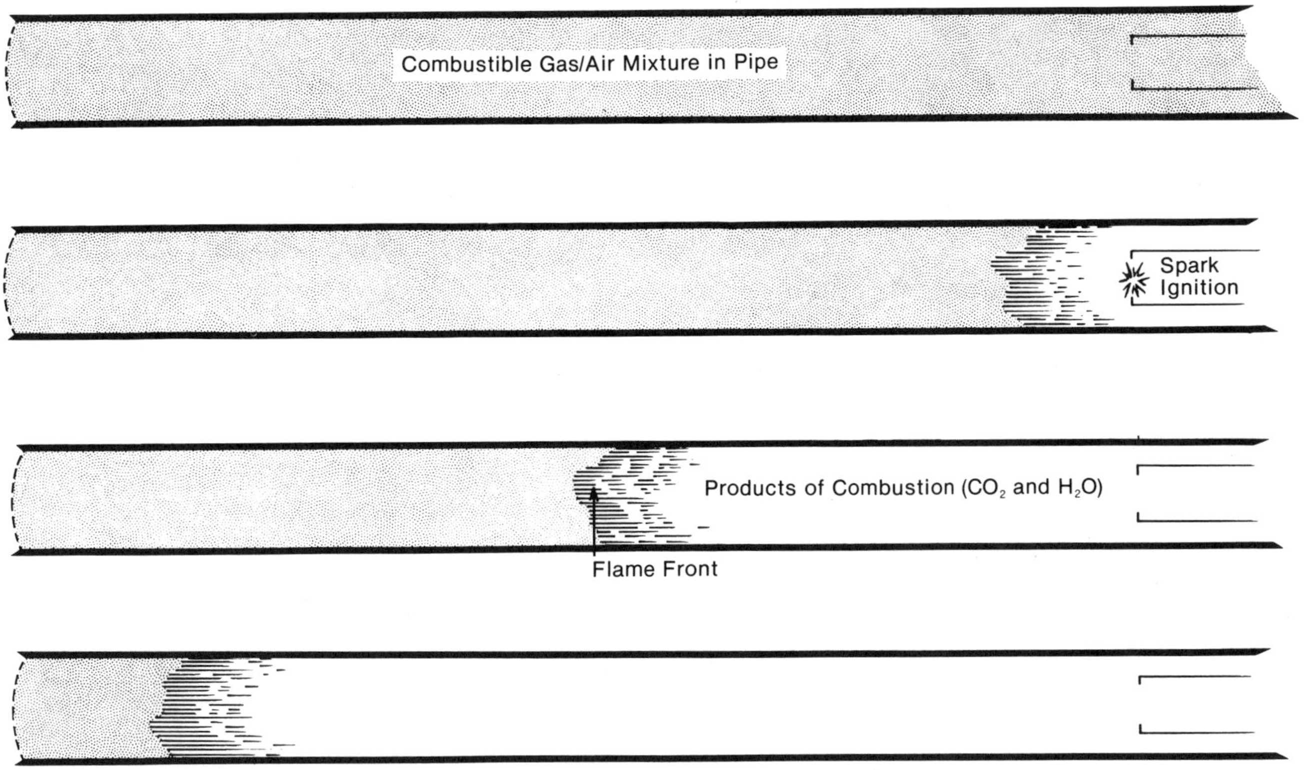

Figure 5-3. The concept of flame velocity. One foot per second is a typical flame speed.

istic velocities at which flames propagate through them. If a long tube is filled with a flammable mixture of a fuel gas and air, and the mixture is ignited at one end, the flame front (combustion zone) moves to the other end (Figure 5-3) at a speed called the *flame velocity*. Gas burners are designed so that the gas velocity is less than the flame velocity. The flame stays fixed in one place with respect to the burner while the gas moves through it. If the gas velocity was too high, the flame would not stick to the burner, but would be blown away or out. Flame velocities of some gases commonly given off by burning wood are given in Table 5-1. Flame velocities of common gaseous fuels are about 1.5 feet per second, or about 2 miles per hour.

This is not very fast, which is why it is so easy to blow out a candle flame (Figure 5-4). If you blow very gently on a candle flame (not so hard as to blow it out), the wax vapor molecules near the wick are swept along with about the same speed as your breath. Air, and especially oxygen, still diffuses into the vapor, and the mixture has no trouble burning because the flame front easily can travel fast enough upstream (toward the

wick) to remain stationary near the wick. In general, as you blow harder, the velocity of the burning gases exceeds the maximum flame velocity, and the flame front gets pushed away from the wick and to the tip of the combustible gases, and there it disappears. The shape of the flame while being blown out can be quite irregular due to turbulence, nonuniform gas composition, and irregularities in the wick. But the basic idea is simple—blowing can force the air and gaseous fuel away from the fuel source (the wick) faster than the flame can move through the gases. For similar reasons, an explosive set off near a gushing and burning oil or gas well can sometimes blow out the flame.

There are "trick" birthday cake candles which are very hard to blow out. I suspect they have imbedded in their wicks small grains of a substance which, like explosives, has the oxygen needed for its combustion built into the material. When the ignition temperature of the substance is reached, it all burns at once, producing intense heat for a short time. After the candle flame is blown out, the wick is consumed downward by smoldering until one of the special grains is encountered; the

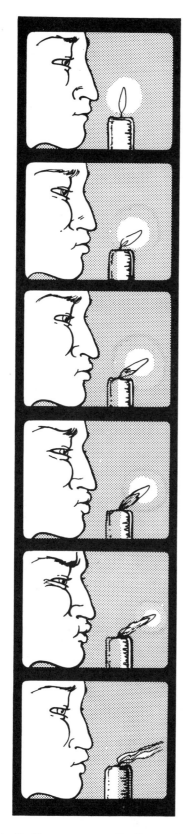

Figure 5-4. To blow out a candle flame, the air velocity must exceed the flame velocity.

temperature of the glowing wick (burning wick "charcoal") ignites the grain which then ignites the unburned gases (smoke), and the candle flame is back again.

THE PRINCIPLES OF SOLID FUEL COMBUSTION

All solid fuels go through the same processes in combustion. For simplicity and concreteness, the following discussion and that of the next section will concern the combustion of wood only.

The combustion of wood is more complicated than the combustion of a candle or any common liquid or gaseous fuel. When wood is heated, it is not all converted to vapors; solid charcoal remains behind. Both the vapors and the charcoal can burn. The vapors burn as flames, and the charcoal glows. Parts of each piece of wood are reduced to ash while other parts are just beginning to warm up. The heating necessary to get wood to the temperature at which it can burn causes very complex chemical reactions in the wood.

The major chemical components of wood are cellulose and lignin, both of which are made of carbon, hydrogen, and oxygen atoms. When wood is heated, as described in more detail in Appendix 3, many chemical compounds come out of the solid wood in the form of gases and vapors. Most of these compounds are formed by the application of heat. Wood, itself, in natural "raw" form, has essentially no combustible volatile component.

For laboratory study of these compounds, the wood is placed inside an almost closed container which is heated from the outside. The generated gases and smoke fill the container and come pouring out. The container prevents oxygen from getting in, making combustion inside the container impossible. This process of thermal degradation is called *pyrolysis*. Pyrolysis is the chemical break-up induced by high temperatures in the absence of oxygen. (The terms "carbonization," "gasification," and "distillation" are used also.) The final solid product after all the chemical changes have taken place is charcoal. Any severely overcooked food also undergoes pyrolysis, emitting smoke and charring. Pyrolysis occurs inside every piece of wood in all fires regardless of the amount of oxygen available to the fire because oxygen from the air does not penetrate inside the wood.

Total yields from the pyrolysis of wood depend

critically on the conditions under which pyrolysis takes place. Generally, the quicker the wood is heated, the larger is the yield of gases and tars, and the smaller is the yield of charcoal. Slow heating, taking about 24 hours to complete the pyrolysis reactions, yields about 50 percent charcoal (by weight, on a dry wood basis). Very quick heating of small pieces of wood, taking 1 minute, yields only about 10–15 percent charcoal. The remainder, in both cases, is in the form of gases and tars; the tar fraction is largest for quick pyrolysis, and smallest for slow pyrolysis. Taking into account the above yields, and the known energy contents of dry wood and charcoal (8,600 Btu. per pound, and about 12,500 Btu. per pound, respectively), one can estimate that in typical wood fires between one-third and two-thirds of the energy content of wood is in the gases and vapors, which are normally burned in the long flames. The rest is in the charcoal. The fraction will be different in different fires, depending on the intensity of the fire and the type of wood, its moisture content and the sizes of the pieces. Tests of stoves with smoldering, flameless fires indicate that 20–40 percent of the energy of the wood can be lost as unburned smoke going up the chimney.

As wood undergoes pyrolysis, combustion takes place if oxygen is available. When a new piece of wood is added to a fire, it is heated by its hot surroundings. Since wood is a moderately good thermal insulator, the heat cannot be conducted quickly to the wood's interior; only a very thin layer at the surface is affected initially. It dries very quickly and starts to char. The gases evolved as the temperature rises to about 540° F. do not ignite because the concentrations of noncombustible carbon dioxide and water vapor are too high. Most of the water vapor passing out through the surface of the wood at this stage originates in deeper layers where the temperature is warm, but less than about 212° F.

As the surface layer temperature of the new piece of wood is raised above 540° F., pyrolysis becomes much more vigorous. The gases and vapors evolved are ignitable partly because the water vapor contribution from deeper layers is now a smaller part of the total. However, the gases will not ignite and burn unless both adequate oxygen is available and the ignition temperature of the gases is reached or exceeded. Both these conditions must be met at the same location and at the same time. The ignition temperature of the gas

mixture is typically between 1,000° F. and 1,200° F. It cannot be ignited by the surface layer of charred wood out of which it has emerged since it is not yet hot enough. If flames are to appear, they must be lighted either by other flames or by burning charcoal from a nearby piece of already ignited wood.

If the gases do ignite, they usually burn at a temperature of about 1,600–2,000° F. Much higher flame temperatures are possible in very large fires and in fires in a relatively well-insulated environment. There is no 1 flame temperature. The actual temperatures inside a typical operating stove span a huge range from a low of 100–200° F. for the air near the air inlet to higher than 2,000° F.

Flaming requires oxygen; but the constant flow of gases passing out through the surface of the wood prevents oxygen from getting close to its surface. The base of a flame is rarely in direct contact with the wood from which the gases emanated. But the radiation from a wood-fire flame is substantial—it is estimated that between 10 percent and 30 percent of the energy released in a "luminous" (yellow) flame is radiant energy. Some of this radiation is absorbed by the surface of the wood. Flames from one log also may lick against another. By these mechanisms the wood is heated, which increases the evolution of gases, which increases the size of the flames, and so on. This is how fires grow bigger and spread.

The charred surface itself usually does not burn for some time. Since charcoal does not vaporize at any temperature achievable in a wood fire, it can only burn when, and to the extent that, oxygen comes to it, and oxygen can get to the surface only when the flow of gases coming out of the wood has subsided. Charcoal starts to glow visibly when its temperature reaches about 900° F. The glow is not necessarily an indication that it is burning, just as the glow of a fire poker or an electric heating element is not evidence that it is burning. However, the glowing of charcoal usually indicates that its surface is burning.

Charcoal burns with little or no flame. Charcoal is mostly carbon, but also contains hydrogen, oxygen, and minerals. Oxygen that wanders onto the surface combines with carbon to form carbon dioxide and carbon monoxide. Carbon monoxide is a combustible gas. Small quantities of other gases from the charcoal, particularly hydrogen, also may be emitted. The quantities are small; if ignited, only small flames result. The flames from charcoal are hard to see because they usually emit only a faint blue light.

As the carbon (and hydrogen and oxygen) are burned out of the charcoal, those few elements that do not form gaseous compounds are left as ash. The ash layer that is always forming (or being left) on burning charcoal is very light, and is easily blown off by the natural air movement in the fire. Some is carried up the chimney and some falls and accumulates under or around the fire. Ash that goes up the chimney is called *fly ash*.

Charcoal is a better thermal insulator than raw wood. The thicker the charcoal layer is on a piece of wood, the slower the penetration of heat to the deeper portions of the wood. This contributes to the steadiness of the burning of wood, especially in larger pieces.

When wood is heated by a flameless source, such as radiant energy, it is observed to ignite spontaneously around 400–500° F.[2] This temperature is below the spontaneous ignition temperatures of the gases coming out of the wood (which range from 800° F. to 1,200° F.). The ignition temperature of tar from pine is about 670° F., but the ignition temperature of charcoal is still lower — perhaps as low as 300° F. for fresh charcoal. In the commercial manufacture of charcoal, great care must be exercised to avoid its spontaneous combustion. After pyrolysis is complete, the charcoal must be cooled and kept in an oxygen-free atmosphere for a few days before it is safe for shipment. Probably, spontaneous flaming ignition of wood starts with the spontaneous glowing ignition of charcoal. The temperature of the burning charcoal then becomes high enough to ignite the gases.

All of these various stages of pyrolysis and burning often take place simultaneously in each piece of wood (Figure 5-5). At sharp edges of the wood, charcoal can be burning and ash accumulating in less than a minute after adding the piece to the fire; oxygen reaches these areas relatively easily. On the other hand, the center of a large log may not start getting warm for half an hour or more. The transitions between various stages of pyrolysis and burning are not sharp but continuous. For example, small amounts of pyrolysis doubtless continue to about 900° F.; and some surface charcoal may be burned even while the gas flow through it is quite strong.

---

2. For a more complete discussion of spontaneous ignition of wood, see J. W. Shelton, *Wood Heat Safety* (Charlotte, Vt: Garden Way Publishing, 1979).

Figure 5-5. Wood combustion process.
1. *Radiation from the flames and the glowing charcoal heats the wood.*
2. *Heated wood decomposes into smoke and charcoal.*
3. *Smoke mixed with air burns in the flames.*
4. *Charcoal and air combine in glowing combustion.*

WOOD FIRES

The ignition of wood fuel and maintenance of the resulting fire require both oxygen and adequately high temperatures. Lack of either will cause a fire to subside and eventually die. Lighting a fire built with closely packed, straight pieces of wood is difficult because not enough air can move between the pieces. On the other hand, wood dust (such as is generated from sanding wood) can burn explosively if suspended in air at a concentration where each particle is surrounded by enough oxygen to burn it, and the particles are close enough together for the combustion of a few particles to ignite others.

It is principally the thinness of a piece of solid fuel that makes it possible for burning to spread over and through the whole piece. A thin wood sliver burns readily; however, if a large log is ignited in a fire and then pulled out and put on the ground (or suspended in the air), the fire will almost always die out. The difference is in how effectively the combustion of one part of the fuel can heat the other parts to the temperature where they, too, burn. For any size piece of wood, self-sustaining combustion is more likely if the wood is ignited at the bottom rather than at the top, for then the flames lap against other parts of the wood.

A large single piece of wood cannot burn completely, no matter where it is lighted, because the heat spreads out (is conducted) into a large amount of wood, and hence cannot warm the surface to a high enough temperature to help the fire grow. A wood sliver has much less mass in proportion to surface area, so the same intensity of surface heating will result in a much greater rise in temperature of the wood. The largest size of normal dry wood which can sustain combustion if lighted at the bottom, is roughly about ¾ inch thick. Both moisture content and resin content are important factors.

Blowing on a single piece of wood does not help it to support a flame. The convective cooling of the wood and increased distance of the flame from the wood are more important effects than the increased oxygen supply. Blowing can extinguish the flame if the velocity of the air exceeds the flame velocity of the burning gases, just as in the case of a candle flame.

In practice, the most important principle in laying and sustaining a fire, particularly an open fire, is to position the pieces of wood near enough to each other so that they will keep each other hot, but far enough apart to allow an adequate oxygen supply to move between them. Straight and smooth pieces of wood may need to be crossed to create adequate air passages among them; crooked or bumpy firewood often need not.

Wood can burn without any flames, but not very efficiently. Typically, this is what happens in stoves operated at very low powers with large charges of wood. The low powers are achieved by severely limiting the air supply. Under this circumstance it can happen that none of the gases and tars emitted from the wood burn, yet combustion of the charred wood can proceed until it is all consumed. Flameless combustion of the surface layer of charcoal is the main source of heat

which pyrolyzes the next layer, but radiant energy from nearby pieces of similarly smoldering wood is usually necessary to sustain the process.

A smoldering piece of wood usually does not ignite its own combustible gases; it does not burst into flame. The glowing charcoal layer of smoldering wood is very hot—about 1,200-1,600° F. judging from the color of its glow. This is above the ignition temperature of the gases and tars. Ample oxygen is available—typically much less than half of the oxygen in the air passing through a stove is used by a smoldering fire. In open smoldering fires, such as in a fireplace, the oxy-

When too close together, not enough air can get in; the fire smolders or goes out.

With appropriate spacing, there is enough air and mutual heating to sustain good combustion.

When too far apart, too much heat is lost to sustain pyrolysis, and the fire goes out.

Figure 5-6. The space between 2 wood slabs affects the possibility of sustained combustion on the facing surfaces.

gen supply is clearly not preventing flaming. And yet, the smoke often does not ignite. The reason is that the oxygen and the high temperatures are not at the same place. The combustible gases are very hot as they pass through the glowing charcoal layer, but have cooled substantially by the time they are a little distance from the surface. Oxygen is abundant some distance from the surfaces, but because of the constant gentle "wind" of gases evolving from the wood, the concentration of oxygen close to the glowing charcoal is low. Thus, it can happen that at no single place in the fire is there both adequate temperature and adequate oxygen for the gases to ignite.

Blowing into a normal multipiece fire heats it up—it can even make a smoldering fire burst into flame. Yet blowing on a burning match blows it out. The main differences are the radiative feedback and the multiple ignition sites in the normal fire. Blowing on glowing charcoal always has the initial effect of making it glow more brightly due to the increased oxygen supply. This is even true for a smoldering match stick. But in the case of a match stick, the small glowing tip by itself does not get the smoke up to its ignition temperature. In a multiple-piece smoldering fire, some of the radiation from each glowing location is absorbed by other parts of the fire. Thus, higher temperatures are achieved. In addition, smoke gets mixed with the air and together they are hurled against hot charcoal where ignition can occur. The blowing improves the mixing of air with the smoke and forces the mixture in close to the hot coals. Since there are many hot coals, smoke from one smoldering ember can be ignited by being blown into another ember. Blowing on a match stick can only sweep the smoke away from the 1 possible ignition source. For similar reasons, large, multipiece fires usually cannot be blown out; even if the flames are blown away (as in blowing out a candle), the intense charcoal temperatures usually will reignite the smoke when the blowing stops.

Green (moist) wood needs more coaxing to burn than dry wood, for a number of reasons, all relating to its water content. The moisture must be evaporated as the wood burns. Since this consumes sensible heat (and converts it to latent heat), the temperature rise of the wood is slowed, which tends to inhibit combustion. The temperature rise at the surface also is slowed because moist wood is a better conductor of heat than dry wood; a larger amount of the energy received at the surface conducts and spreads out into the interior of the wood. As the water vapor passes through the hot charcoal on the surface of the wood, some of it reacts with the carbon, producing carbon monoxide and hydrogen; this reaction uses heat energy and has a cooling effect on the charcoal. Finally, the water vapor that does leave the wood dilutes the combustible gases coming out of the wood, making them more difficult to burn. All these effects slow the combustion rate; this is why some people like to use some green wood in a fire when a low combustion rate is desired (such as overnight in a stove). But a fire of only very moist wood usually is difficult to sustain.

FLAME COLOR

Most, but not all, organic fuel gases are capable of burning *either* with a mostly blue flame *or* an orange-yellow flame. The latter are called "luminous" flames because they are so much brighter. In early and mid stages of combustion, wood-fire flames are long and luminous, while pure charcoal at the end of the combustion cycle may burn partly with a small blue flame. Flame color is not a good indicator of completeness of combustion.

With adequate oxygen throughout a flame, the flame color is predominantly blue. The blue light does not come from the ultimate combustion products: CO_2, CO, and H_2O; it originates from intermediary and unstable molecules, such as CH, C_2, and OH. Most natural gas and propane flames in appliances such as ranges, clothes dryers, water heaters, and furnaces are predominantly blue because air is thoroughly premixed with the fuel before it burns, ensuring adequate oxygen throughout the flame.

The yellow light in luminous flames comes from very small glowing carbon particles which form in fuel-rich regions. They are glowing simply because they are hot, just as a hot fire poker or any hot solid does.[3] The yellow light is so much brighter than the blue associated with the chemical reactions that the human eye and brain sense only the yellow; in fact, the fainter blue light is

---

3. The spectrum of the carbon-particle emission is not quite that of a black body radiator at the same temperature, but is shifted a little towards the blue—the color temperature is about 100° C. higher than the true temperature. This is because the particles are much smaller than visible wavelengths. See A. G. Faydon and H. G. Wolfhard, *Flames* (London: Chapman and Hall, 1960), p. 231.

still there. The glowing particles are much too small to be seen by the unaided eye as individual sources of light (they are about a millionth of an inch in size). One piece of evidence that the particles really exist is that they can be collected as a deposit of soot on a cool surface inserted into the luminous region of a flame.

Most wood-fire flames are luminous or have luminous regions. Since the oxygen is not premixed with the fuel, there will be regions where the temperature is high enough to induce particle-forming reactions, but where the oxygen concentration is too low to allow complete combustion. This is the case for the long luminous flames in wood fires. The blue flames from charcoal do not form carbon particles because the gaseous fuel which is burning is mostly carbon monoxide and hydrogen. Hydrogen gas, of course, cannot form carbon particles, and carbon monoxide is one of the few exceptions among carbon-containing fuels in that it cannot burn with a luminous flame under any circumstances.

The yellow color of a flame is not necessarily a sign of incomplete combustion. Whenever a flame is yellow, there are carbon particles, but the particles are highly combustible. If they pass through a region near the outside of the flame with sufficient oxygen and at a high enough temperature, they will be burned. This is largely the case in a nonsmoking candle flame and in wood-fire flames under many circumstances.

Similarly, blue flame color does not indicate that combustion is complete, only that soot particles are not present inside the flame. Blue flames can emit carbon monoxide. And seeing a blue flame in a wood burner does not preclude the existence of considerable smoke elsewhere in the system which is not burning at all.

SMOKE

The term "smoke" is sometimes used to designate all airborne by-products of combustion, sometimes just the combustible ingredients, but usually just the visible components.

The total chimney effluent from a wood fire contains hundreds of different components. Nitrogen is by far the most abundant ingredient—about 55-75 percent by volume. Its source is the combustion air, which is approximately 78 per-

cent nitrogen. Unused oxygen from the air is usually the next most abundant, constituting about 5-20 percent of the volume. Pure air is about 21 percent oxygen. The 2 products of complete combustion, carbon dioxide, at typically 3-15 percent, and water vapor, come next. All other ingredients taken together rarely amount to more than a few percents of the total volume. They include combustible gases, solid particles, and liquid droplets. The gases and liquid droplets include the airborne products of pyrolysis mentioned previously, plus derivatives which may be produced from them in the fire and chimney after they have emerged from the wood. The number of identified chemicals is large; the actual total number is doubtless still larger.

There are 2 kinds of solid particles in wood-fire effluents—carbon and fly ash. The carbon particles are some of the same ones that made the flames yellow-orange in color, although, often, the small particles in the flames amass to form larger particles in the flue gases. They are called soot when deposited on surfaces such as fireplace, stove, or chimney walls. Suspended in the gases, they are part of the smoke.

The visible part of wood-fire emissions consists of only the liquid and solid particles. If the particles are all very small (smaller than can be seen in an ordinary microscope, which means smaller than the wavelengths of light), the smoke looks white or bluish-white. The color, in this case, is due mostly to light scattering, the same effect that makes the sky blue. If the particles are big enough to be seen individually, they appear either black if carbon particles (soot) predominate, or yellow-brown if tar and other condensed vapors predominate. Smoke with a sufficient number of larger particles to appear dark is rare from wood fires. The visible smoke from a flameless (smoldering) fire is mostly tar droplets, and they are usually so small that they appear white or bluish. Cigarette smoke, which is also in a smoldering process, also is high in tar content.

In cold weather, droplets of condensed water can make any flue effluent look quite dense as it emerges from the chimney. When the water vapor, which is always in flue gases, encounters the cold air outside the chimney, it condenses into droplets, essentially forming a cloud. As this cloud drifts away, it mixes with air, and soon the water droplets evaporate, making the water invisible again.

CHAPTER 6

CHIMNEYS

Chimneys are just as important as heaters for successful heating with solid fuels. If a heater lets smoke into the room, an inadequate chimney is the most likely reason. If you are not getting enough heat from a stove, the problem could be inadequate draft from the chimney. Creosote problems are not exclusively related to chimney design but can be lessened by good chimneys. Heating efficiency is affected by chimney location and type. A large fraction of house fires associated with wood heating is due to unsafe chimney installations. The practicality and installation cost of solid fuel heating is significantly affected by whether or not an existing chimney can be used; new chimneys can cost as much as new stoves. For all these reasons and more, the chimney is a critical part of any solid fuel heating system.

DRAFT FUNDAMENTALS

A chimney carries the undesirable combustion products (smoke, and so on) out of the house and supplies the draft necessary to feed air to the fire. The force necessary for both functions comes from the tendency of hot air to rise, an effect called *buoyancy*. The flow up a chimney is restrained by resistance from chimney walls, bends, dampers, and the appliance itself. It is the balancing of buoyancy and flow resistance that determines the smoke velocity in a chimney.

Because of the buoyant effect of the hot stack

gases, air pressure outside the stove is greater than air pressure inside the stove. If the air inlets of a stove are open, or if there are any cracks or holes in the stove, air will be pushed (or drawn) in. The term used to describe and quantify this effect is *draft*.

Draft is a measure of the force making gases flow. At a place where the draft is high, air will be drawn hard into any opening, but if the opening is small, not much air will get in. Thus, draft and actual air or flue-gas flow are not the same.

Measuring Draft

A common device used to measure draft is a transparent U-shaped tube partly filled with water (Figure 6-1). One end is left open and "senses" the pressure of the atmosphere outside the chimney or stove. The other end is connected to a metal tube that is inserted into the chimney or stove perpendicular to the flow; it senses the pressure inside the chimney. If there is some draft, the pressures are unequal, as indicated by a difference in level of water on the 2 sides of the U-tube. In America, the common unit used to measure draft is "inches of water," referring directly to this height difference. Drafts in residential chimneys are usually between .01 and .15 inches. Because this difference in water level is too small to be seen easily in a U-tube, many draft gauges use different geometries to amplify the visual effect or use different principles entirely.

The draft is zero at the top of the chimney; the

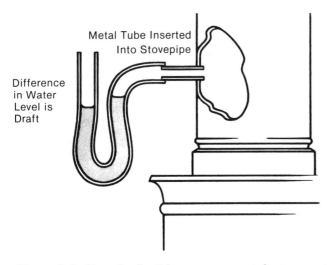

Figure 6-1. *Use of a flexible, transparent tube to measure draft, or chimney suction. Practical draft gauges are often built quite differently.*

pressure of the flue gases as they emerge essentially equals that of the surrounding air. The draft is usually highest at the bottom of the chimney or in the stovepipe connector. If there is some draft (suction) everywhere in the system, no smoke can leak into the house even if there are cracks; air will be pulled into the cracks rather than smoke pushed out. This is usually the case in solid fuel heating systems.

Chimney Flow Capacity

The ultimately critical question about chimneys is not draft, but flow. Draft is the cause – force or push – behind the flow. The flow is the net effect – the number of pounds or cubic feet of flue gases that pass up the chimney in a given time. If a chimney has an inadequate flow capacity for a particular stove, the stove will not be able to operate at its maximum heat output rate. Even with its air-inlet damper wide open, the suction of the chimney will not draw combustion air into the stove at a high enough rate.

An inadequate chimney may cause smoke to leak out of cracks in the heater and/or stovepipe when the air-inlet damper is opened beyond a certain point or when the heater door is open.

Chimney capacity has 2 approximately equivalent definitions: the maximum flue-gas flow (in units such as pounds per hour or standard cubic feet per minute) under given typical flue-gas temperature conditions,[1] or the largest normal type of heating appliance that can be safely vented through the chimney. For this purpose, appliance size is usually rated in terms of fuel consumption in units such as Btu. per hour.

Flow in a venting system is constrained by friction between the moving flue gases and their surroundings. Stovepipe elbows, chimney rain caps, stove air-inlet dampers, stovepipe dampers, and even straight sections of pipe or chimney all offer resistance to the flowing gases. (Gases here means air or smoke or the combination.) If a chimney could be made hot suddenly, the initially stationary gases would start rising and increase

---

1. The theoretical expression for flue-gas mass flow, M, is
$$M = (A/k) \sqrt{2ghd_f (d_a - d_f)}$$

where A = flue cross-sectional area
      k = system resistance coefficient
      g = gravitational acceleration
      h = chimney height
      d_f = density of flue gas
      d_a = density of outdoor air

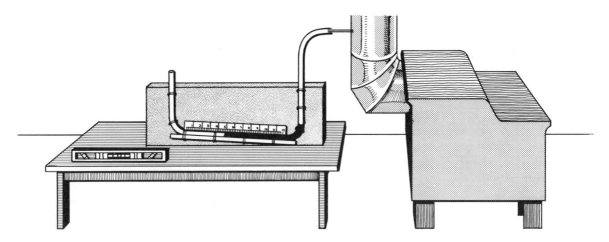

Figure 6-2. *A schematic drawing of a homemade, slanted draft gauge.*

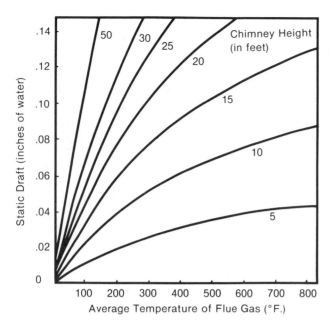

Figure 6-3. Static (theoretical) draft as a function of chimney height and average flue gas temperatures, assuming an outdoor temperature of 40°F. Actual drafts are usually less due to flow resistance. Height is measured from the draft-measurement location to the top of the chimney, and the flue gas temperature is an average over that portion of the chimney.

Static draft has dimensions of pressure and is expressed by gh $(d_a - d_f)$, where: g = gravitational acceleration; h = chimney height above the location where draft is being calculated; d_a = density of the outdoor air; and d_f = density of the flue gas.

in speed until the frictional resistance just balanced the buoyance effects.

In practice, the flow of gases through a chimney is determined by 4 factors.

- The temperature difference between the flue gases and the outdoor air.
- Chimney height.
- Chimney diameter.
- The whole-system resistance coefficient.

Flue-Gas Temperature

Conventional wisdom says the hotter the chimney, the better it draws. For practical purposes, this is true. But as indicated in Table 6-1, sometimes chimney performance may run counter to this traditional wisdom.

A chimney need not be very hot to perform at a substantial portion of its maximum possible capacity. Even when it is a warm 70° F. outside, a

chimney is at 45 percent of its maximum capacity when the flue-gas temperature is only 100° F.; at 200° F., the capacity is already 80 percent of maximum. (These flue-gas temperatures are averages over the entire vertical extent of the venting system).

TABLE 6-1
INFLUENCE OF TEMPERATURE ON CHIMNEY CAPACITY

Average Temperature of Flue Gases (°F.)	Relative Chimney Capacity Ratings
70	0
75	19
80	27
100	45
200	80
400	97
600	100
800	99
1,000	96

The influence of flue-gas temperature on chimney capacity. The capacity at 600° F. is arbitrarily given a rating of 100. The flue-gas temperature is the average over the height of the chimney from the flue collar to the top of the chimney. An outdoor temperature of 70° F. is assumed.

In many systems, an average flue-gas temperature as low as 100° F. provides adequate draft. This tends to be true of closed solid fuel burners (not fireplaces or fireplace stoves) because their chimneys tend to be oversized. But a serious disadvantage of such low flue-gas temperatures usually is considerable creosote accumulation.

By and large, a chimney should be moderately well insulated and be located as much as possible in heated interior parts of the building to keep flue-gas temperatures reasonably high.

Chimney capacity actually decreases for temperatures higher than about 600° F. Although buoyance, the driving force, is always greater for higher temperatures, the resistance to flow increases even faster. This is because hot gases are less dense and so must have a larger velocity for the same mass flow. However, the effect is very small; practically speaking, the capacity of any given chimney is essentially the same for any average flue-gas temperature over about 300° F.

Chimney Height

To estimate drafts and capacities of venting systems, chimney heights are measured from the flue collar on the appliance to the chimney top. Vertical sections of stovepipe connector contain buoyant gases just as does the chimney; so they also contribute to chimney capacity.

Increasing chimney height improves flow capacity, but not in direct proportion to the change in height. If all other things remained unchanged, a doubling of chimney height would result in a 41 percent increase in capacity; a 10 percent height increase would improve performance by 5 percent. (Chimney flow is proportional to the square root of its height, other things being equal. See footnote 1.)

Other variables do change with height alterations. The higher the chimney, the more the flue gases cool before emerging from the top, which has a negative effect on chimney capacity. There is always an improvement in chimney capacity with increased height, but the amount of improvement is diminished if heat is conducted out through the chimney walls. Insulated chimneys benefit the most from height increases. Table 6–2 illustrates the average *net* effect of height on chimney capacity.

In practice, height is determined by the need to end the chimney above wind-pressure influences and by architectural considerations. In mobile homes and other flat-roofed, single-story struc-

TABLE 6-2
INFLUENCE OF HEIGHT ON CHIMNEY CAPACITY

Chimney Height (feet)	Relative Chimney Capacity Ratings
6	60
8	69
10	76
15	88
20	100
30	115
50	135

The influence of height on chimney capacity. Capacities are relative to a 20-foot-high chimney, which is arbitrarily given a capacity rating of 100. Heat loss effects are included. The numbers are approximately valid for any type of, and any diameter, chimney, but are specifically for a masonry chimney with an internal area of 38–50 square inches (6–7-inch diameter, if round). Data is adapted from Table 17 in Chapter 26 of The ASHRAE Handbook and Product Directory, 1975 Equipment Volume (New York: American Society of Heating, Refrigerating and Air-Conditioning Engineers, 1975).

tures, chimney heights are often only 10 feet (measured from the stovepipe collar to the chimney top), and they usually work well. For such short chimneys, adding 2 or 3 feet to the height can make a noticeable improvement. But for more typical installations, where the venting system height is 15–30 feet, adding a few more feet is not likely to solve calm-weather draft problems.

Chimney Diameter

Chimney diameter affects the flow capacity more than chimney height. Assuming the whole venting system (chimney and chimney connector) is of the same diameter, the capacity of the system is approximately proportional to its cross-sectional area. (The proportionality is not exact since heat loss through chimney walls and the resistance to flow also are affected by chimney diameter.)

Table 6–3 illustrates the relationship between chimney diameter and capacity. An 8-inch chimney has about 90 percent more capacity than a 6-inch chimney, and a 6-inch system has about 55 percent more capacity than a 5-inch system. In other words, diameter is *critical*.

In practice, it is usually safe to take the recommendations made by the appliance manufacturer concerning minimum chimney diameter. Lacking

TABLE 6-3
INFLUENCE OF DIAMETER ON CHIMNEY CAPACITY

Chimney Diameter (inches)	Relative Chimney Capacity Ratings
3	20
4	38
5	64
6	100
7	139
8	192
10	330
12	506

The influence of diameter on chimney capacity. Capacities are relative to a 6-inch-diameter chimney, which is arbitrarily given a capacity rating of 100. The numbers are approximately valid for any type of, and any height, chimney. Data is adapted from Tables 13 and 14 in chapter 26 of The ASHRAE Handbook and Product Directory, 1975 Equipment Volume (New York: American Society of Heating, Refrigerating and Air-Conditioning Engineers, 1975).

explicit instructions, a chimney of the same size as the flue collar on the appliance is usually adequate. Substantially larger flues are *not* desirable.

Flow Resistance

Do the number of elbows and the length of stovepipe affect chimney performance? Yes.

Building codes usually require appliances to be located close to their chimneys. Some codes specify no more than two 90-degree elbows in the connector. Some metal fireplace manufacturers specify no more than four 30-degree elbows. But, in practice, many stoves have been used successfully with very long stovepipe connectors (50 feet or more) and more than two 90-degree elbows. This was common a century ago in churches and schoolhouses in New England.

Such long and complex additions to normal venting systems are not recommended, but they will work—if the original system had sufficient excess capacity. This is usually the case with closed burners, such as stoves, furnaces, and boilers; it is usually *not* the case with open burners, such as fireplaces and fireplace stoves. The air inlet on closed burners offers so much more resistance to flow than does any other part of the system that a few more stovepipe lengths or elbows are just not very important to the performance of the whole system. On the other hand, in open burners most of the resistance to flow is due to

the venting system itself—since the appliance is open, it offers little flow resistance. In this case, extra elbows, extra lengths of horizontal stovepipe or, in some cases, even a chimney cap can be disastrous—smoke will spill out of the appliance into the house because it does not all get out through the stovepipe and chimney.

Long lengths of stovepipe inside a building increase the heating efficiency of the system; but on balance, they are not recommended. Although stovepipe is not inherently unsafe, the sloppy way in which it often is installed and maintained results in a serious hazard. In addition, such installations usually result in very high rates of creosote accumulation.

Excess Capacity

Too large a flue results in excess capacity, which means less draft and more creosote. The larger the flue, the more chimney surface area there is through which the flue gases will lose heat, and the slower the flue gases will rise, allowing more time for heat to be lost. Cooler temperatures mean less draft and more creosote accumulation.

Excess capacity occurs most often when a large fireplace flue is used for an insert or a stove. If the chimney runs up the outside of the house, the cooling of the smoke is even greater. In extreme cases, cold outdoor air can descend down the chimney from the top at the same time smoke is rising. Recommendations for installing inserts and stoves in fireplaces are given in chapter 12.

To avoid the problems of excess capacity, do not use a chimney flue with a cross-sectional area greater than twice the area of the flue collar on the appliance. See Table 6-4.

House-Chimney Interactions

Many draft and smoking problems are not due to inadequate chimneys. Tight houses can restrict the air supply to the appliance, other appliances can depressurize the house, and the "stack effect" of the house itself can compete with the draft of the chimney.

What happens if a house is *too* tight? The effects depend on the type of appliance. For a closed heater, such as a stove, furnace, or boiler, the consequences may be annoying but are rarely dangerous. When less air enters a stove, the fire is less intense, and so the heat output is less. But smoke may spill out of an open burner, such as a fireplace or fireplace-stove. A minimum flow of

TABLE 6–4
CHIMNEY SIZES

Appliance Type	Collar Size (inches)	Suggested Inside Diameter for Round Chimneys or Flue Liners (inches)	Suggested Rectangular Flue Dimensions (nominal exterior (inches)
Small stoves	4	4–6	4×8
	5	5–7	8×8
Medium and	6	6–8	8×8
large stoves	7	7–8	8×8
Fireplace stoves,* small fireplace inserts,* furnaces,	8	8–10	8×12
boilers	9	10	8×12
Medium fireplaces*		10–12	12×12
Large fireplaces*	12	12	12×16
	14	14	16×16

* A common rule of thumb for fireplaces and fireplace-stove chimneys is that the cross-sectional area of the flue should be about $\frac{1}{10}$ ($\frac{1}{8}$ for chimneys less than 15 feet tall) of the area of the fireplace opening.

Here are the suggested chimney sizes for residential solid fuel heating equipment lacking manufacturer's instructions. Bigger is not always better. Oversized flues tend to create less draft and accumulate more creosote or soot. Adapted from Table 5–4 in J. W. Shelton, Wood Heat Safety (Charlotte, Vt.: Garden Way Publishing, 1979).

air, about 0.8 foot per second averaged over the opening, is needed to keep the smoke eddies inside the combustion chamber. Typical air consumption rates of various wood burning appliances are indicated in Table 6–5.

All houses have some air leakage both into and out of the structure, even with all doors and windows closed. Much of this occurs around doors and windows. A certain amount of this *air exchange*, or infiltration plus "exfiltration," is necessary to keep the humidity from getting too high. Typical air exchange rates are from ½ to 2 air changes per hour, although in extremely tight houses the rate can be as low as ¼ change per hour.

In homes with typical natural air leakage, only open appliances are likely to run short of air. Typically, *closed* burners need roughly 10 percent of the air that is naturally entering and leaving the house anyway; but *fireplaces* may need more air than the house can provide.

This air shortage is most likely to occur in electrically heated houses, energy conservative

houses, mobile homes, and earth-sheltered (partly underground) houses. Opening a window, decreasing the area of the fireplace opening, and ducting outdoor air to the fireplace (see chapter 9) may make it possible to use open burners in houses that are too tight. In very tight houses, or even in closed rooms, even closed burners may perform better with outside air.

Another reason for poor draft can be the slight depressurization of a house caused by other appliances. Kitchen and bathroom exhaust fans are common culprits. Perhaps more important in practice are other vented fuel-burning appliances, such as central heaters and other solid fuel heaters. For instance, in many houses with more than 1 fireplace, only 1 can be used at a time. Even then, air is pulled down the unused flue, bringing with it creosote odors and sometimes even smoke from the other fireplace flue. A strong draft in 1 large chimney literally sucks on the house, making it harder for any other chimney to operate properly. The suction can be enough to pull outdoor air down any unused (cool) flue, to cause smoke to spill into the house from operating open appliances, to result in sluggish operation of closed burners, and to cause poor combustion and fume spillage in gas-burning and oil-burning equipment. The only practical solution is to supply outside air to one or more of the larger appliances.

Sometimes the house itself acts somewhat like a chimney. On a cold day, the warm air inside the house is relatively buoyant. The warm air pushes

TABLE 6–5
CONSUMPTION AND AIR EXCHANGE RATES

Appliances and Houses	Air Consumption Cubic Feet per Minute (at 70° F.)	Grams per Second
Small stoves	10–50	5–25
Large stoves, furnaces, and boilers	20–100	10–50
Fireplace stoves with doors open		
Small fireplaces without doors	100–300	50–150
Medium and large fireplaces (without doors)	200–1,000	100–500
House air exchange rate	100–600	50–300

Typical air consumption of solid fuel heaters and air exchange rate for houses. For open burners the flow is roughly 50 cubic feet per minute for every square foot of the opening (for example, the fireplace face).

Neutral

Strong Draft

**INTERIOR
CHIMNEY**

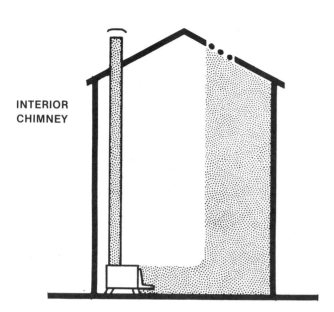

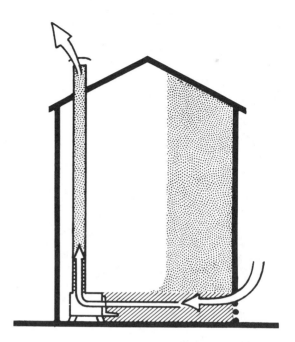

Severe Draft Problems

Neutral

**EXTERIOR
CHIMNEY**

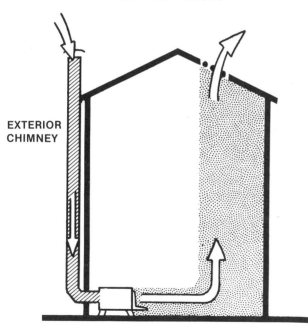

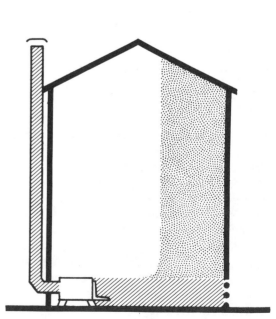

Figure 6–4. House-chimney draft interaction on a cold, calm day.

out any available cracks in the upper portions of the building, and cold outdoor air is sucked in through openings in the lower portions of the building. In very tall buildings, the resulting wind coming into open street-level doors can be very fierce unless special countermeasures are taken.

But the more important effect on chimney performance is the slight changes in house pressure caused by this buoyant house air. If the house has more cracks in its upper portions than near ground level, warm air can get out more easily than cold air can get in, and the result is a slight depressurization of the whole house. Lower pressures inside the house mean less tendency for air and smoke to go up the chimney. In essence the house itself is acting like a competitive chimney (Figure 6-4). In fact, building engineers call this pressure effect in buildings the *stack effect*. In a 2-story or 3-story house, the effect can be as large as 0.03 inches of water.

At the opposite extreme is a house that is predominantly leaky in its lower portions—as would be the case with a ground-level door open. Since cold air can then move in more easily than the warm air can get out, pressure at the bottom of the house tends to equalize with the outdoor pressure at ground level. This pressurization encourages the flow of smoke up the chimney.

When the stack effect of the building is working against the chimney's own draft, not only is the chimney's draft generally decreased, but the flow in the chimney can even reverse! This is most likely with an exterior chimney starting on the ground floor of a multistory house. Exterior chimneys run cooler, so the smoke is less buoyant. A multistory house has a stronger stack effect. If the average temperature of the smoke in the chimney is less than the average temperature of the air in the house, the flow is likely to reverse. Cold air descends down the chimney, feeds the fire, and all the smoke comes into the house through the air inlet of the appliance (Figure 6-5).

Since a contributing cause of flow reversal is cool smoke in the chimney, chimneys with high heat loss contribute to the problem, as does operating the appliance at a low firing rate (often done overnight with stoves). Also colder outdoor temperatures make reversal more likely.

Correcting reverse flow in a chimney can be difficult. The first thing to do *any* time smoke is spilling into the house from a wood burner is to open a door or window on the ground floor and on

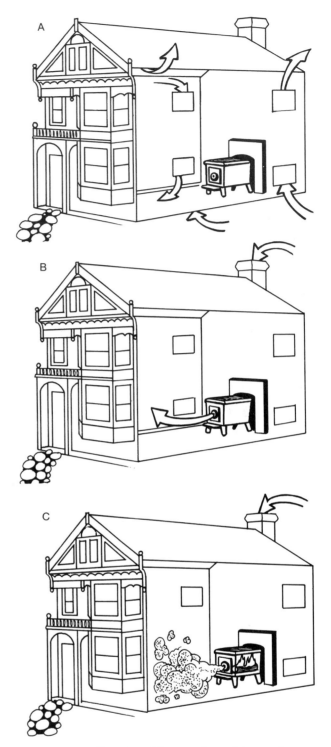

Figure 6-5. Exterior chimneys in multistory houses often have poor draft. A: The normal air flow into and out of a house on a calm winter day. The air tends to be drawn into the house on the ground floor. B: As a result, many exterior chimneys are non-self-starting; air flows down the chimney when it is cold. C: Even when the chimney is in use, smoke can become sufficiently cool that flow reversal occurs.

the upwind side of the house if it is windy, and *close* all other openings. This maximizes the pressure in the house, which may by itself correct the situation. If a window fan is available, use it to force outside air into the room; close all other windows and all doors, both exterior and interior. The resulting pressurization of the room will almost certainly correct the chimney's reverse flow. Directing a strong portable fan at the air inlet or into the open combustion chamber is *not likely* to help and can cause sparks to be blown into the room. In chimneys with easily accessible clean-out doors or unused breachings, it sometimes helps to insert some newspaper inside the chimney and light it; the additional warmth increases the buoyance enough to get the chimney operating properly.

Interior chimneys are rarely susceptible to flow reversal because their warmer environment keeps the smoke at least as warm as the house air temperature.

You can test a chimney for its flow-reversing tendency before actually using it. On a very cold, calm day when the chimney has not been in use for a day or two, susceptible chimneys are likely to have cold outdoor air descending and entering the house. If the air flow is not obvious by its force, temperature, or its chimney odor, you can use cigarette or incense smoke, or tissue paper, or tinsel to see which way the air is moving. Chimneys that run backwards when cold are called *non-self-starting*. The non-self-starting problem itself is traditionally solved by lighting newspaper stuck up into the chimney.

Non-self-starting is only an annoyance. But flow reversal is dangerous because it can lead to asphyxiation. It is best not to use such chimneys, especially for airtight stoves and inserts. If such chimneys are used, smoke detectors in the house are essential.

Weather and Altitude

Windy weather can affect chimney performance adversely by reducing or even reversing the flue-gas flow—making the fireplace or stove smoke. Properly locating the chimney top relative to the roof can alleviate some of the problem, and chimney caps can be very effective.

There are 3 kinds of wind effects. If the wind direction is into an opening through which smoke is trying to come out, the smoke flow is impeded. For example, a low, open (uncapped) chimney at certain locations downwind of a building may not perform well if the wind blows too hard (Figure 6-6). This first effect is caused by what is called the "velocity pressure" of the wind.

There are 2 ways to alleviate this problem. If the chimney extends high enough, the wind generally will blow across it, not into it. This is 1 reason why it is usually recommended that the chimney top be 2 feet higher than any portion of the roof within 10 feet of it and 3 feet higher than the roof area through which it penetrates (Figure 6-7). (The other reason is to prevent hot flue gases or sparks from igniting the roof.)

Installing a chimney cap is the other way to alleviate this wind back-pressure effect. Wind caps make the calm-weather smoke discharge omnidirectional. Then, with wind blowing from any direction, some smoke can still always get out on the other side. A flat plate on posts is a common and simple wind cap for masonry chimneys. Many kinds of metal caps are used, especially on metal chimneys (Figure 6-8). Chimney caps that rotate or have moving parts are sometimes unsatisfactory for use with wood fuel; creosote deposits can foul up the caps to the point where they become stuck.

A second wind effect always helps chimney performance. Air exerts what is called *static pressure*, which, in still air, is the same as atmospheric pressure. But in moving air, the static pressure is less.[2] In principle, a barometer in moving air would read a lower pressure than in nearby still air. Thus, if a chimney top is in a region of strong winds, the general (static) pressure level all around it is reduced. This effectively applies suction on the gases in the chimney, and so it draws better—the wind aspirates smoke from the chimney. The stronger the wind, the larger is the effect. This effect is in addition to the velocity pressure effect previously discussed.

A third effect, related to the viscosity of the air and its turbulent motion, is the creation of a large low pressure eddy region on the downwind side of a house (and a much smaller high pressure region on the upwind side). A chimney terminating in this low pressure region performs better, since the low air pressure tends to suck the smoke out.

When all the effects are combined, and if the

---

2. This is approximately described by Bernoulli's equation. Specifically, the pressure decrease in moving air compared to still air is $\frac{1}{2} dv^2$, where d is the mass density of the air, and v is its speed.

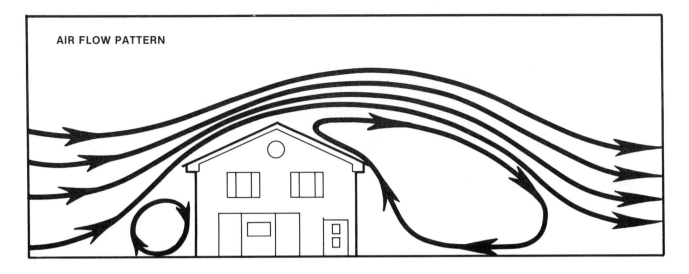

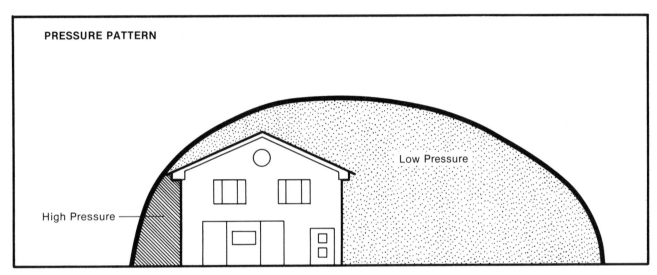

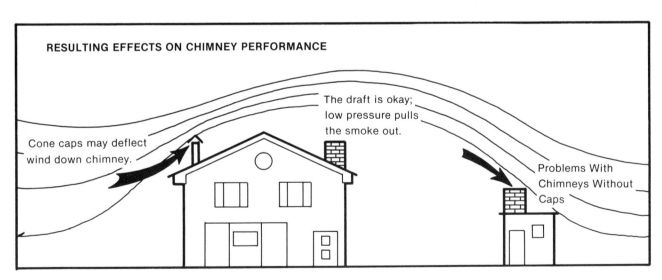

Figure 6-6. Typical wind effects around a building. Similar patterns are caused by trees and hills.

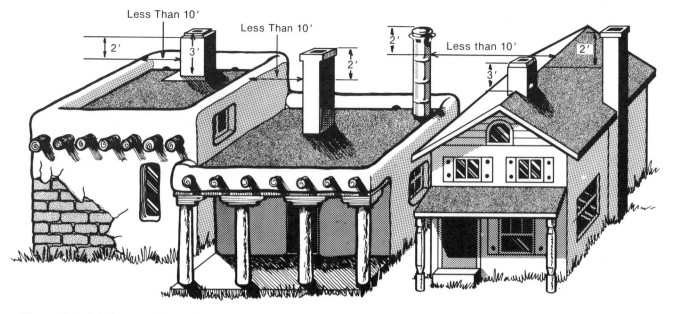

Figure 6–7. Minimum chimney height recommendations according to NFPA and most building codes.

chimney has a good cap, winds usually do not cause serious problems. For wind velocities up to the flue-gas velocity, there is very little effect. (Typical stove flue-gas velocities are 1–6 miles per hour.) For wind speeds 1–3 times the flue-gas velocity, there is a slight reduction in chimney flow. At wind speeds greater than about 3 times the flue-gas velocity, chimney performance is better than in still air. With an effective chimney cap, wind speeds between roughly 5 miles and 20 miles per hour may cause a slight decrease in chimney capacity, slower winds will have little ef-

fect, and faster winds often will improve performance.[3]

Gusty winds can cause slight smoke spillage into a house with any kind of chimney and cap, but less with taller chimneys. The currently recommended chimney heights, illustrated in Figure 6–7, are shorter than the chimneys on most eigh-

3. These quantitative estimates are from R. L. Stone, "The Role of Chimneys on Heat Losses from Buildings" (Paper delivered at the Annual Meeting of ASHRAE, Boston, Mass., June, 1975).

Figure 6–8. Chimney cap designs.

teenth and nineteenth century houses in New England. Taller chimneys are recommended as long as there is still good access for chimney cleaning.

Wind can change the pressure inside a house, which then affects chimney performance. If the house has some open windows or doors on the side facing into the wind, the house pressure increases, which helps the chimney; openings on the downwind side decrease the pressure inside the house, hurting chimney performance. A temporary solution to a badly smoking stove or fireplace on a windy day is to open a downstairs window or two on the upwind side of the house.

Outdoor temperature can affect chimney performance because the draft depends on the difference between the flue-gas temperature and outdoor temperature. The colder the weather, the better the draft in interior chimneys. As the temperature drops from 50° F. to 0° F., chimney capacity can double. Exterior chimneys also perform better in cold weather if their insulating value is high enough and if the flue-gas temperature and flow rate is high enough to prevent excessive cooling of the smoke.

Barometric pressure also has a small effect on chimney capacity. When the pressure is high, the combustion air and the flue gases are a little denser; more smoke can get through a chimney. An increase in pressure of 1 inch of mercury increases chimney capacity by about 4 percent.[4] Typical variations in barometric pressure span no more than 2 inches of mercury, which causes an 8 percent change in chimney capacity.

It is unlikely that humidity in the air has an effect on stove performance. Even when the relative humidity is 100 percent, water vapor can never constitute more than 1.5 percent of air by weight (at 68° F. or less); so changes in humidity can affect the composition of air by only this small amount. The maximum effect of humidity on the oxygen content, density, thermal conductivity, and specific heat in air is less than 1.5 percent, which is negligible.

None of these weather effects is very big by itself, but they frequently act together. When barometric pressure is high in the winter, the weather is often cold and calm; when the pressure drops, the weather is often windy and not as cold. Adding all the effects together suggests that weather can affect chimney capacity by as much as 20 percent, the higher capacity often coinciding with high barometric pressure, making it easier to have hotter fires.

Weather effects will be most noticeable in systems that are marginal to begin with. In many cases, there is enough excess chimney capacity that these weather effects will never be noticed.

Altitude also affects chimney performance. The thinner air at high altitudes decreases chimney capacity just as does low barometric pressure. For every 1,000 feet of elevation above sea level, chimney capacity is decreased by about 4 percent relative to the same chimney at sea level.[5] At 5,000 feet, a chimney would have to be designed with 20 percent extra capacity compared to a chimney at sea level serving the same appliance. At 10,000 feet, the effect is about 40–45 percent.

Although elevation effect is much larger than the barometric pressure effect, it is usually of little practical consequence for solid fuel heaters because chimneys usually have generous excess capacity. Also, the thinner air at high altitudes can mean that less oxygen gets to the fire; if the fire is not as intense, the chimney need not have as much capacity.

SAFETY FUNDAMENTALS

Chimneys are frequently implicated in wood-heating-related house fires, but exact causes are often difficult to determine. Certainly, improper installation and, in particular, inadequate clearance between a chimney and combustible parts of houses are common causes. In dry climates, sparks coming out of chimney tops can start fires. Some chimneys are unsafe due to improper construction or manufacture. Chimneys must be able to safely endure chimney fires—the burning of accumulated creosote deposits inside the chimney.

Regardless of the quality of the chimney, chimney fires are always potentially dangerous. Chimneys must have convenient access for inspection and cleaning of creosote.

Stovepipe (single-wall sheet metal pipe) is *not* a safe material for the chimney itself, but it can be used for the *chimney connector*, the connection between the appliance and the chimney.

4. *ASHRAE Handbook and Product Directory, 1975 Equipment Volume*, (New York: American Society of Heating, Refrigerating and Air-Conditioning Engineers, 1975) p. 26. 13.

5. Ibid.

Stovepipe has very little insulating value. As a result, the exterior can get very hot, which is why most building codes require 18 inches of clearance to combustible materials (see chapter 10 for details), an inconveniently large clearance for a chimney. Also, the flue gases can get quite cool, resulting in decreased draft and the possibility of very rapid creosote accumulation. Stovepipe is relatively flimsy. Such thin-gauge ordinary steel can rust out in a matter of months, particularly when used outdoors.

There are 2 basic traditional types of potentially safe chimneys for solid fuel burning appliances: prefabricated (also called factory-built) metal chimneys and masonry chimneys. I use the phrase "potentially safe" because chimney safety also depends on proper construction and installation, proper maintenance, and nonabusive use of the chimney.

Prefabricated Chimneys

Prefabricated metal chimneys are usually easier to install and less expensive than masonry chimneys. There are 3 common types of factory-built chimneys (Figure 6-9 and 6-10).

- Double-wall insulated (also called "solid" or "packed" or "mass" insulated).
- Triple-wall air-insulated.
- Triple-wall air-cooled (also called "thermosyphon").

There has been considerable confusion about the 2 triple-wall types. Air-cooled chimneys intentionally draw in outdoor air at the top between the outer 2 layers, circulate this air down the entire length of the chimney and back up between the inner 2 layers, and discharge the air back to the outdoors. The moving cold air contributes to keeping the outside of such chimneys from getting overly hot. The triple-wall air-insulated chimneys do not draw in outside air intentionally, nor do they have such an organized air flow pattern. Rather they intend to utilize the insulating value of relatively trapped air.[6]

But air-insulated chimneys are not tight enough to trap the air. There is an air flow; typically 5-20 cubic feet per minute is drawn into the chimney air spaces (some also is drawn into the

---

6. In practice, these conceptual distinctions between chimneys may not be as clear. See Appendix 3 of Jay Shelton, *Wood Heat Safety*.

TABLE 6-6
HEAT LOSS COEFFICIENTS
OF CHIMNEYS AND CONNECTORS

Type	Heat Loss Coefficient*
Heavy steel, clay, or iron pipe	1.3
Dark or weathered steel stovepipe	1.2
Shiny, galvanized or stainless stovepipe	1.0
Brick, tile lined	1.0
Double-wall metal, air space	0.4
Double-wall metal, insulation in between, stainless steel inner	0.4

* Suggested design value—actual measured values span a considerable range. Units are Btu. per hour per square foot of inside area per °F. temperature difference (flue gas to exterior).

Heat loss coefficients of some common chimney and connector types. From Table 6 in chapter 26 of The ASHRAE Handbook and Product Directory, *1979 Equipment Volume (New York: American Society of Heating, Refrigerating and Air-Conditioning Engineers, 1979).*

flue) and is vented out the top. This flow is less than in thermosyphon chimneys; but in many air-insulated chimneys, it contributes significantly to keeping the outside of the chimney at a safe temperature. (In a sense, the mass-insulated chimneys are more deserving of the name "air-insulated" because it is the trapped air between the fibers or powder particles that makes the insulating materials good insulators.

Looking at the ends of a triple-wall chimney

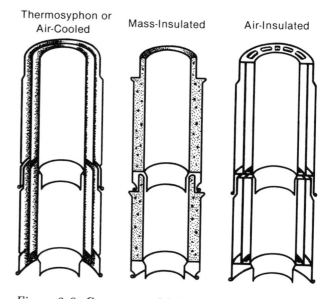

Figure 6-9. Common prefabricated chimneys.

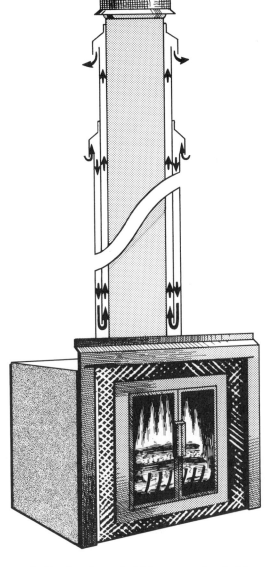

Figure 6-10. An air-cooled (thermosyphon) chimney. The air flow is driven by the buoyance of the air between the inner 2 metal layers.

section will not readily reveal which type chimney you are examining. Most air-insulated types do not really seal in the air. The only reliable way to tell what kind of chimney you have is to examine the topmost chimney section. A thermosyphon chimney has special openings and/or drip collars where the outdoor air is drawn in and where the heated air is discharged.

Triple-wall air-cooled chimneys are not generally intended or recommended for use with solid fuel stoves, furnaces, and boilers. They are most commonly used to vent factory-built metal fireplaces, and usually fireplaces from the same manufacturer as the chimney. The reason they often are not recommended for stoves is that the air-cooling system works so well that the flue itself is also cooled. In very cold weather, with airtight appliances, the result can be substantial creosote buildup. With factory-built fireplaces this is less of a problem because the appliances tend to be less airtight, which results in less creosote. These comments apply only to air-cooled (or thermosyphon) chimneys, and not to all triple-wall chimneys.

The insulation used in most double-wall mass-insulated chimneys is an amorphous silica powder plus a little fiberglass; it does not contain asbestos.

Some new factory-built chimney designs are becoming available, partially in response to the facts that a few factory-built chimney lines have warped during chimney fires and others have suffered corrosion, particularly when used with coal appliances.

The safety standards for solid fuel chimneys are evolving. Both the Underwriters Laboratories and Underwriters Laboratories of Canada have either proposed or finalized changes involving higher flue gas temperatures (to better simulate chimney fires), and they require either a corrosion test or using materials known to be more corrosion resistant. Chimneys that pass these new requirements are sometimes called "super chimneys." Although such chimneys offer improved safety, they are also more expensive. The past generation of prefabricated chimneys has *very* rarely been inadequate. The vast majority of chimney-related house fires in the past have been caused by not following the manufacturer's installation instructions. A contributing cause has been not keeping the chimney relatively clean of creosote. These problems will continue to exist with the "super chimneys."

Design features in these new chimneys may include:

- More corrosion-resistant stainless steel alloys (such as 304 instead of 430).
- Heavier gauge metal.
- More or better insulation in mass-insulated chimneys.
- Refractory linings in prefabricated metal chimneys.
- Redesigned chimney section ends to accom-

modate more thermal expansion and to retard heat leakage.

Appearances can be dangerously deceptive. There are some superficially look-alike factory-built double-wall chimneys that melt when used with wood stoves (Figure 6-11). "Type B" vents are intended for use with gas appliances only. They have aluminum liners that cannot take wood-appliance flue-gas temperatures. There are also Type L vents intended for oil appliances. They have steel liners which will not melt, but with only 2 metal layers and no insulation, they can become very hot on the outside. Type L vents are permitted by most codes as chimney connectors but not as chimneys for solid fuel appliances.

Buying a Prefabricated Chimney Or Using an Existing One

To be sure that your chimney is intended for use with wood or coal appliances, and that it meets nationally recognized safety standards, look for a label declaring the chimney to be a "Class A" or "All Fuel" or "Solid Fuel" chimney. The chimney should be *listed*. Listed means included in a list published by a recognized testing laboratory or inspection agency, indicating that the equipment meets nationally recognized safety

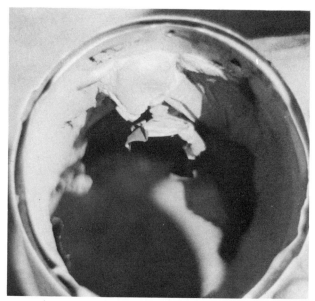

Figure 6-11. A melted liner in a B-vent chimney that was misused as a chimney for a wood stove.

standards. Two major chimney listing organizations in North America are Underwriters Laboratories (UL) and Underwriters Laboratories of Canada (ULC). The "UL Listed" or "ULC Listed" label means the chimney design satisfies nationally recognized standards for its intended use. These laboratories also check the adequacy of the instructions as part of the listing procedure.

Installation instructions must be followed very carefully. Insulated chimneys are not cool on the outside; *most prefabricated chimneys require a minimum 2-inch clearance to any combustible material*, except where following the instructions results in smaller clearances. It is unsafe to mix different brands of factory-built chimneys. They usually do not fit together anyway; but even if they do, the resulting chimney may be unsafe.

Masonry Chimneys

Masonry chimneys are usually built of brick or concrete block, but stone, adobe, and other materials are used also (Figure 6-12). Since masonry chimneys are traditional, their safety is rarely questioned; but they are by no means immune from potential problems.

A liner is essential. Traditional fireclay liners are considerably more resistant to flue-gas corrosion than are bricks or concrete blocks. Codes specify that the small space (roughly ½ inch) between the liner and the masonry should be left as an air space. The liner plus air space adds to the insulating value of the structure, keeping flue temperatures warmer and the outside of the chimney cooler. The space also allows for thermal expansion of the liner without the masonry structure itself cracking, which is important during chimney fires. The liner is kept centered in the chimney by the mortar overrun both between the bricks or blocks and, preferably, between the liner sections (Figure 6-13).

There appears to be a trend towards requiring more clearance to combustible materials. Many codes still permit direct contact between the outside of a standard 4-inch-thick masonry chimney and wood. However, the Canadian code does not, and neither do the codes in some states. Typical requirements in the newer codes are:

> The clearance between standard (4-inch wall) masonry chimneys and combustibles should generally be at least 2 inches, but exterior chimneys may be as close as ½ inch to the outside of the

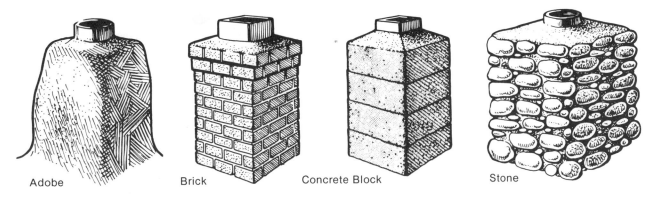

Adobe   Brick   Concrete Block   Stone

Figure 6-12. Masonry chimney types.

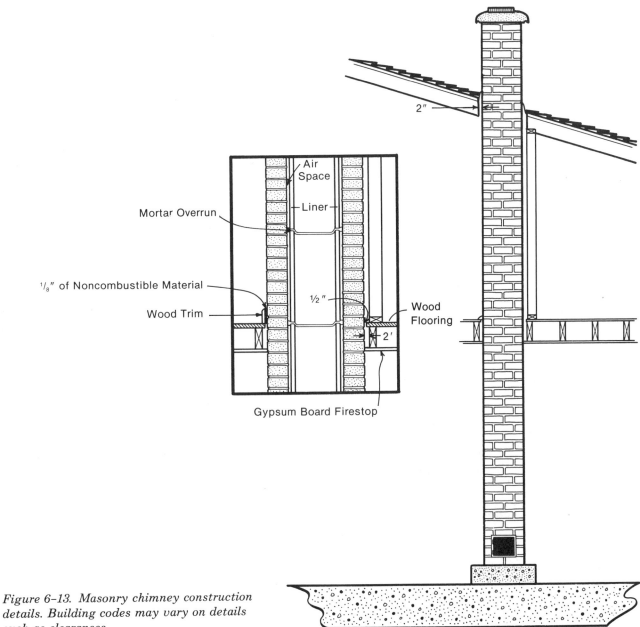

Air Space

Liner

Mortar Overrun

¹/₈″ of Noncombustible Material

Wood Trim

½″

Wood Flooring

2′

2″

Gypsum Board Firestop

Figure 6-13. Masonry chimney construction details. Building codes may vary on details such as clearances.

house, flooring may be ½ inch from the chimney, and wood trim can have only ⅛ inch of any non-combustible material separating it from the chimney. Nowhere may there be direct contact between the chimney and anything combustible.[7]

It is not clear how important these clearances are. It is likely that a large part of the current problems with masonry chimneys is not code inadequacies but rather that so many chimneys are not built in conformance to even older codes. However, I favor the increased clearances.

Why does a masonry chimney need these clearances when usually such chimneys are only warm to the touch? Because "usually" is not good enough. Most masonry chimneys now in existence were built for use either with gas or oil appliances, or with an open fireplace. Under sustained hard use, stoves, inserts, and solid fuel furnaces and boilers can generate hotter flue gases. Of course, during a chimney fire, temperatures can be higher still. Many house fires have started in ceilings, attics, and roofs where wood is in direct contact with a masonry chimney. The fact that a chimney has been used for years with no problems does not imply that it will be safe when used to vent a different kind of appliance.

The space between the chimney and nearby joists should not be filled with insulation. Convection in the relatively open air space usually does a better job keeping the wood cool than does insulation in this situation. All kinds of insulation should be kept away from chimneys and from recessed lighting fixtures in ceilings.

There is a new type of masonry chimney that is partly prefabricated and has passed the new, tough Canadian chimney standards. It uses pre-cast pumice aggregate outer casing and liner sections, and a special poured filler between the two, which sets up to provide strength and some additional insulating value. The liners and filler mix may also be used to reline an old chimney. Special steel collars are used to help position and join the liner segments.

Maintenance Accessibility

The best way to avoid a chimney fire is to keep the chimney clean enough so that a significant fire cannot occur. A buildup of ¼-inch or more

creosote indicates cleaning is advisable. The *only* way to know when cleaning is necessary is to inspect the chimney and stovepipe connector.

Chimneys must be easy to inspect. If inspection is awkward, it is less likely to be done, and chimney fires will be more likely. Chimneys also need convenient access for cleaning.

Although creosote can accumulate anywhere in the venting system, inspection should not require time-consuming or awkward disassembly of the system. Stovepipe inspection and cleaning can be made easy by installing a T either in place of an elbow or just in a straight portion of pipe (Figure 6–14). All chimneys should have a cleanout door or a cap at their bottoms; if located conveniently, a small mirror can be used at this opening to look up the chimney. When installing an insert in an existing masonry fireplace, it is often wise to install a cleanout access door just above the fireplace damper (or throat); otherwise, inspection and cleaning may require taking the insert out, which in some cases can be very awkward. Inspection from the chimney top may be reasonable if the roof is easily accessible, not steeply sloped, and rarely covered with snow or ice.

Lightning Protection

Where lightning protection for buildings is recommended, chimneys should be protected also. Usually, this involves a metal rod sticking slightly above the chimney, connected to an electrical ground with fairly heavy cable (about ¼-inch-diameter solid copper or AWG 2-stranded copper, or equivalents).

Metal chimneys themselves are good electrical conductors; they do not need rods, but they do need grounding. The cable should be wrapped snugly around the chimney at any convenient location and safely grounded.[8]

Spark Screens

Some chimney caps have screens (Figure 6–8) to keep birds and other animals out. Screens also can be important (and are sometimes required by codes) where the fire hazard is very high—to keep sparks and burning material from getting out and landing on dry vegetation. The screen size usually

7. Freely paraphrased from the "Canadian Heating, Ventilating and Air-Conditioning Code 1977," NRC No. 15560 (National Research Council of Canada, Ottawa, 1977).

8. For more details, see Jay Shelton, *Wood Heat Safety*, pp. 15–16; and NFPA 78, "Lightning Protection Code 1975" (Quincy, Mass., National Fire Protection Association, 1975).

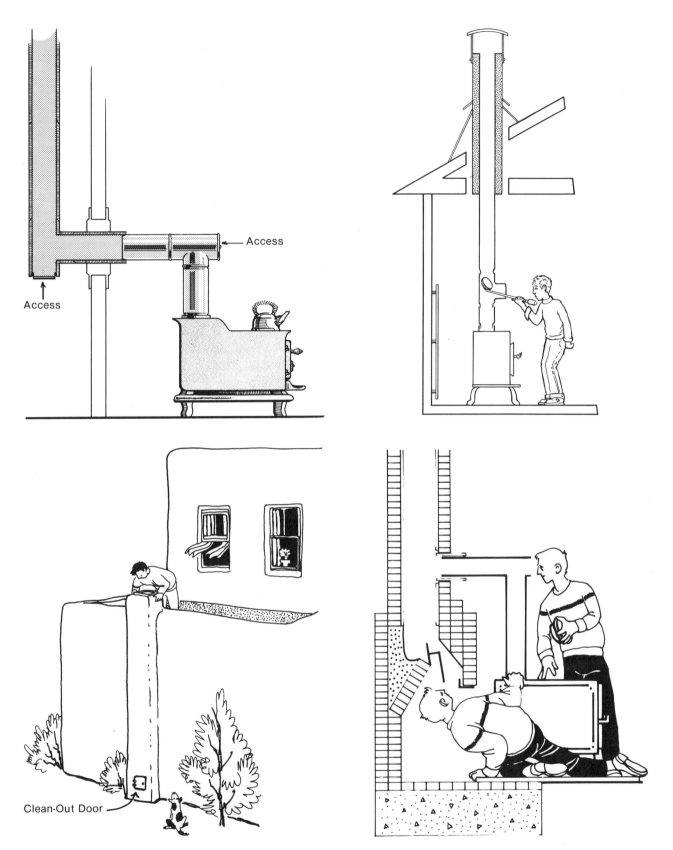

Figure 6–14. Some installations with good accessibility for inspecting and cleaning.

Access

Access

Clean-Out Door

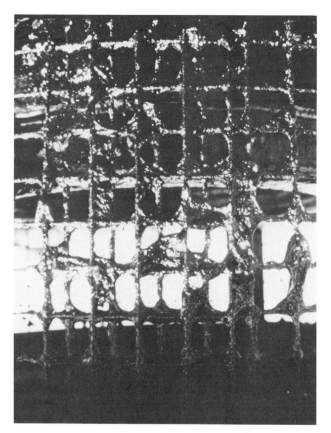

Figure 6-15. It is important to clean spark screens as they can become clogged with creosote.

recommended is ½-inch mesh. Larger holes may let out too much burning material; smaller holes easily clog up with creosote. Screens need frequent inspection and cleaning to keep them clear of creosote. It is usually recommended that the screen area be 4 times the area of the flue.

PRACTICAL CHOICES FOR SOLID FUEL INSTALLATIONS

When planning a solid fuel burning appliance installation, 5 broad questions concerning the chimney often arise.

- Is the existing chimney safe?
- Can a liner be added to an existing chimney?
- What is the best way to install a stove or insert using a fireplace chimney?
- Is it safe to vent a solid fuel appliance into the same flue used by a gas or oil appliance?
- When installing a new chimney, what is the best type and location?

The particular problems of using fireplace chimneys are discussed in chapter 12. The rest of these issues are discussed below.

Inspecting Old Masonry Chimneys

There is no such thing as an absolutely safe chimney, especially an existing masonry chimney. There is often no practical way to give such chimneys a really thorough examination.

An important first step is to clean the chimney. The cleaning itself may reveal weaknesses—the feel of the brush may signal a defective liner.

Here are some important things to look for (see Figure 6-16).

- General mechanical soundness. There should not be loose or missing masonry units (bricks, blocks, and so on). The mortar should be hard. If an awl can be forced into the mortar, it is not in good shape. Creosote stains on the outside of the chimney indicate either improper construction or decay. Some stains are related to excessive rain water entering the chimney.

If these problems are localized, they can be fixed. If these conditions exist over much of the chimney, the whole chimney may need rebuilding.

- Intact liner.
- Obstructions in the flue.
- Clearances. Unfortunately, most masonry chimneys have inadequate clearances. It is very difficult to assess the seriousness of this. There is no doubt that very many masonry chimneys without clearance to combustibles have been used for years with no problems. My personal feeling is that such chimneys, if otherwise sound, can be used with reasonable safety *if* the chimney is kept sufficiently clear to avoid all significant chimney fires, *and* the temperature of the flue gases entering the chimney is kept below 500° F. most of the time.[9] This is difficult to assure. A better (and more expensive) alternative is to add another liner.

If clearances are not quite what they should be but are greater than zero and accessible, some additional protection can be gained by putting shiny sheet metal between the chimney and the wood, ideally with open air spaces on both sides.

9. This is an estimate based on data in G. T. Tamura and A. G. Wilson, "Fire Hazard Tests on Small Masonry Chimneys," Internal Report No. 202 (Division of Building Research, National Research Council of Canada, Ottawa, 1960).

Any insulation should be pulled back a few inches away from the chimney.

• Unused and improperly sealed breachings. Before the advent of central heating, it was common to have many small stoves, each one in different rooms, for heating. Today, in many older houses, there can be many unused breachings in a chimney—places where appliances were connected once. If the house has been remodeled so that the chimney is covered, it may be very hard to find these locations. Unused breachings ought to be sealed with bricks and mortar to make them as tight and secure as the rest of the chimney. Pie-plate metal covers have very little insulating value and can get blown out by a pressure surge inside the chimney.

• Easy accessibility for creosote inspection and cleaning. Cleanout doors can be added, but if the job involves cutting through a tile liner, an experienced mason probably should do it.

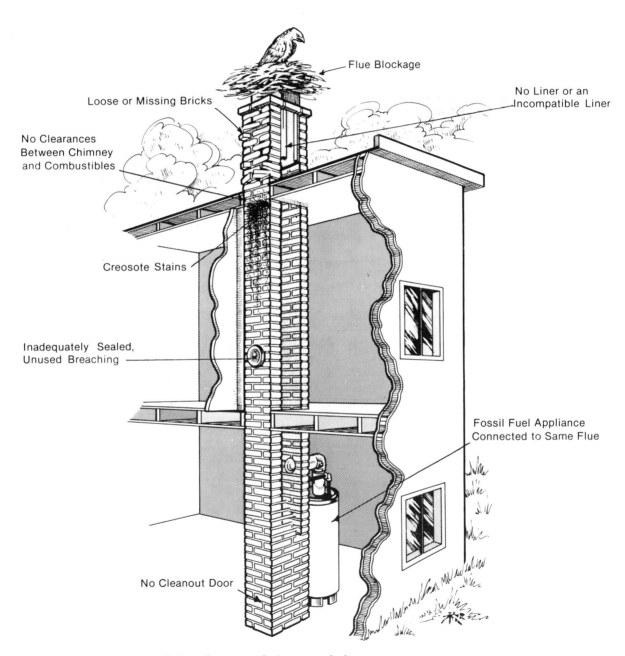

Figure 6-16. Some things to look for when considering an existing masonry chimney for use with a solid fuel appliance.

Lining Existing Chimneys

Unlined masonry chimneys must be lined to make them reasonably safe. Adding a second liner to a chimney gives an extra margin to safety, and can be particularly useful where the masonry chimney may not have desirable clearances to wood or may have undiscovered, poorly sealed breachings. Another reason for adding a liner to a chimney can be to reduce the size of the flue; many fireplace flues are oversized for stoves and inserts. Added liners usually reduce creosote and improve draft.

Chimneys should be cleaned as well as possible before relining to minimize the chance of the old creosote igniting.

Liner systems should withstand surface temperatures up to roughly 1,500° F., resist corrosion from wood and coal smoke, contain dripping creosote, and have a useful life expectancy of decades.

Adding a normal tile liner to an existing chimney is very difficult and not generally advisable without the use of special systems to assure the tiles are seated properly. Some sweeps and masons have the necessary equipment. The flues must, of course, be straight.

Ordinary steel stovepipe is not a suitable chimney liner. Although it performs satisfactorily while intact, it is likely to corrode or burn out quickly. Annual replacement would be necessary.

Stainless steel stovepipe is probably a suitable chimney liner. It is lightweight enough to handle easily. The sections can be fastened together with sheet metal screws as they are lowered into the chimney. Any chimney that is not straight can be difficult to line, but elbows are available. Ts are useful at the breaching (where the stovepipe enters the chimney). The crimped ends of each section should point down so that dripping creosote is contained inside the liner.

Make sure the liner is strong and well supported. It must be cleanable with a stiff wire chimney brush.

Stainless steel stovepipe liners are increasingly available. Complete systems are available, including the components for bringing the liner down through the damper and fireplace throat all the way to the appliance.

Stainless stovepipe liners may be hung from the top of the chimney or supported from below. The top of the chimney should be sealed around the liner with sheet metal to keep precipitation

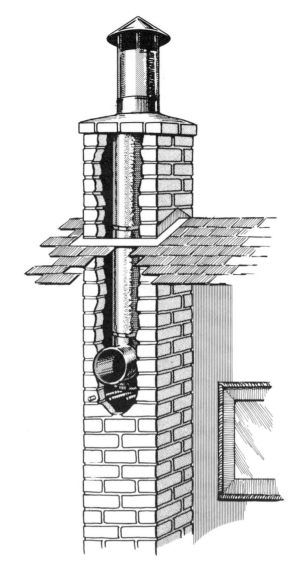

Figure 6–17. A liner of stainless steel stovepipe installed in a masonry chimney.

out and to keep the warm air trapped inside. In some instances, manufactured chimney flashing and drip collars can be used.

Virtually all materials expand when they become hot. Since a liner, especially if surrounded by insulation, will get much hotter than the original chimney, the installation must allow for some give. Stainless steels expand 2–3 times as much as ordinary steel. A 20-foot stainless liner will become longer when hot by inches! In some cases, lateral movement of the liner can accommodate the expansion, but where insulation prevents this, the installation should allow the top of the liner to slide up and down through its flashing.

Stainless steel is not indestructible. Some alloys and gauges may be inadequate. But because stainless liners are relatively new, it is not perfectly clear what is adequate. Most likely, the alloys and thicknesses used for the liners in factory-built chimneys are appropriate. Coal smoke is more corrosive than wood smoke; type 304 and 316 stainless are better than type 430 for coal.

Flexible stainless tubing is also available. It eases the problem of lining a chimney with offsets—chimneys with bends.

Another suitable liner is porcelain-coated, heavy-gauge steel.

Insulation can be added between these liners and the chimney to reduce creosote and increase draft. But insulation also reduces heat transfer to the house from interior chimneys. Since interior chimneys have less creosote and draft problems to begin with, adding insulation in these cases is not worthwhile usually.

Ordinary fiberglass insulation should not be used to insulate between liners and chimneys. It is not sufficiently durable at high temperatures. Perlite (expanded volcanic material) and vermiculite (expanded mica) are suitable. As with all masonry chimneys, the top should be carefully flashed to keep precipitation out. Also, the liner joints must be sufficiently tight to prevent the insulation from leaking inside the liner and accumulating, causing blockage.

An insulation that hardens after it is poured

Figure 6–18. Flexible stainless steel pipe for lining or relining chimneys.

will not gradually leak into the liner through joint cracks and can give the system a useful life beyond the time when the liner may start to deteriorate. Up to roughly 1/3 cement by volume and enough water to make the mix appropriately fluid has been tried with apparent success, although few, if any, such systems have been in use for decades.

Obviously you will not want to redo a relining, especially after adding insulation. Bear in mind that insulation helps hold the heat in during normal operation and during chimney fires. So use a high quality liner material that will last a long time and withstand extreme heat.

If a chimney flue is very large, it may be possible to use a standard factory-built metal chimney as a liner. This is a more expensive liner than some, but certainly results in a relatively safe chimney. However, it will essentially eliminate heat transfer to the house from an interior chimney.

Another relining method involves lowering an inflatable tube into the chimney. The tube has spacers to keep it centered. The tube is inflated, and a fluid insulating mixture is pumped down the chimney from the top. After the material hardens, the tube is deflated and removed, leaving an insulated liner bonded to the chimney walls.

This system has long been used in England, apparently with success. It is expensive compared to some other lining systems but has the important advantage of accommodating chimneys with offsets. Ideally, it would be good to have more information on durability. The thermal shock of chimney fires tends to crack masonry materials. How well does this type of liner and its chimney stand sudden extreme temperature changes? How well will the liner withstand the corrosiveness of wet creosote? In England, wood is not a common fuel; hence creosote fires are relatively rare.

I am aware of no problems with this system, but again, as with most relining systems, there has not been a lot of experience in this country.

Many of these insulated relining systems can be used in new masonry chimneys as a way to keep the flue gases hotter. In fact, an ordinary tile liner can be used with a few inches of poured insulation around it, if the chimney structure is large enough for the size liner needed. The liner can be held centered using plumber's strapping or the equivalent, and the insulation poured around it.

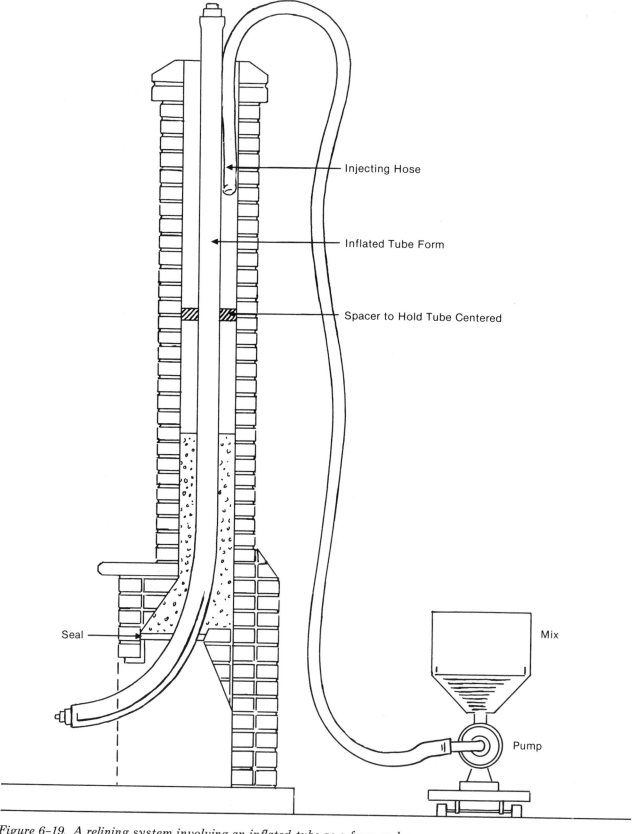

Injecting Hose

Inflated Tube Form

Spacer to Hold Tube Centered

Seal

Mix

Pump

Figure 6-19. A relining system involving an inflated tube as a form and an injected insulating mix that hardens.

Shared Flue Installations

Are shared-flue installations safe? Although many building codes allow it, the trend is towards prohibiting it. NFPA 211 prohibits it except for appliances so listed.

A flue is a single passageway. Many chimneys have more than 1 flue built into the same overall chimney structure. There is no controversy about the safety of having more than 1 well-built flue in a chimney. The issue here is whether it is safe to vent a stove into a flue already used by an oil or gas appliance.

What are the problems? First, creosote buildup can cause lethal fumes from the fossil fuel appliances to enter the house. Fallen creosote in the chimney bottom, if it rises high enough to start blocking the fossil fuel breaching, can have the same effect. Both these blockages also can cause soot deposits inside the fossil fuel appliance itself, substantially decreasing its energy efficiency.

Second, should a chimney fire occur, it is likely to be more intense because more air can get into the shared flue to feed the fire. It is also more difficult to put out a chimney fire by depriving it of air.

One *potential* benefit is reduced creosote. If the fossil fuel appliance is used often, it will help keep the flue warm and thus reduce creosote accumulation.

Overall, I recommend against multiple use of a single flue. But if it is done, here are the most important aspects to remember.

• Be sure the chimney has adequate capacity to handle all connected appliances. A common rule of thumb for multiple connections to a single flue is that the cross-sectional area of the flue must equal or exceed the sum of the area required by the largest appliance plus 50 percent of the areas required by the other appliances. Required areas usually are equal to flue-collar areas.

• Be sure the chimney and cleanout area are kept clean so that blockages and chimney fires cannot occur.

• If both appliances are on the same floor, try to put the two breachings a few feet apart (vertically) and put the wood-appliance connection below the fossil fuel appliance. Separating the connections by a few feet helps assure that each flue gas stream enters the chimney uninhibited by the other. Having the solid fuel smoke enter *below* assures that the solid fuel appliance will suffer first if fallen creosote is building up at the bottom of the chimney, resulting in decreased heat output from the solid fuel appliance, but not a hazardous situation.

SELECTING NEW CHIMNEYS

Selecting a chimney involves compromises. No one chimney type performs best at all desired chimney functions. Chimney location also affects the choice. The usual choice is between standard masonry (brick or concrete block) and the 3 types of prefabricated chimneys—insulated (with powder and/or fiber), air-insulated, and air-cooled. Let's look at the location first.

Locating a New Chimney

Wherever possible, particularly in multistory houses, new chimneys should run up inside the building. *Interior* chimneys have 4 distinct advantages over exterior chimneys, 3 of which are a consequence of the warmer environment, resulting in warmer temperatures inside the chimney.

• Better draft generally.
• Less creosote accumulation.
• Much less chance of chimney flow reversal.
• Added heat gain to the building.

Figure 6-20. Multiple flues inside 1 chimney.

The heat gain from interior chimneys can be substantial, particularly in a 2-story or 3-story house. For maximum heat gain, the chimney should be fully exposed to the living spaces, and not boxed in, or placed in closets, or covered with paneling. Among standard chimney types, masonry conducts flue-gas heat best (Table 6–6). I estimate the available improvement in overall energy efficiency of the system to be roughly 5–20 percentage points in 2-story and 3-story houses.[10]

Unfortunately, interior chimneys take up living space and are usually more expensive because of the additional finish work needed where ceilings and floors are penetrated and, in the case of masonry chimneys, the extra masonry needed to reach the foundation if the house has a basement. There is also the need with interior chimneys to *maintain* clearances—furniture, rugs, clothing, and so on should be kept at least 2 inches from the chimney. For safety, some homeowners *do* enclose the chimney, but install large registers in the enclosure at floor and ceiling level to allow the chimney heat to get out. However, the chance of a fire with unenclosed chimneys is not large, particularly with reasonably careful operation and maintenance of the system.

In single-story houses, or top-floor installations, chimney location is much less important. Not much chimney can be exposed to the living space for additional heat, a large portion of the chimney is inevitably exposed to outdoor temperatures, and flow reversal is rarely a problem in these installations.

However, a problem can arise in very cold weather with exterior chimneys, even in 1-story houses. Ice plugs can form, a problem most common in small masonry chimneys. Each pound of wood burned generates more than half a pound of water vapor. This water vapor often condenses with the creosote. When chimney interior surfaces are very cold, the volume of condensed water can be very large, and it will freeze on surfaces that are sufficiently cold. With exterior chimneys, the condensed water often will not freeze until it drips down below the breaching. As additional water freezes, a ring of ice gradually grows towards the center of the flue. When a full plug forms, the ice starts growing upwards. When it

10. This estimate is based in part on data in W. S. Harris and R. J. Martin, "Heat Transmitted to the I-B-R Research Home from the Inside Chimney," *ASHRAE Transactions* Vol. 59 (1953), pp. 97–112.

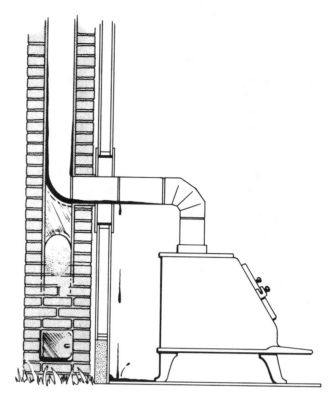

Figure 6–21. An ice plug in an exterior chimney.

reaches the breaching the water starts flowing into the house through the connector. Ice plugs, or at least constrictions, also can form at the top of any chimney, including interior chimneys.

The Pros and Cons Of Masonry Chimneys

Here are some of the advantages and disadvantages of standard masonry chimneys.

• Because the chimney is massive, it takes some time for it to warm up. Thus, it is common for a small amount of creosote to be deposited inside masonry chimneys every time a fire is started.

• On the other hand, this very same massiveness is beneficial during a chimney fire. The mass can temporarily absorb some of the heat from a chimney fire. This keeps the exterior of the chimney from getting as hot as it otherwise would.

• Massive interior chimneys provide beneficial heat storage. Substantial heat is stored in the masonry during hot fires. As the fire cools down, much of this heat can enter the house so the house does not cool off as quickly. Generally, mass increases the steadiness of the heat output.

Figure 6-22. A cracked masonry chimney.

• Masonry materials are poor thermal insulators. Since in steady conditions heat conducts through the chimney walls so easily, masonry chimneys are ideal in interior exposed installations for obtaining extra heat from flue gases.

• This same heat loss characteristic makes masonry chimneys more susceptible to creosote accumulation, especially as an exterior chimney.

• An intense chimney fire can crack a masonry chimney (Figure 6-22). Dangerously large cracks are rare but do occur. They are most likely in improperly built chimneys where the narrow space between the liner and the masonry wall has been filled with mortar, and does not allow the liner to expand when heated. Small hairline cracks are common, and a chimney can have a good many of them without being seriously weakened or leaky. But repeated chimney fires cause gradual deterioration.

• Masonry chimneys, even new ones, are rarely up to code, and even the codes themselves may not specify adequate clearances. Solid fuel heater use can be harder (hotter) on a chimney than oil or gas appliances. However, with reasonably careful inspection of a chimney before use, and most importantly, reasonably careful operation and maintenance of the system (not overfiring the appliance, and keeping the chimney clean), problems are very unlikely to arise.

The Pros and Cons
Of Prefabricated Chimneys

The standard types of prefabricated metal chimneys also have pros and cons.

• Prefabricated chimneys are usually less expensive.

• Some types of prefabricated chimneys, particularly the air-cooled, seem to accumulate large creosote deposits in cold climates because of excessive cooling of the flue. This can be true even with interior locations since it is usually outdoor air which is used for the cooling.

• The mass-insulated and air-insulated types of prefabricated chimneys seem to be the best of all standard chimney types at keeping the flue gases hot. Their use results in the least creosote, and they are, therefore, a good choice for an exterior chimney.

• No type of prefabricated chimney is particularly good as a supplemental heat source in a house. The mass-insulated type holds the heat in; the air-cooled type vents much potential heat gain up the cooling passages to the outdoors; and the air-insulated type holds most of the heat in and vents a little of it.

• Some types of prefabricated chimney apparently can be damaged in an intense chimney fire (Figure 6-23). Warping of the liner can allow insulation to settle, leaving an uninsulated gap where the outside of the chimney can then get very hot. I do not consider this distortion of liners to be a

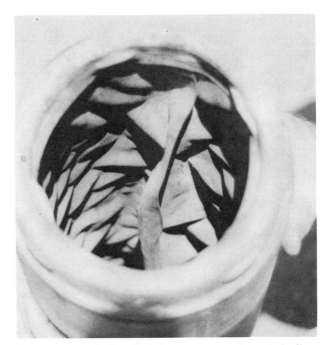

Figure 6-23. A distorted liner in a prefabricated all-fuel chimney. Such distortion occurs only very rarely, during intense chimney fires.

major problem. The probability of liner distortion is *very* low, especially if chimney fires are avoided, as they easily can be. The greatest danger of a house fire may be during a second chimney fire. Chimneys should be inspected after a chimney fire, and any damaged sections should be replaced.

• There has been some concern about the durability of the stainless steel liners when coal is burned. Coal fumes are potentially more corrosive than those from wood. Some chimney manufacturers have switched to types 304 and 316 stainless, which should help.

Concluding Thoughts On Selecting Chimneys

Conclusion? First, some general comments. All code-approved solid fuel chimneys are reasonably safe. All must be carefully installed with particular attention to clearances. None should be subjected to intense chimney fires. All chimney fires can be prevented by appropriate operation of the heater to prevent creosote buildup, or by cleaning the chimney whenever necessary.

The best kind of chimney depends on which performance characteristic is deemed most important. Interior masonry is the best for heating efficiency and for heat storage. Interior mass-insulated or air-insulated prefabricated chimneys are probably the best for minimizing creosote. Some of the new chimneys are probably best at enduring intense chimney fires. Masonry chimneys are probably more resistant to corrosion from burning coal.

DIAGNOSING AND SOLVING DRAFT PROBLEMS

If a solid fuel heater or fireplace is letting smoke into the house, the best ways to solve the immediate problem are to make sure any flue-gas dampers (bypass dampers and dampers in the collar, throat, or stovepipe) are fully open; and to open windows and/or doors in the first floor or basement of the house and close any openings in the upper parts of the house. If the weather is windy, the open windows and/or doors should be on the windward side of the house.

If a stove or fireplace tends to smoke only at the beginning of a fire, opening a window or door for a few minutes during light-up can be an easier and safer solution than shoving a lighted newspaper up the chimney.

If the solid fuel appliance only smokes when the weather is *windy*, 3 possible remedies are installing a chimney cap if there is none, installing a better chimney cap if there is one, and increasing the height of the chimney. Changing the height of a chimney is relatively easy with prefabricated metal chimneys.

If chronic smoking is a problem even in *calm weather*, check the venting system for obstructions and clean if necessary. Birds' nests and creosote are possible causes of blockage. Check for an open clean-out door, for unsealed breachings, and any other sources of air leakage into the chimney, especially in its lower half. If an oil or gas appliance is connected to the same flue, considerable air will always enter the chimney through the connector.

Another cause of *calm weather* draft problems is improper chimney diameter. Both undersized and oversized chimneys result in poor draft. A less likely cause is inadequate chimney height, but if increasing the height is not too difficult, it is worth a try.

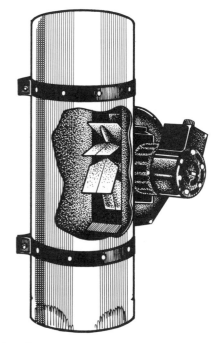

Figure 6–24. One type of draft inducer. Before purchasing a draft-inducing fan, make certain it is designed for the high temperatures of solid fuel flues.

If opening a window or door in the same room as the appliance improves performance substantially in calm weather, either the house was too tight or the stack effect of the house was too strong for the exterior chimney to fight. In either case, supplying outside air to the appliance can solve the problem.

A smoking fireplace often can be cured by making the fireplace opening smaller. This can be done by raising the hearth or installing a canopy hood extending down from the top of the fireplace opening. Alternatively, glass doors can be used to stop open fireplaces from smoking.

If all else fails, a draft-inducing electric fan, with its propeller inside the flue, can be used to force the smoke up the chimney.

CHAPTER 7

WOOD STOVE DESIGN AND PERFORMANCE

The ideal wood heating appliance should be able to do many things well. The appliance should perform with a heating rate both high enough and low enough to be useful as a heater; it should have high energy efficiency, low creosote potential, and be easily operated and durable. These characteristics can be affected by how the stove is designed (as well as how it is operated and even how it is installed). In many cases there are conflicts — a stove designed to be the best in one area may, as a consequence, perform poorly in another. In order to choose and operate a stove wisely, it is necessary to understand the interrelations among all the performance design characteristics of stoves.

Most of the principles to be discussed apply to all solid fuel appliances used in the home — stoves, fireplace inserts, furnaces, boilers, water heaters, and cookstoves — but they are discussed principally in the context of wood stoves in this chapter.

POWER OUTPUT RANGE

Heating is the fundamental job of a stove. The rate at which an appliance puts out heat is its *power output*. In the United States, heating appliances are usually rated in Btu. per hour; in Europe, watts are used (1 watt equals 3.41 Btu. per hour).

Rating solid fuel appliances is considerably more complicated than rating oil or gas heaters for 2 reasons. First, the maximum heat output of a stove is affected by many more variables, such as the amount of fuel per loading, its moisture content, the piece size, the frequency of reloading, and the draft developed by the chimney. All these factors can affect the rate at which the fuel burns. Most oil and gas burners inject and burn fuel at one constant rate.

The second complication is that stoves really need 2 ratings — a maximum and minimum power. Most oil and gas heaters, in a sense, have only 1 power; the device is either on or off. When the burner is on, heat is delivered at the maximum rate, and ratings are based on this maximum output. Usually less than the maximum output is desired, and the burner continuously cycles on and off, controlled by a thermostat or aquastat. The *average* power varies from a maximum value all the way down to zero. Thus, rating conventional heaters by their maximum powers tells essentially the whole story.

But a solid fuel fire cannot be turned on and off so easily. Low outputs are often needed in the fall and spring when the weather is cool but not cold. Not all stoves can sustain a slowly burning fire easily. Testing and rating ought to address the low as well as the high end of the power output range of an appliance.

How does stove design affect the minimum and maximum power output capability of a stove? For low power output, the single most important design feature is airtightness. An airtight or "controlled combustion" stove is a stove in which the fire can be extinguished by shutting the air inlet(s), leaving unburned fuel in the stove.

How much the air supply can be restricted in an airtight stove before the fire goes out depends mostly on the properties of the fuel. The drier the fuel is, the less likely it is to go out. Power outputs as low as 3,000 Btu. per hour can be obtained (Figure 7-1) and not much less is ever needed – 3,000 Btu. per hour is equivalent to the body heat output of about 6 resting adults.

Large power outputs require high rates of combustion and good heat transfer, and this requires a large amount of fuel and air, and a large stove. Heat outputs of 50,000 Btu. per hour can be achieved with most medium and large stoves. This is enough to heat many homes entirely, even in the coldest weather.

Matching the power output range of a stove to the heating needs of a house is discussed in chapter 9.

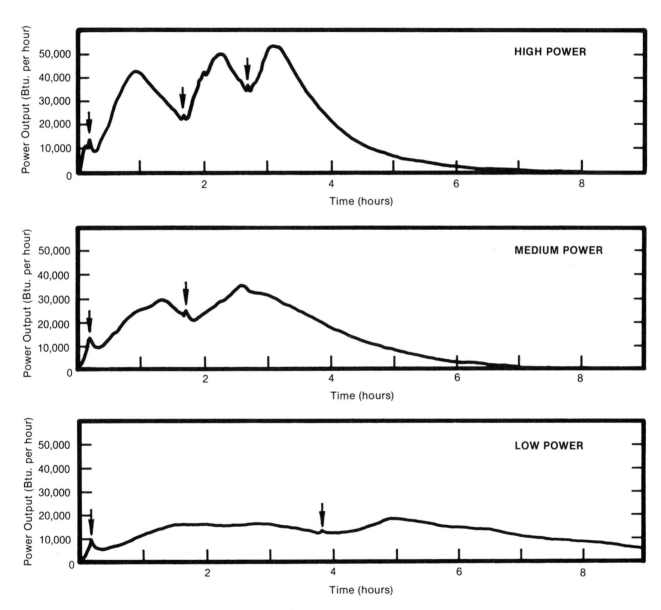

Figure 7-1. Power curves for the same stove operated at 3 different air settings. The arrows indicate when fuel was added. Aside from the kindling at the beginning, all fuel loads were about the same size. Higher powers require more frequent refuelings. Less frequent refueling results in lower average power output.

TIME BETWEEN REFUELINGS

Having to refuel a stove frequently is an inconvenience. Sometimes you will want to just keep the fire going, even if only slowly, so that when you get up in the morning or return home from work, you can make a hot fire quickly by adding wood, without having to start from scratch with kindling, paper, and a match. Again, the most critical design feature here is airtightness—almost any airtight stove with a full load of fuel and a restricted air supply can hold a fire for at least 10 hours.

Sometimes you will want to keep the house warm during the long duration burn. This requires a large fuel capacity. A small airtight stove that can hold 20 pounds of wood will average only 7,000 Btu. per hour over a 10-hour burn (assuming that the wood energy content is 7,000 Btu. per pound, moist basis, and the overall energy efficiency of the stove is 50 percent). To obtain 30,000 Btu. per hour output over a 10-hour burn would require a fuel load of about 85 pounds, which in turn requires a large fuel chamber.

Maximum power output and maximum time between refuelings cannot occur simultaneously. A manufacturer's brochure which claims both an 80,000 Btu. per hour output and a 24-hour burn is misleading, if it does not also mention that these 2 kinds of performance cannot occur at the same time. Maximum power requires a wide open air inlet, which results in the wood burning quickly. Maximum burn durations require restricted air settings, and this results in very low power output.

Each equal fuel load in a given stove will produce roughly the same amount of heat. If it is burned quickly, the heating will be at a high rate but of short duration. If it is burned slowly, the heating will be at a low rate but will last for a longer time.

ENERGY EFFICIENCIES

The term "efficiency" as commonly used has many meanings. An "efficient" stove can be one that has a high power output, or that is easy to operate, or can hold a fire for a long time, or is a good buy, or almost any other desirable attribute.

The term "energy efficiency" has a precise meaning. It is the fraction (or percentage) of the chemical energy in the fuel that becomes heat in the house:

$$\text{Overall energy efficiency} = \frac{\text{heat energy output to house}}{\text{solid fuel energy input}}$$

The principal advantage of an energy efficient system is that less solid fuel is consumed to produce the same amount of useful heat.

Any appliance with energy as the principal output can be described by an energy efficiency. Incandescent light bulbs have electric energy as their input and light energy as their main functional output; the conversion efficiency is 3–8 percent depending on bulb type and size. (Over 90 percent of the electrical energy leaves the bulb as heat, not light!)

We can compare the efficiencies of conventional heating systems to wood stoves, which typically have overall energy efficiencies of 50–60 percent. Electric space heating is essentially 100 percent efficient—all of the electric energy entering the house heating system is converted into useful heat in the house. Electricity *generation* is less efficient; in fossil fuel and nuclear plants, only

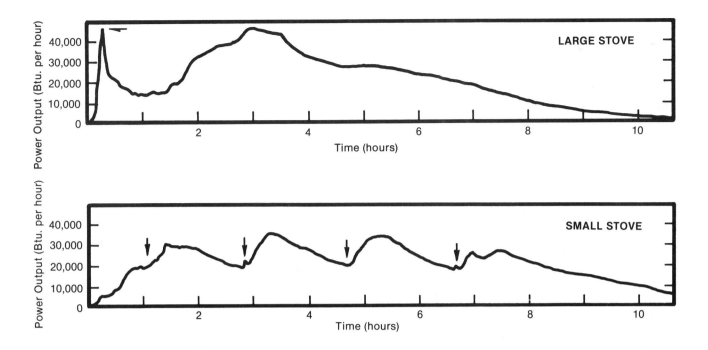

Figure 7-2. *Equal average power output from a large and a small stove. The arrows indicate when the main fuel loads were added. The small stove required 4 fuel loads during the same time the large stove consumed 1.*

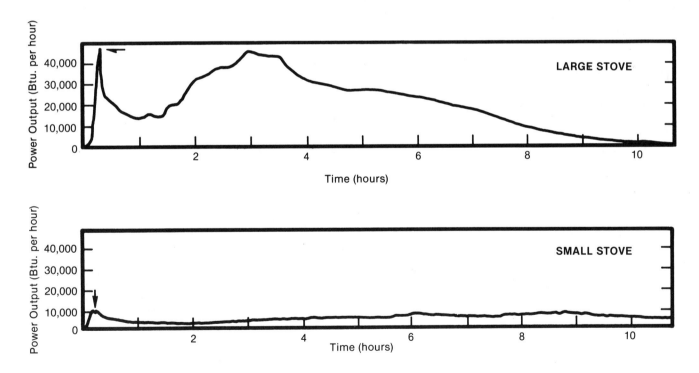

Figure 7-3. *Power curves for a large and a small stove, each burning a single full load of fuel with the air inlets set for achieving approximately the same burn duration. Both achieved a long duration burn, but the average power outputs were very different.*

about 30–40 percent of the energy in the fuel is converted into electrical energy. The "waste" heat is usually dumped into nearby water, or dissipated in the air. The net heating efficiency of electric space heating, including generation, pump storage, and transmission losses is about 30 percent (the amount of heat energy delivered into the house is about 30 percent of the corresponding fuel energy consumed at the electric plant). Energy efficiencies of gas fired and oil fired furnaces and boilers in homes range from about 30 percent to 75 percent.

For a solid fuel heater to have a high overall energy efficiency it must do 2 jobs well. First, the fuel must be burned as completely as possible so that very little smoke goes up and out the chimney (smoke is unburned, gasified solid fuel). Second, the stove must transfer the heat from inside the stove to the room, minimizing the heat lost up the chimney.

Combustion and heat transfer are quite distinct processes; different stove design features are beneficial to each. It is easier to understand stove design and performance by considering each of these processes separately. Each process has its own energy efficiency. The product of the 2 com-

---

Energy Efficiencies of Oil and Gas Central Heaters

Typical efficiencies of gas-fired and oil-fired furnaces and boilers in homes vary from 30 percent to 75 percent. The most important factors in determining this rate are the design of the furnace, maintenance of the furnace, and the sizing of the furnace relative to the building's needs, or equivalently, its duty cycle or fraction of "on" time. The kind of heat-distribution system (steam, hot water, or air) matters some, but usually not as much as the above 3 factors.

Basic design principles for fossil-fueled furnaces are the same as for wood stoves, and are discussed at length in this chapter.

Furnace maintenance is critical in 2 areas: keeping the fuel-to-air ratio optimal, and keeping heat-exchange surfaces clean of soot. The latter is especially important in oil furnaces since oil, a relatively nonvolatile liquid fuel, tends to burn less completely, generating soot.

Efficiencies as high as 80–85 percent are often claimed for gas and oil central heating systems. Most of these claims are exaggerated. The measurement conditions on which they are based may include a perfectly adjusted burner, clean heat transfer surfaces, properly adjusted draft control (for oil), complete warm-up of the system (continuous furnace operation), and that all heat losses through the sides of the unit and from the ducts or pipes are in fact useful gains to the house.

In practice, jacket and distribution system losses are not all always useful house gains. Not all the heat losses into a basement space get upstairs to the living spaces. Hot air rises, but radiation travels in all directions.

Fossil fuel heaters have their highest energy efficiency when on continuously. In practice, furnace capacities are chosen conservatively, so that they exceed by a comfortable margin the maximum expected heating needs of the buildings in which they are installed. As a result, furnaces are constantly cycling on and off. The fraction of "on" time rarely exceeds 75 percent and, of course, approaches 0 percent at the beginning and end of the heating season.

Furnaces have lower average efficiencies when cycled on and off than when operated continuously. There are many reasons for this, some of which may not apply in particular situations, depending mostly on the type of fuel (gas or oil), the type of heat distribution system, and whether or not you want to heat the basement of your house.

Some of the possible reasons for the lower efficiencies are the following. Gas pilot lights are on throughout the heating season, but may not be contributing useful heat to the house, except when the main burner is also on. During the first few minutes of "on" time, when the furnace and its flue are not hot, combustion is less complete since the combustion region is not warmed and since normal stack draft has not developed. Some of the heat that is stored in the furnace walls and chimney, and in the heat-distribution fluid, ducts, or pipes, is lost each time the system cools off—some goes up the chimney, and some may leak out into parts of the house where heat is not desired or needed.

When all factors are taken into consideration by measuring the actual heating efficiencies of fossil fuel systems in people's homes, the net result is efficiencies in the range from about 30–75 percent.

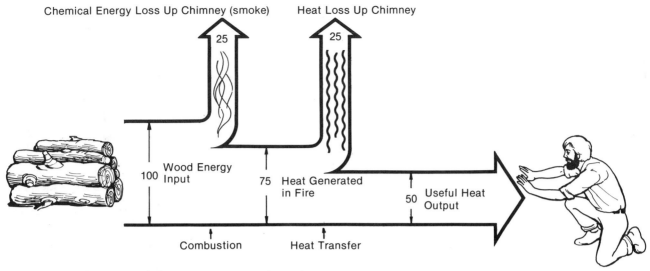

Figure 7-4. Energy flow diagram for a typical wood stove.

ponent efficiencies (expressed as fractions) is the overall energy efficiency.

Combustion efficiency =

$$\frac{\text{heat generated in combustion}}{\text{fuel energy input}}$$

Heat transfer efficiency =

$$\frac{\text{useful heat energy output}}{\text{heat energy generated in combustion}}$$

Overall energy efficiency = (combustion efficiency) × (heat transfer efficiency) =

$$\frac{\text{useful heat energy output}}{\text{fuel energy input}}$$

The maximum possible energy efficiency is achieved when both combustion and heat transfer are complete. Complete combustion occurs when all the carbon, hydrogen, and oxygen atoms in the fuel are converted to carbon dioxide and water vapor. Under these ideal conditions, there is no smoke, no odor, no creosote, and no air pollution, and all possible chemical energy in the fuel is released as heat in the fire.

Heat transfer essentially is complete if the flue gases have cooled down to room temperature before leaving the interior of the house. If both complete combustion and complete heat transfer were achieved in a heating system, its efficiency would be essentially 100 percent. (If the European system for reporting energy efficiencies, which is based on the *low* heat value for fuels, were used, efficiencies in this ideal case would exceed 100 percent. See chapter 3.)

Although complete combustion is desirable, complete heat transfer usually is not desirable for 2 reasons. In some installations, room temperature flue gases would not induce adequate draft.

TABLE 7-1
CONVERSION CHART FOR
ENERGY EFFICIENCIES

Energy Efficiencies Using Lower Heating Value or European Basis	Energy Efficiencies Using Higher Heating Value or American Basis
0	0
10	9
20	18
30	27
40	36
50	45
60	54
70	63
80	72
90	81
100	90
111	100

Conversion chart for energy efficiencies. European efficiencies generally must be decreased by about 10 percent (not percentage points) for fair comparison to efficiencies reported by American laboratories. The assumptions in deriving this table are that the hydrogen content of the oven-dry fuel is 6 percent, and that its moisture content is 20 percent (moist wood basis). See Appendix 6 for more information on heating values.

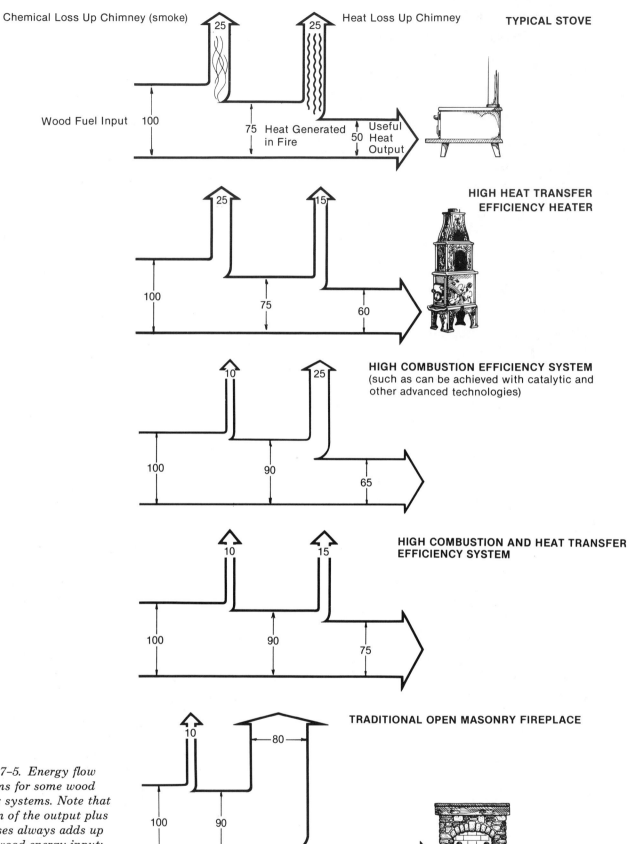

Chemical Loss Up Chimney (smoke) 25 25 Heat Loss Up Chimney **TYPICAL STOVE**

Wood Fuel Input 100 75 Heat Generated in Fire 50 Useful Heat Output

HIGH HEAT TRANSFER EFFICIENCY HEATER

100 25 15 75 60

HIGH COMBUSTION EFFICIENCY SYSTEM
(such as can be achieved with catalytic and other advanced technologies)

100 10 25 90 65

HIGH COMBUSTION AND HEAT TRANSFER EFFICIENCY SYSTEM

100 10 15 90 75

TRADITIONAL OPEN MASONRY FIREPLACE

10 80 100 90 10

Figure 7-5. Energy flow diagrams for some wood heating systems. Note that the sum of the output plus the losses always adds up to the wood energy input; energy is conserved.

Then smoke likely would enter the room, and/or it would be impossible to get very high rates of combustion due to the small air flow in the appliance. In addition, considerable condensation would occur inside the stovepipe and chimney. Since in practice combustion is never complete, the condensate would contain tar, acids, and other ingredients in creosote. This can create a mess, a fire hazard after the water has evaporated away, and increase chimney corrosion. This conflict between maximizing heat transfer and minimizing creosote deposition is a fundamental problem in stove design and use.

Combustion Efficiency

Combustion efficiency is the percentage of the chemical energy in the fuel consumed that is converted to heat (sensible, latent, and radiant) inside the stove. If combustion is complete, there is no chemical energy loss (smoke) up the chimney, and there is no charcoal in the ash.

Generally, the charcoal residue is negligible. In all my years of testing wood burning stoves, I rarely have found more than 1 percent of the energy content of the wood left as charcoal after a day of stove operation. In addition, particularly in stoves without grates, most of the charcoal is left on top of the ash bed where it is burned in the next fire. In most cases, it is safe to neglect the residual charcoal and equate poor combustion in wood stoves only with chemical losses—smoke—up the chimney. This is often not the case in coal burners. Coal ash can contain more unburned fuel than wood ash.

In a fire, pyrolysis always converts a substantial amount of "raw" wood energy content into smoke—30–70 percent, depending on wood type, its moisture content, the piece sizes, and how quickly the wood heats up in the fire. Obviously, substantial energy is lost if none of the smoke is burned. The yellow flames in a typical wood fire are the burning gases. Most fires include flames, indicating that at least some of the gases are being burned. Only in a totally flameless (smoldering) fire is most of the energy in the combustible gases lost—roughly 30 percent in this case.

Conditions necessary for complete combustion are easy to state and difficult to achieve in practice (Figure 7-6 and Table 7-2). Adequate oxygen in the right places and high temperatures, both for a sufficient time, are the essential ingredients.

Excess combustion air usually is necessary to assure an adequate supply to the regions in the stove where combustion is taking place, since some air usually bypasses the fire, and the mixing

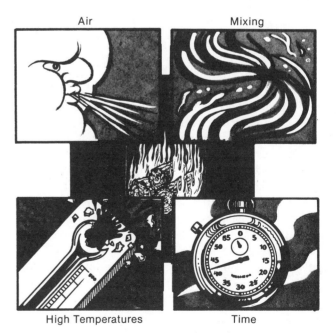

Figure 7-6. The necessary ingredients for the complete combustion of any fuel.

TABLE 7–2
TOWARDS COMPLETE COMBUSTION

Fundamental Requirements	Conditions Conducive Towards Meeting Fundamental Requirements
Adequate oxygen supply	Wide open air inlet Secondary air Air leaks Open appliance (e.g., fireplace)
Good mixing of oxygen with combustible gases	High velocity combustion air Air directed at fire, or distributed uniformly within and around fire Appropriate baffling Recirculation of smoke back through fire
High temperatures in combustion regions	Insulated combustion chambers (e.g., liners of firebrick) High combustion rate Preheated combustion air Recirculation of smoke back through fire Minimum excess air
Enough time at high temperature for combustion reactions to be completed	Large combustion volume and/ or long flame path Insulated combustion chambers (to avoid flame quenching)
Lowering the "activation energy" for the combustion reactions	Presence of a catalyst

Design fundamentals and techniques for achieving high combustion efficiency.

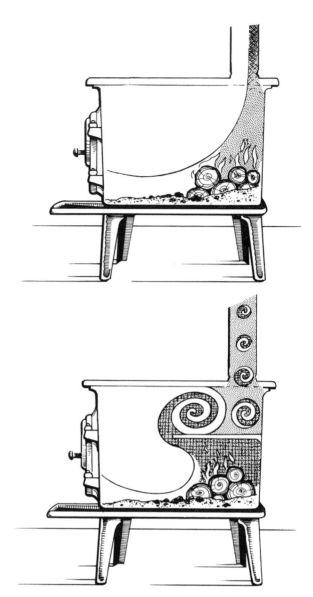

Figure 7-7. Baffles induce turbulence to mix the air and smoke, promoting better combustion. The figures are schematic only; real flow patterns are more complex, particularly when flaming, which causes turbulence, is present.

of air with combustible gases coming out of the wood is not ideal.

Thorough mixing of the air with the combustible gas is important, for fuel molecules can burn only if oxygen molecules are right next to them. Oxygen naturally wanders or diffuses into fuel-rich regions but not usually far enough or fast enough to assure complete combustion. Much better mixing can be achieved by inducing turbulence in the mixture by having the air enter the combustion region at relatively high velocity, or by forcing the burning gas mixture to make abrupt turns as it travels through the stove (Figure 7-7).

The importance of high combustion-zone temperatures has been clearly demonstrated experimentally.[1] A small wood burning furnace was

---

1. C. B. Prakish and F. E. Murray, "Studies on Air Emissions from the Combustion of Wood-waste," *Combustion Science and Technology* 6 (1972), pp. 81–88.

built and equipped with electric heating elements in its walls so that any temperature could be maintained regardless of other conditions. Completeness of combustion was assessed by measuring carbon monoxide, total hydrocarbons, particulates, and smoke density. The variables investigated were temperature in the combustion region, total combustion air flow, the ratio of secondary to primary air (discussed later), moisture content of the fuel, type of fuel, fuel size, preheating of secondary air, and type of grate.

Of the variables tested, the combustion-region temperature was found to be the single most important factor affecting combustion efficiency. As long as the temperature around the fire was about 1,100° F. or higher, combustion was very nearly complete. At lower temperatures, the amount of incompletely burned materials (smoke) was always significant.

In practice, based on tests at Shelton Energy Research (SER), combustion efficiencies in most stoves range from about 70 percent to 90 percent. A combustion efficiency of 70 percent means 30 percent of the wood energy was not burned, but went up the chimney as smoke.

The dominant factor affecting combustion efficiency is how the stove is operated, not how it is designed. For most of the relatively tight stoves tested at SER, a full load of fuel and a restricted air supply results in poor combustion. Whenever you strive for the maximum burn duration, poor combustion results.

Most stoves can operate with relatively good combustion efficiencies. What is required are smaller fuel loads and more air—perhaps only a quarter of a full load, and a wide open air inlet, particularly during the first half hour or so of the burn, when most of the smoke is generated.

Stove Design and Combustion Efficiency

While stove operation is the primary factor in combustion efficiency in most stoves, stove design does have some effect. SER has investigated 3 standard design features in some detail: secondary air, firebrick liners, and a particular baffle system.

Secondary Air. Some stoves have 2 distinct air inlets. The *primary air inlet* feeds air to the wood, usually near the bottom of the fire. In many wood and coal stoves with grates, the primary air is admitted under the grate and is appropriately called "underfire" air. *Secondary* (or "overfire" in some cases) *air* is admitted either above the wood or into a separate secondary combustion chamber located above, beside, or even below the primary combustion chamber (Figure 7–8).

The intended function of secondary air is to assure enough air for complete combustion of the smoke. It can happen in many stoves that much of the oxygen in the air which has entered through the main or primary air inlet is consumed in the part of the fire near this inlet (Figure 7–9). So little oxygen is left further back in the stove that the smoke cannot burn. In some cases there is a *secondary combustion chamber* where the secondary air helps the smoke to burn; in other cases there is no additional specialized chamber.

Why not just let in more primary air and eliminate the secondary air? Primary air usually stimulates combustion of the wood, making the main fire hotter. If the fuel load is fairly fresh, the increased temperature can increase the rate of pyrolysis (smoke generation) without necessarily

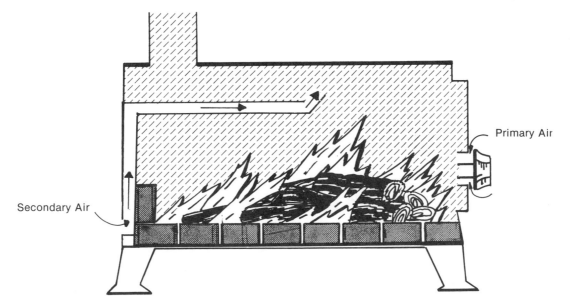

Figure 7–8. A stove with a secondary air inlet near the back of the stove.

Figure 7–9. The rationale for secondary air. The oxygen content of the air is decreased as it passes through the burning fuel. If smoke and unburned gases are also leaving the fuel bed, there may not be enough oxygen left to burn them.

burning any larger fraction of the smoke. So, admitting more primary air can, in some cases, *increase* the amount of smoke leaving the stove. Thus, the oxygen concentration in the back of the stove can *decrease* when more primary air is admitted. In addition, admitting more primary air almost always will increase the heat output; unless more heat is needed, this is not a useful solution. Also, primary air must remain restricted if you are striving for a long-duration burn.

SER has tested the secondary air systems in 5 representative conventional stoves. A minimum of 8 day-long tests were run on each stove, half with the secondary air inlet blocked and half with the inlet operating normally. The tests spanned a full range of power outputs from low to high.

The results were that the secondary air systems on these stoves worked to some extent, but were not very effective when needed most. I would expect that the same is true of most other typical stoves with secondary air. Specifically, it was found that when the stoves were operating at medium or high power, letting in secondary air improved combustion efficiency a few percentage points.

At low power, secondary air had no effect on combustion efficiency. This is unfortunate because this is when combustion is at its worst and

needs the most help. But in addition to not helping combustion at low powers, adding secondary air decreased heat transfer efficiencies.

The effect of secondary air on *overall* energy efficiency was a slight decrease at low powers and a slight increase at high powers for an average effect of about zero. (This is an average result for all 5 stoves—individual stoves had slightly different behaviors.)

Why doesn't secondary air help more? Because good combustion requires more than just ample air. If that was all that was needed, the smoke would burst into flame when it came out the top of the chimney into the atmosphere. The other important requirements are good mixing of the air with the smoke and, most important, high temperatures. If smoke is not already burning, temperatures of roughly 1,100° F. are needed to ignite it.

The principal reason secondary air systems are often ineffective is that the air is admitted at a point where the temperature of the smoke is already below the ignition temperature; therefore, no additional combustion takes place. The net effect is the same as an air leak at that location, which increases the amount of excess air and decreases the efficiency of heat transfer.

A rough rule of thumb is that secondary air can

help only if admitted into a region where flames already exist. This is likely to be the case for any kind of secondary air inlet in a stove operated at very high firing rates, for then flames fill the combustion region and may even extend up the stovepipe a little way; in this case most air leaks can be effective secondary air supplies. Unfortunately, secondary air is least effective for low power, smoldering burns, when combustion is the poorest.

Keeping the smoke as hot as possible enhances combustion. Very few stove designs attempt to do this. If the smoke/flames are routed through a relatively small volume passageway, and this passageway is highly insulated, such as with insulating firebrick, then combustion of the smoke will be more complete generally, including when secondary air is added (Figure 7-10).

There is another reason why secondary air is usually ineffective in stoves operated with low power outputs. There is already ample air in the stove—secondary air is not needed! Although visible dense smoke coming out a chimney *is* an indicator of high chemical energy loss, it is not a good indicator of oxygen concentration or whether secondary air could, in principle, be useful. Measurements of oxygen concentrations at the flue collar in a variety of stoves indicate that most stoves operated at low powers with smoldering smoky fires have substantial oxygen in the smoke. Without active flaming in a stove, actual combustion

takes place only at the solid wood (that is, only the charcoal burns, not the gases). Much of the primary air bypasses the fire since very little of it usually comes in contact with the charcoal surface.

People speak of smoldering, low-power-output fires as being air-starved or air-limited. Such fires are air-starved in the sense that if more air were admitted, the combustion rate would increase, and the stove would get hotter. But such fires usually are not air-starved in the sense that all available oxygen in the stove has been used up by the fire.

In principle, preheating the secondary air can help combustion efficiency, but the heating must be substantial (at least a few hundred degrees Fahrenheit) to be very significant. Most stoves with this feature cannot preheat the air sufficiently to be of much use.

If the secondary air is added to already burning, fuel-rich gases, then the preheating can make the secondary combustion a little more complete by increasing the average temperature of the burning mixture. But the effect of preheating is unlikely to be large since the burning gases are already so hot. Flame temperatures are roughly 1,500-2,000°F.; preheating by even a few hundred degrees of the relatively small secondary air flow is unlikely to have much effect.

If the secondary air is added to flue gases that are not burning because the temperature is below

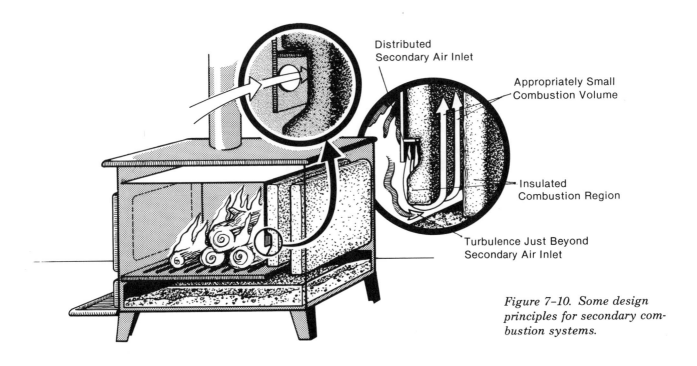

Figure 7-10. Some design principles for secondary combustion systems.

the ignition point, the secondary air would have to be heated to well over 1,100°F. to help initiate ignition. This is not practical.

Finally, if the gases are not burning because there is not enough oxygen (but they are hot enough), ignition will usually occur without pre-heating the secondary air.

Another reason many secondary air systems do not perform well is that the air they admit does not mix adequately with the unburned gases. A localized single inlet that admits air smoothly into the flue-gas stream usually is not optimal. Mixing, of course, eventually takes place, but often not until the gases have cooled below their ignition temperature. The inlet should distribute the air as evenly as possible in the flue gases, and/or turbulence should be high. Experiments with coal heaters indicate that a good system is to admit air through an essentially continuous slot across the region where secondary combustion is desired.[2]

The amount of secondary air admitted is critical. If too much air is admitted, the cooling and diluting effects may inhibit combustion, and excess air is detrimental to heat transfer. The amount needed is not very predictable and not constant. It depends on how much of the oxygen in the primary air is left and on how much of the smoke is unburned. These variables in turn depend on many details, such as the exact location of the primary air inlet; the amount, direction, and speed of the air as it enters the interior of the stove; how close the wood is to the primary air inlet; how much wood is in the stove, which, of course, changes constantly during each fuel cycle; the average size of the pieces; and how densely the wood is piled.

On the other hand, gases can burn to some extent over a wide range of fuel-to-air ratios (Table 5-1), and the mixing of the air with the gases is never uniform. Thus, secondary combustion is theoretically possible over a moderately wide range of secondary air flows. In practice, however, the secondary air flow will not be optimum much of the time, if for no other reason than that wood fires are so variable over time—the optimum amount of secondary air near the beginning of a fire is unlikely to be optimum at a later time because composition and concentration of the smoke will be different.

Secondary air by itself will never be a panacea because it works least well when needed the most—during low-power, long-duration, smoldering burns. But I believe secondary air systems could be worthwhile—they could be engineered to work better at lower powers than most now do by providing the proper amount of air, well-mixed with smoke, and, most often neglected, in an environment that stays hotter longer through the use of heat storage and insulation, such as with refractory materials, once flaming is established. In fact, a few solid fuel heaters are available which now make a serious engineering effort towards more effective secondary combustion (Figures 7-11 and 13-2). It is likely that these units have better combustion efficiencies over a range of burn rates from high to medium low.

Firebrick and Metal Liners. Stove liners (Figure 7-12), especially firebrick liners, are supposed to make the combustion chamber hotter, which is sometimes claimed to make combustion more complete. Not only should the combustion chamber as a whole be warmer, but the firebrick surface itself should be hotter, thereby reducing the amount of flame quenching at the surface.

However, a liner makes the combustion chamber warmer by inhibiting the heat from getting out. Thus, one would expect heat transfer efficiencies to be lowered by the presence of a liner.

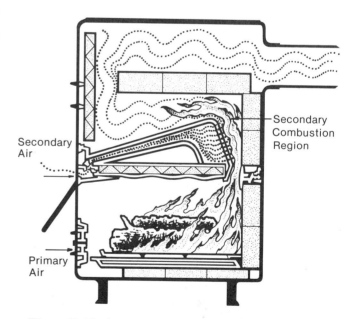

Figure 7-11. A stove (the Jøtul 201) with substantial insulation around the secondary combustion region, and with a distributed air inlet.

2. B. A. Landry and R. A. Sherman, "The Development of a Design of a Smokeless Stove for Bituminous Coal," *Transactions of the ASME* (January, 1950), p. 9.

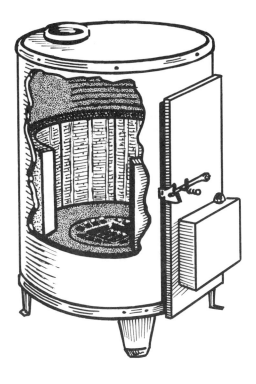

Figure 7-12. Stoves with liners made of refractory material, such as firebrick (left) and cast iron (right).

On the balance, do liners hurt or help overall efficiency? SER's experiments on liners have involved 2 stoves. Since I expected (on theoretical grounds) that the effects of typical liners would be small, we first studied an extreme case, comparing the performance of a stove without any liner to the same stove with a specially built 4-inch-thick insulating (soft) firebrick liner covering the full height of the combustion chamber. Fifteen day-long tests were run with a variety of fuel loads and air settings.

How did the presence of the brick liner affect the stove's performance? It depended on how the stove was operated.

• When plenty of air was supplied, so that the fire was relatively clean-burning, the presence of the liner increased the combustion efficiency slightly, by 2-3 percentage points (Figure 7-13). Only when oxygen is available can increasing the combustion zone temperatures help.

• When the fire was relatively smoky due to a restricted air supply, then the liner decreased the combustion efficiency by 2-9 percentage points.

How can a firebrick liner in the primary com-

bustion chamber hurt combustion efficiency? First, if there is not any spare oxygen available in the stove, increasing the temperature in the combustion region cannot help—it takes both high temperatures and adequate oxygen to achieve good combustion. But perhaps more importantly, increasing the temperature of the wood in the stove causes smoke to be generated at a faster rate since the smoke-generating reactions are driven by heat. When oxygen is already in short supply, only a certain amount of smoke can be burned. Increasing the rate of smoke evolution out of the wood means a smaller fraction of the smoke will be burned. The basic problem here is that the same chamber in the stove is used both to burn the wood and to store wood to be burned over the 4-12-hour period between fuel loadings.

Even if oxygen is not in short supply during smoldering burns, which is more often the case, a firebrick liner can still decrease combustion efficiency. The problems of inadequate mixing of air with the smoke, and of the mixing not occurring until the mixture is too far from the glowing coals to be ignited by their heat, can together result in excess air plus poor combustion. In this case also, adding a firebrick liner will increase the rate of evolution of smoke. If the same problems preventing the smoke from burning remain, then the liner results in more smoke going up the chimney. (Note that these liabilities of firebrick liners do not exist for liners around *secondary* combustion chambers, which contain only the gases to be burned, and no solid fuel. Liners around secondary combustion chambers can only improve combustion.)

The expected effect of firebrick liners on heat transfer was also observed in our experiments. The liner caused a decrease in heat transfer efficiency of 2–5 percentage points.

Combining the effects of the 4-inch-thick liner on combustion efficiency and heat transfer, we found that with relatively clean-burning fires (small fuel loads and ample air) the overall energy efficiency was approximately unchanged—combustion being helped slightly and heat transfer being hurt slightly. However, with smoky fires (large fuel loads and restricted air supply), both combustion and heat transfer were lowered by the liner, and the overall energy efficiency was decreased by about 5–10 percentage points. This result was for an unusually thick, extensive, and highly insulating liner.

We also tested a stove with and without its normal factory-supplied liner, over a range of power outputs. On the average, the liner improved combustion efficiency by a few percentage points, decreased heat transfer by a few percentage points, and thus had little effect on overall energy efficiency (Figure 7–14). I expect this is typical for

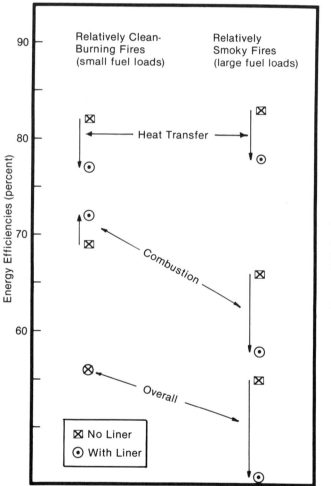

Figure 7–13. *The effects on energy efficiencies of a very extensive and highly insulating liner in a wood stove.*

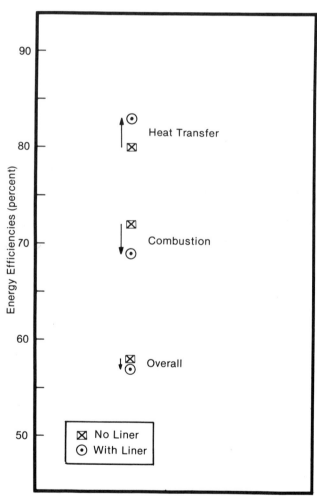

Figure 7–14. *The effects on energy efficiencies of a typical firebrick liner in a stove, averaged over low, medium, and high firing conditions.*

most stoves with firebrick (or other refractory material) liners. I would also suspect the effect of metal liners to be similar but smaller. As seems to be the case for secondary air, liners are least effective at improving combustion efficiency when needed the most—during low power smoldering burns.

Primary combustion chamber liners (both brick and metal) can be useful despite the fact that the effect on energy efficiencies is often quite small. A liner keeps the main stove structure from getting as hot as it otherwise would. Excessive heat causes both steel and iron to oxidize (leading ultimately to burning out) and can cause cast iron to crack, steel and cast iron to warp, and paint to turn white. Depending on wall thickness and other design aspects, not all stoves need liners; but the principal purpose liners *can* serve is protection of the stove body against heat-caused deterioration. Extensive liners can also contribute to heat storage which improves the steadiness of the heat output from wood heaters and can make it easier to burn green wood.

Baffles. Some baffle systems in stoves (Figure 7-15) are claimed to increase combustion efficiency. Such baffles are interior partitions in or near the combustion region around which the flames or smoke must flow. Baffles could promote combustion by bringing smoke and air closer together, creating turbulence to mix smoke and air, and creating higher temperatures in the primary combustion area by acting as a barrier between the fire and the relatively cool exterior wall or top of the stove.

It is risky to generalize about the effectiveness of baffles because so many design variations are possible. I have done enough testing of baffles to demonstrate that the effect of baffles is unpredictable.

Two Scandinavian stoves, a large and a small model of the same design (Figure 7-15), were tested for the effect of the baffles. Tests were run with and without the horizontal baffle plate present, and over a range of power outputs. We found that the baffle in the larger stove had little effect (less than about 2 percentage points) on any of the 3 energy efficiencies—combustion, heat transfer, and overall. In the work on the smaller stove, we found its overall energy efficiency to be improved by a few percentage points, particularly at medium and high power outputs, by the presence of the baffle. (At the time the latter experiments

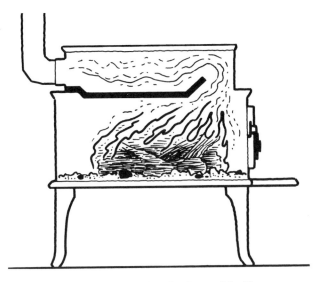

Figure 7-15. A stove with a horizontal baffle.

were conducted we did not have adequate instrumentation to distinguish between combustion and heat transfer efficiencies.)

Thus, it is not possible to generalize about the effectiveness of even similar-looking baffles on energy efficiencies. It may be that a difference in leakage around the baffles made the difference in our tests. The baffles in the stoves tested are not welded or sealed in place, so this is a plausible explanation.

Even when baffles do not improve combustion, they can be useful for other purposes. In particular, the baffle system in the Scandinavian stoves tested appears to improve the steadiness of the heat output. Baffles may also improve heat transfer efficiency.

Baffles *can* have many effects, and the effects certainly depend on the design of the rest of the stove. I believe that the baffles in many stoves are beneficial. But, since the effects are so varied and complicated, more testing is necessary before generalizations are justified.

Preheated Combustion Air. One way to increase combustion-zone temperatures is to preheat the incoming combustion air by ducting it through channels or pipes in the stove or its chimney (Figure 7-16). The heating must be substantial to be effective; 1,100° F. is the critical temperature of the smoke/air mixture for relatively complete combustion to occur.

The flames in fires are always hotter than 1,100° F., and the smoke near the stove walls is

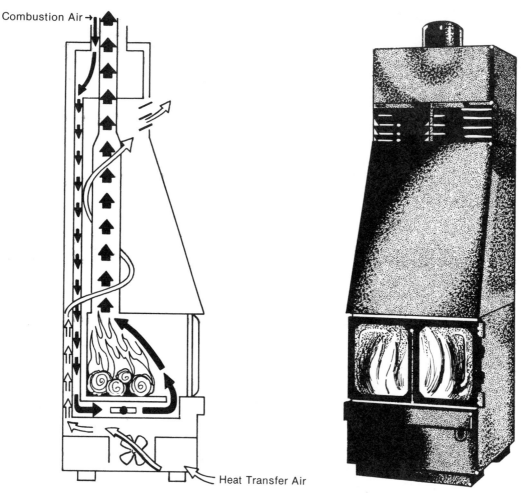

Figure 7-16. A stove with combustion air preheating (the Combitherm).

usually cooler. When it works, preheating the combustion air improves combustion probably by increasing the size of the region in the stove over which temperatures are above 1,100° F.

However, increasing temperatures in the combustion region does not necessarily lead to better combustion. Sufficient oxygen must also be available, and ignition of the smoke must be achieved. Without these conditions, the effect of preheating combustion air can be negative, due to increased rates of smoke generation.

Not much testing has been done on preheating combustion air. J. Allen at Battelle Columbus Laboratories has conducted related studies.[3]

Smoke Recirculation. Another way to improve combustion is to recirculate some of the flue

gases back through the combustion region, giving the combustible gases a second chance to burn as they pass through or near the hot fire (Figure 7-17). Although the basic principle of flue-gas recycling for more complete combustion is valid, I am not aware of any careful measurements on stoves verifying the effectiveness of such designs.

I suspect many stoves achieve some smoke recirculation unintentionally. In relatively tight burners with glass doors, one can see that the flame and smoke path is not straight up and out the flue collar; the flames rise and then roll back down the perimeter of the combustion chamber to the fuel bed. This is also a typical pattern in downdraft stoves.

Who's in Control Here?

If stove design features, within traditional limitations, are not very effective, what determines

3. John M. Allen, "Techniques for Achieving More Complete Combustion in Wood Stoves," *Proceedings of Residential Wood and Coal Combustion Specialty Conference*, Air Control Association (1982).

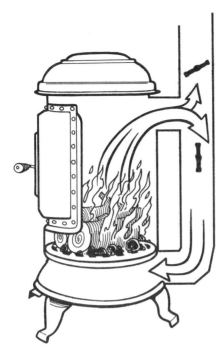

Figure 7-17. A stove design in which there can be some recycling of gases through the combustion chamber, contributing to more complete combustion.

- Large exterior surface area.
- Long residence time for smoke and/or hot gases.
- Convection and turbulence both inside and outside the system.
- No insulating materials in walls.
- An exterior finish with high radiating efficiency.

In tests at SER, we have found heat transfer efficiencies in stoves range from about 60 percent to 80 percent. Let's consider the various factors in approximate increasing order of importance in practice.

Stove Finishes. As indicated in Table 2-4, typically 60-70 percent of a (radiant) stove's total energy output is in the form of radiation. The amount of radiation given off by a hot surface depends not only on the temperature of the surface, but also on the nature of the surface. Most stove

whether a stove is operating at a 70 percent or a 90 percent combustion efficiency? You do! You, the operator, have more control over combustion efficiency than the designer of the stove.

To achieve better combustion, do not starve the fire of air. In practice, this is accomplished by using smaller fuel loads and maintaining a more open air setting. Refueling will be more frequent, but preliminary testing at SER indicates that the same total amount of fuel is burned to produce the same amount of useful heat. In this way, the combustion efficiency can exceed 90 percent in most stoves. Improved combustion means less smoke and, hence, less creosote and less air pollution.

HEAT TRANSFER EFFICIENCY

The second major task an energy-efficient stove must do well is to transfer the heat released in the fire to the living spaces of the house. In contrast to the combustion process, the net effects of design features are much more predictable. The fundamental requirements are outlined in Table 7-3.

In practice, the following features usually improve the heat transfer efficiency of a stove.

TABLE 7-3
TOWARDS COMPLETE HEAT TRANSFER

Fundamental Requirements	*Conditions Conducive Towards Meeting Fundamental Requirements*
Large surface area	Large exterior surface area
High thermal transmittance:	
From fire to stove wall, by radiation	No baffles in stove which block radiation; radiation-absorbing inner wall surface
From fire to stove wall, by convection	Baffles for increased convection
Through stove wall	Wall construction: single metal layer (any reasonable thickness) No liners Clean inside surface—no soot or creosote
From outer surface of stove to room	High emissivity surface finish Fan blowing on stove for increased convection
High temperatures of fire and gases just downstream	High combustion rate Minimum excess air
Long travel time for hot gases	Minimum excess air Low combustion rate Large stove volume Baffling for hot gases

Design fundamentals and techniques for achieving high heat transfer efficiency.

finishes are good radiators with a radiating efficiency (or infrared emittance) of 80–95 percent (of the maximum possible). This includes paints, enamels, and porcelain of *any color* except metallic colors. Black is a traditional color for stoves; but, in fact, green, blue, red, yellow, and white are equally good. There are differences, but the differences may depend more on the type of paint used than on the color. More measurements are needed in this area to sort out the small differences that possibly exist.

Dark surfaces, especially black ones, are better absorbers and emitters of *solar* radiation. But solar radiation is mostly in the visible part of the spectrum; radiation from stoves is in the invisible infrared part of the spectrum. Thus, the color of a surface as perceived by the human eye is not a meaningful indicator of the infrared properties; that is, the radiating potential of the surface.

Neither metallic paints nor bare clean metal surfaces (see Table 2–2) are efficient radiators. Galvanized pipe is only about half as efficient a radiator as black pipe. Nickel-plated or chrome-plated stove parts or stovepipe radiate very poorly. You can sense the radiation from hot black pipe from many feet away, but with nickel-plated pipe at the same temperature you would almost have to touch the pipe to sense it was hot. Fortunately, little wood heating equipment is offered with poorly radiating finishes.

Since it is only the *surface* that affects radiation, a thin coat of ordinary, nonmetallic paint over a bare, polished metal surface will transform it from a poor to a good radiator.

Some stoves are finished in (or even built of) soapstone, tiles, and other masonry materials. The emissivities of these materials are high—roughly comparable to other more common finishes, such as paint. Claims of *far* superior radiating efficiency from such materials cannot be valid. Common stove finishes have radiating efficiencies between 80 percent and 95 percent; you cannot get much higher.

Stove Construction Materials. To maximize heat transfer, the insulating value of materials and constructions must be minimized. There can be a conflict between keeping the heat in to improve combustion efficiency and letting it out to improve heat transfer efficiency. If maximum heat transfer were the only objective, plain metal walls would be favored over firebrick-lined or metal-lined walls.

Neither the thickness nor type of metal (cast iron or plate steel) has a large effect on how much heat gets through a simple single-layer wall. Heat comes through a thin wall more quickly; thick walls store some heat for later use. But, on the average, or under steady conditions, nearly the same total amount of heat gets through.

Soapstone, tile, brick, and other masonry or refractory materials have more insulating value than do steel or iron. This can impede the outward flow of heat slightly. However, such materials can be useful for other reasons—improving heat storage, appearance, combustion efficiency, durability (in some cases), and lowering cost (in some cases).

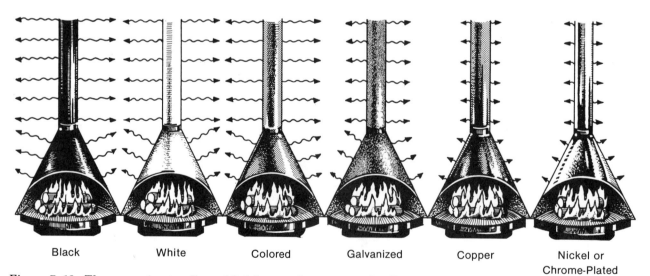

Figure 7–18. *The approximate effect of finishes on the amount of radiant energy given off. Arrow length is proportional to radiant heat output.*

The claim is often made that the air in a house is drier if wood or coal is used as the heating fuel. Consequences could be increased nasal and throat discomfort, increased frequency of colds and other respiratory illnesses, increased chapping of skin, and increased stress on the wood structure of the house and furniture.

But, in fact, use of the most efficient wood or coal heating systems does not cause a house to be unusually dry. The reputation for dryness most likely is based on heating with a traditional open fireplace (without an outside air system). Here the very large air consumption of the fireplace causes significant extra infiltration of outdoor air into the house. Since cold air is almost always dry (because cold air cannot hold as much water vapor as warm air), the house becomes dry. But it is the increased infiltration of air that causes the dryness, not the solid fuel itself. A comparable size gas-log fire would have the same effect.

In fact, *only* through this mechanism of increased infiltration can a wood heating system be the cause of dryness of air in houses. Tight wood heaters, such as most contemporary stoves, furnaces, and boilers, cause only a small, if any, increase in house air-exchange rates. Any system that draws most of its combustion air directly from the outdoors (and is reasonably airtight) cannot adversely affect humidity.

Not all outside-air systems are favorable in this regard. Some factory-built metal fireplaces draw in outside air, heat it, and deliver it to the inside of the house. This air is bound to be very dry. Thus, outside air used for transferring heat from the fire to the house will have a drying influence on house air, whereas outside air used for combustion only and/or which is sent up the chimney will not.

Drying of *nearby* furniture and structures may be caused by any localized (or "room") heater. For instance, wooden furniture near a radiant stove will be slightly warmer than room air temperature and hence will be drier than otherwise would be the case. But the same would be true of a radiant gas or electric heater. Thus, *wood* heat is not the cause, but localized heat is.

Whether or not a particular solid fuel heating system is the cause of excessive dryness in a house, a traditional way to humidify the air is to keep a pan of water on the wood stove, although in many cases the amount of the humidification from such systems is very small. (Typically 2–12 gallons per day is needed to humidify a house.) Another humidification system which works well with *any* kind of heating system is indoor plants.

Convection and Turbulence. About 30–40 percent of the heat output from a typical radiant-type stove is in the form of hot air that *convects*, or moves, up and away from the stove. If a medium-size household fan is used to blow air at a radiant stove (*forced* convection), the convected heat output increases somewhat. The fan keeps sweeping away the layer of hot air next to the stove, making it easier for more heat to emerge from the stove wall. This also cools the stove surface, reducing the radiant output, but the net effect is an increase in total heat output by a few percent.

Some stoves come with blowers that force air through internal passageways. Tests at SER indicate that such blowers sometimes help and sometimes do not. There are 2 reasons such blowers may be ineffective.

• Some systems achieve good air flow by natural convection alone. A blower cannot help unless it results in a significant increase in air flow. Some blowers supplied by manufacturers are too small to significantly increase convection.

• In a few units, the pipe or passageway inside the stove is so small in surface area that not much heat could be extracted in any case, even with a large blower.

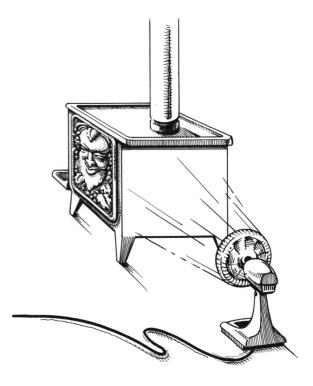

Figure 7-19. A fan blowing on a radiant stove can increase the total heat output somewhat.

How can you estimate if a blower is significant? First, check the area of the passage inside the stove through which the air moves. Very roughly, if this surface area is less than 10 percent of the exterior surface area of the stove, not much additional heat can be gained, whether or not a blower is used. Second, when the stove is hot, compare the force of the air on your hand with and without the blower. If the blower does not increase the force (air speed) significantly, it cannot do much good. (Since the blower when turned off may impede the natural air flow, it is best to physically dismount the blower for the no-blower test.)

Keep in mind that stoves with large blowers and large interior surface areas through which the air is forced may suffer a decrease in combustion efficiency due to the resulting cooling of the fire's surroundings. The air passageways themselves may impede heat transfer by other means. For instance, many circulating (or jacketed) stoves have blowers which circulate air between the jacket and the combustion chamber body. The jacket provides a large passageway for air. However, the jacket also blocks direct radiation from the main stove body. The net effect, according to tests at Auburn University,[4] is a lower overall energy efficiency for typical circulating stoves compared to typical radiant stoves.

In some cases, even though a blower may improve the efficiency of a particular stove, it is possible that the stove would be more efficient still if the air passages and blower were removed. This is by no means always the case, but it certainly complicates generalizations about blowers.

Increasing the turbulence of the smoke and hot gases inside the system can increase heat transfer. For example, when smoke rises up in a straight vertical segment of stovepipe, the smoke nearest the pipe walls becomes relatively cool, while the smoke in the middle remains quite hot. A difference in temperature of a few hundred degrees is possible. If a stovepipe damper in the pipe is rotated to a partly closed position, this central hot core will be forced over to the pipe wall where its heat will get out into the room more easily. Introducing any kind of turbulence tends to help by mixing the smoke with itself,

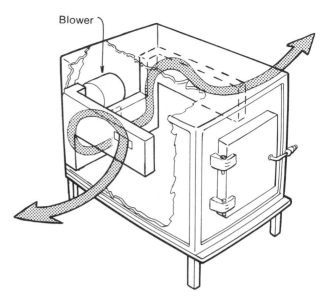

Figure 7-20. A stove with a blower that forces air through internal passageways for additional heat transfer.

making the smoke near the pipe wall hotter, making the pipe wall itself hotter, and causing more heat to enter the room.

Elbows also increase turbulence inside a stovepipe. However, this also decreases draft and so is not recommended. Closing a stovepipe damper also decreases the draft, but the damper has the clear advantage that full draft can be restored at any time merely by opening the damper.

Turbulence inside stoves also aids heat transfer. Turbulence generally is increased by baffles. However, baffles also block radiation from the fire to the stove wall, and this inhibits heat transfer. The net effect of baffles on heat transfer is unclear. SER's tests on the 2 stoves with baffles showed no net effect on heat transfer in one case, and a probable slight increase in the other. More testing is needed.

Smoke Residence Time. The longer the time smoke spends in the stove and stovepipe, the larger will be the fraction of its heat that will get out into the house, other things being equal.

Residence time is affected both by stove design and by how the stove is operated. Residence times tend to be relatively large in stoves with either a smoke chamber, or extensive internal baffling, or both. The baffling in this context is not

---

4. *Design Handbook for Residential Wood-burning Equipment,* Department of Mechanical Engineering, Auburn University (1981).

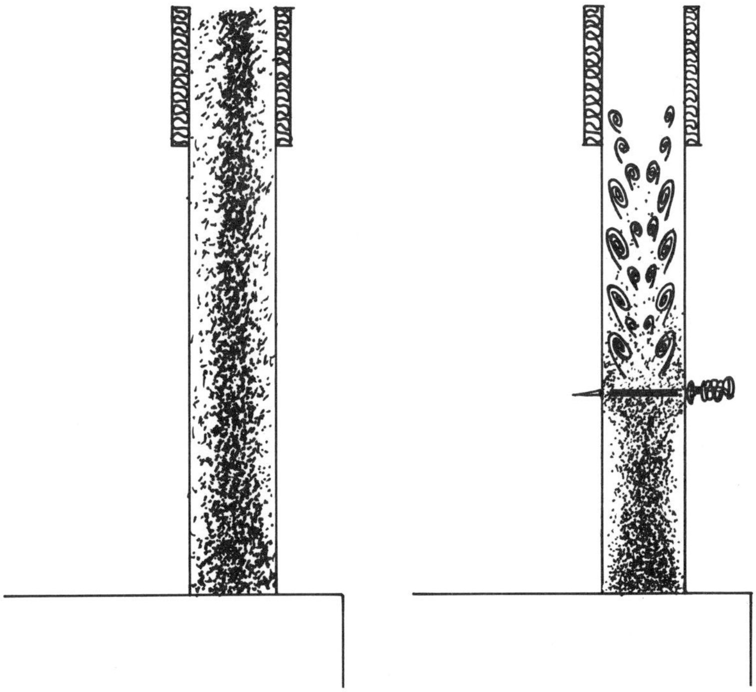

Figure 7-21. A closed stovepipe induces turbulence. The damper can force the hotter smoke in the center of the pipe to come into contact with the wall. It also causes general mixing of the hot and cooler smoke. Both effects tend to assist the movement of heat from the smoke into the room.

to slow down the smoke in terms of its speed or velocity, but to ensure it does not bypass portions of the stove as it travels through. The smoke will take a longer time to get out because it has traveled further.

The other major factor affecting residence times is the amount of smoke and air moving through the system. The less flow there is, the longer will be the time the smoke takes to get out of the stove. Stoves usually have their highest heat transfer efficiency when operated at lower power outputs—with small fires. Heat transfer tends to be the worst with the largest fires, when the hot gases, smoke, and air are racing through the stove.

For a given power output, the flow depends on the stove design and on how it is operated. The critical factor is the amount of excess air.

Excess Air. Air that enters a stove but whose oxygen content is not actually used in the combustion reactions is called *excess air*.

A certain amount of excess air is desirable to promote more complete combustion of the smoke. Since mixing of air and fuel is not perfect, some extra air must be provided to assure there is enough air where it is needed.

But too much excess air is very detrimental to heat transfer. Unused air has 2 negative effects. It dilutes and cools the hot gases and smoke. This decreases the temperature of the stove and the heat output. Excess air also decreases the residence time for the smoke—the speed of the smoke/air mixture must increase to accommodate the extra air. This means less time is available for the heat in the mixture to be transferred into the house.

How Airtight Is an Airtight Stove?

No stove is literally airtight. Even with the air inlet shut, it is inevitable that some air will leak in—through the air inlet and around the door.

Does it matter? No, not as long as the stove is tight enough so that you can control the combustion to any desired rate, including suffocating the fire. In some ways the term "controlled combustion" is more accurately descriptive than "airtight." However, airtight is more commonly used, and it serves the purpose of distinguishing between air-limited and fuel-limited modes for controlling combustion. So do not worry if your stove has some leakage—as long as you can always keep the heat output down to a comfortable level.

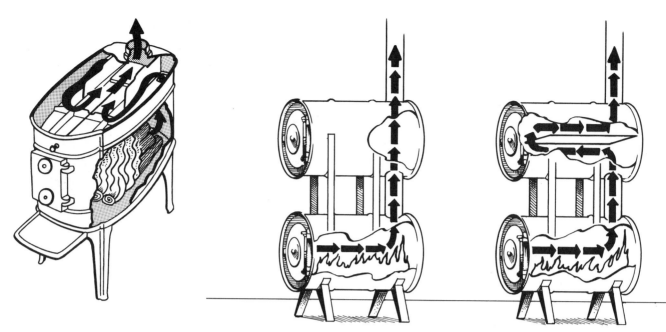

Figure 7-22. Baffles in a stove can increase the smoke residence time, partly by slowing the velocity, but primarily by forcing the smoke to travel a longer path.

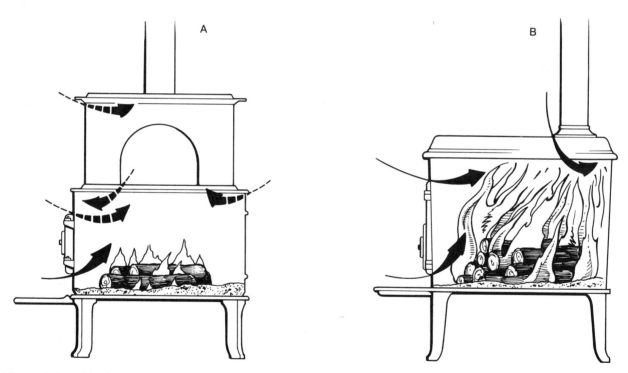

Figure 7-23. Air that leaks into a stove may either constitute excess air (reducing heat transfer) or additional combustion air (improving combustion), depending on the location of the leak and the size of the fire. In Stove A, with a small fire, most leaked air becomes excess air. In Stove B, with a large fire, most leaked air is used in combustion. If there is flaming where the air enters, the air likely will be used in combustion.

Figure 7–24. Fire size affects the amount of excess air, even in airtight stoves. Most oxygen is used up in large fires. With relatively small fires, considerable air bypasses the fire. Shading indicates oxygen-depleted air. These diagrams are simplified—there is much more turbulence and mixing in real fires.

One potential source of excess air is air leaks in nonairtight stoves (Figure 7–23). But not all leaked air is necessarily excess air. Particularly with relatively large fires in a stove, most of the air entering the stove, even through leaks, tends to be used in the combustion process. This is because the flames fill the stove and often can use a little more air than enters the air inlets to burn more completely.

But with relatively small fires, leaked air is usually excess air. Even if there are unburned gases in the stove, temperatures tend to be so low that leaked air cannot help stimulate combustion. Leaked air and secondary air are similar in this respect.

Another source of excess air is the air inlet itself. Ideally one would like to have all the air that enters through the inlet be used, with no excess and no incomplete combustion. But wood combustion is so complex—so inhomogeneous in space and variable in time—that this is impossible in any normal stove. Two extreme examples illustrate the point.

With a very small fire in a large stove, it is virtually inevitable that some of the entering air will not be used in combustion—much of the air will bypass it, and because the fire is so small and localized, even some of the air that is directed at it will not be used (Figure 7–24). There is not enough turbulence for good mixing with the com-

bustible gases. Even if the air were adequately mixed with the gases, small fires in large stoves are unlikely to be hot enough to achieve complete combustion, and thus some air likely would be unused.

At the opposite extreme is a stove with a full load of very dry wood on a bed of hot coals with the air inlet wide open. Here the result is an extremely hot fire completely filling the combustion chamber. All the oxygen is used up in combustion reactions regardless of the direction the air was moving as it entered the air inlet. There is no excess air. In fact, combustion in such cases is usually air-starved—the fire would be hotter still and less smoke would go up the chimney unburned, if still more air had been supplied. Secondary air would be beneficial here.

The design of the air inlet itself can have a considerable effect on stove performance. Some stoves have the air directed straight at the wood (Figure 7–25). This tends to promote turbulence and mixing in the area where the air hits the fire, making it easier to start fires and to burn green wood. The air inlet blows on the fire much as the old bellows blower does. The turbulence may improve the heat transfer also. But it is difficult to predict the overall effect.

Some stoves have the combustion air passing up through a grate on which the wood rests. This kind of stove (if it is airtight) is easy to operate

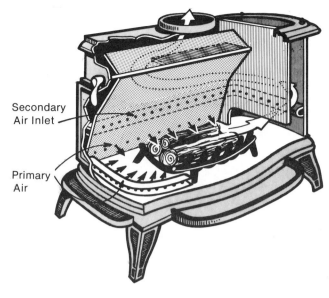

Secondary
Air Inlet

Primary
Air

Figure 7–25. Two types of air inlets. Left: *a stove with relatively directed, high-velocity air inlets (the Tempwood).* Right: *a stove with distributed primary and secondary air inlets (the Defiant).*

with close to zero excess air merely by having a full load of dry wood and any reasonably open air setting. Particularly in designs with small central grates, the air cannot bypass the fire, so nearly all the oxygen in the air is used. However when so operated, the stove may not burn up all the smoke. Thus combustion efficiency may be low, while heat transfer efficiency is high. Introducing secondary air above the wood can improve combustion in hot fires; but with small fires, it is likely to constitute unused or excess air and reduce heat transfer.

A stove design in which most of the combustion air enters above the fire (and the smoke exit is also above the fire) will tend to have low overall energy efficiency. Heat transfer will be poor because inevitably most of the air will be unused. Combustion efficiency generally will be low also because of the difficulty in obtaining a reasonably hot fire. Thus, power output also tends to be low. SER has tested one such stove.

It seems clear that stove design can have a large impact on the amount of excess air going through the stove. Stove operation is also very important. But it also seems clear that considerably more research is needed to determine which designs are best.

Surface Area. One of the most important design features needed for good heat transfer effici-

ency is a relatively large heat transfer surface area relative to the combustion chamber, or more specifically, relative to the combustion rate. For most stoves, the heat transfer surface area is just the exterior area.

An example of a stove with high heat transfer efficiency is the double-barrel stove (Figure 7–22). Barrel stoves are typically owner-built from kits. The kits contain the legs, door, and smoke collar(s). The purchaser supplies the barrel and mounts the kit hardware.

The upper barrel contains no wood and has no effect on combustion—it is a chamber the smoke passes through on its way to the chimney. A double-barrel stove, compared to the more common single-barrel stove, clearly has twice the surface area and at least twice the smoke residence time. SER has measured the heat transfer efficiency of such a stove. It is as high as any we have tested—around 80 percent, including the heat from the first 5 feet of stovepipe—a typical installation.

But despite its high heat transfer efficiency (and high overall efficiency), the particular barrel stove tested would not be a practical heater in most homes. In the tests, we could not easily get power outputs less than 30,000 Btu. per hour, which is more heat than usually needed. Air leaks caused this high minimum power—the air-inlet damper leaked air prodigiously when closed, and the door did not seal tightly against its frame.

Figure 7-26. Stoves with relatively large surface areas (and smoke residence times) by virtue of smoke chambers or passageways. Left: an antique American stove. Right: the Ulefos 172.

Other brands may be tighter. Note that in this case leaked air did not decrease energy efficiency, but decreased some of the control of the power output – lower powers were not achievable.

Many other stoves have similar extra smoke chambers (Figure 7-26). SER has experimentally verified the resulting high heat transfer efficiency for a number of them. Again, these chambers gen-

erally have nothing to do with combustion, but are merely smoke passageways which provide more exterior surface area and longer smoke residence time.

Stoves can achieve high heat transfer without the external appearance of an extra chamber (Figure 7-27). Some have extensive baffles which, among other things, occupy so much space that

Figure 7-27. Stoves with relatively large heat transfer surface areas by virtue of extensive internal baffling (left: the Home Warmer), or interior air passageways (Juca stoves).

the exterior surface of the stove is large relative to the combustion chamber. Air passageways *inside* a stove also can add to overall heat transfer if the surface area and air flow are both adequate.

An interesting but not usually effective way to improve heat transfer is to convolute the stove walls (Figure 7-28). Although one can easily double the exterior surface this way, it is not as effective as adding a smoke chamber. The radiant output is not increased much since so much of the radiation leaving each surface does not get into the room but hits and is absorbed by another part of the stove surface. Since most stove finishes already have an emissivity of around 90 percent, the *most* that convoluting or roughing the surface could do would be to increase radiant output by about 10 percent. In practice, the effect would be considerably less.

Convoluting can cause a larger percentage change in the convected heat output since this kind of heat transfer is more nearly proportional to the surface area. The net effect on overall energy efficiency of convoluted stove walls is probably significant, but I am not aware of any experimental work to measure it.

In Korea, it is common to route smoke from coal stoves through passageways in the floors and walls of houses. If most of the heat gets into the house (as opposed to into the ground, under the house, and into the outdoor air through the outer skin of the walls), this system can have very high heat transfer efficiency. And, in principle, heat transfer losses can be prevented by using appropriate insulation. Access should, of course, be provided for cleaning coal soot from these passageways.

However, these systems have 1 serious problem. A few thousand Koreans die every year from asphyxiation! Carbon monoxide leaks into the house through inevitable cracks in the floor and walls.

Installation Factors. Heat transfer efficiency is dramatically affected by the length of the stovepipe exposed in the house. Stovepipe constitutes additional surface area, and is a factor in the residence time of smoke in the system. In chapter 9, we will look at stove installations more closely, but here, on the subject of heat transfer efficiency, stovepipe lengths must be considered.

Typical stove installations have up to about 6 feet of stovepipe between the stove and its chimney. Measurements at SER indicate the first 6 feet of pipe typically contribute 15-30 percent of the total heat output!

The ultimate in heat transfer can be achieved if 50 or 100 feet of stovepipe are installed between the stove and its chimney (Figure 7-29). With such installations, the smoke will cool down to room temperature before entering the chimney, and this means 100 percent heat transfer—no more heat could be extracted. But, such long runs of pipe are not recommended for 3 reasons.

• Long runs of pipe are illegal where building codes prevail. All codes require stoves to be placed as close as possible to their chimneys to minimize the length of the connecting pipe. Stovepipe is considered (correctly) a weak link in solid fuel heater installations.
• Such extreme cooling of the smoke all the way down to room temperature may result in very poor draft.
• The cooling of the smoke almost always leads to very large amounts of creosote.

It is interesting that such long runs of pipe were common in colonial America, particularly where space permitted—in churches, schools, and Shaker community buildings. So it certainly can be done. What would it take to manage the draft and creosote problems?

It *is* possible to get adequate draft with only room temperature smoke in the chimney. The key

Figure 7-28. The convoluted walls provide extra surface area, but the design is not effective in a radiant stove. This stove is an antique American product. (Photo courtesy Sturbridge Village)

Figure 7-29. The long stovepipe results in increased heat transfer efficiency but causes substantial creosote and draft problems.

is to be sure the smoke does not get any cooler as it rises up the chimney. This essentially requires a chimney that runs up inside the building. If the chimney is kept at room temperature, the smoke cannot become any cooler than room temperature. Having a *multistory* interior chimney is also helpful. The theoretical draft in an interior chimney in a multistory house with no fire in the chimney is a few hundredths of an inch of water on a cool day, which is enough for most stoves. Of course, not every interior chimney has such draft. In the case of a house that is very leaky in its upper portions and relatively tight in its lower portions, the draft will be close to zero.

I had a dramatic experience with this draft. A prefabricated chimney had been installed in my house starting at the ceiling of the ground floor and running up through a second-story room, through the attic, and out next to the peak of the roof. The stove was not yet installed, and it was a cool, calm fall day. Worried about losing heat up

the chimney, I took a piece of aluminum foil and was lifting it up to cap off the bottom of the chimney when it was sucked from my hand and stuck to the chimney bottom. The draft was that strong, with nothing but room temperature air in the chimney!

The creosote problem is not so easily managed. Even if combustion is complete, water vapor condensation is a problem. For each pound of wood burned with 20 percent moisture content, 0.63 pound of water vapor is generated. Even burning wood with zero moisture content results in 0.54 pounds of water per pound of wood burned completely. Much of this water will condense, and since combustion is rarely complete, tar, soot, and vapors will be mixed with the water. If 50 pounds of wood per day are burned, as much as 25 pounds (about 3 gallons) of dirty, acidic, corrosive, smelly water—creosote—will condense.

Stovepipe has never been watertight, particularly in a horizontal position. The large amount of liquid condensate in this kind of long stovepipe installation was apparently "managed" with gutters hung below the pipe and/or pails hung on the pipe or set on the floor. The stovepipes were disassembled and brushed out *frequently*—not once a year as is common in this country today, not 2–4 times a year as is often done in Europe, not once a month as was done in Philadelphia 200 years ago when wood was the only source of heat, but once a week!

There is one other way to achieve 100 percent heat transfer efficiency, and that is to neglect to hook up the stove to a chimney. Then one can be sure *all* the heat generated in the fire enters the house. The problem is that all the smoke enters the house as well—a decidedly unhealthful situation.

Hundreds of years ago such open fires (without chimneys) in houses were fairly common, and they still are in some parts of the world today (Figure 7–31). Asphyxiation was avoided (most of the time) by having a hole in the roof to let the smoke out, and having a very large amount of ventilation of outdoor air into the house. Such high ventilation rates are very costly when trying to keep a house warm. I am an absolute advocate of chimneys.

Figure 7-30. In rural Guatamala, ventilated roofs serve as chimneys. Cooking is done over an open fire on the floor. With no chimney, as such, heat transfer efficiency to the food and the room combined approaches 100 percent. The smoke helps control insects in thatched roofs.

OVERALL ENERGY EFFICIENCY

What is the most energy efficient stove? Overall energy efficiency is affected equally by combustion efficiency and heat transfer. The dominant factor affecting combustion efficiency, in most cases, is not stove design, but stove operation, which is not something you can buy. On the other hand, stove design *does* affect heat transfer efficiency. Therefore, for the average or typical stove user, the most energy efficient stoves are those with high heat transfer efficiency. Some examples were given in Figures 7–22 and 7–26. They tend to produce the most heat for a given amount of fuel consumed.

In my testing, I have observed overall energy efficiencies for stoves ranging from about 45 percent to about 65 percent. These efficiencies are averages over the full range of power outputs from low to high. They also include the heat output from 4 or 5 feet of exposed stovepipe.

A stove with a relatively low overall energy efficiency was a traditional Franklin stove (Figure 7–31) with folding doors, which left large cracks around them when closed. Actually, combustion tends to be more complete in this stove than many because of its air leakage—it is not possible to starve the fire of air as much as one can in an airtight stove. It is poor heat transfer that makes this stove less efficient. The air leakage is one cause. Leaked air (or more generally, excess air) dilutes and cools the hot gases in the stove and stovepipe and increases the total flow of gases and air through the system, thus reducing the residence time. The other cause is the small surface area and residence time relative to the typical fire size. The stove is not very deep, and the fire generally fills the whole volume with flames which sometimes pass up through the collar. Small fires are difficult to maintain because of the lack of control over the air supply.

Most of the stoves SER has tested with overall energy efficiencies in the 60 percent range had large smoke chambers. A smoke chamber, however, is no guarantee of high energy efficiency. We tested a brand available in 2 models, one with and one without a smoke chamber, and did *not* find a big difference in efficiency. I suspect the reason was air leakage—there were many cracks between the various cast-iron parts.

Figure 7-31. Heat transfer and overall energy efficiency tend to be low in traditional Franklin stoves due to excess air leaking in from cracks around the bifolding doors, low smoke residence time, and low surface area relative to fire size.

Is the spread in energy efficiencies—from 45 percent to 65 percent—significant? Is energy efficiency a factor worth considering? Yes!

If a 50 percent efficient stove uses 3 cords of wood to heat a house, a 60 percent efficient stove would need 2.5 cords to do the same heating job. That is a 17 percent fuel savings. If a 45 percent efficient stove uses 3 cords to do a particular heating job, a 65 percent efficient stove would use about 2 cords to do the same job—a savings of a little over 30 percent. These are substantial savings, whether they be in dollars spent to purchase fuel or effort expended to prepare the fuel.

How does the overall energy efficiency of a stove vary as a function of the size of the fire? Figure 7-32 summarizes all 3 energy efficiencies as a function of power output for a particular stove. The overall energy efficiency is about the same at all power outputs. But combustion and heat transfer efficiencies show considerable variation. At high power outputs combustion is at its best because of the ample air supply to, and high temperature of, the fire. Heat transfer efficiency is at its worst because of the low residence time— the flames and hot gases are moving through the system so quickly that much of the heat is swept up the chimney.

At low power outputs, combustion efficiency is at its lowest. Air-starved smoldering fires do not burn wood smoke very well; a lot of chemical en-

ergy is thrown away up the chimney. However, heat transfer efficiency is at its best because the smoke velocity is so low—there is more time available for the heat to get out of the smoke.

Although the details of performance vary from stove to stove, for all conventional stoves we tested, the trends are similar: overall energy efficiency does not change a great deal from low to high power (rarely more than 10 percentage points and usually less), combustion efficiency is always poor at low powers and much better at high powers, and heat transfer efficiency is usually best at low powers and worst at high powers.

An oversized stove will be operated much of the time at the low power end of its capabilities. This generally means that combustion efficiency will be low; hence creosote and air pollution will be problems.

Final Comments on Energy Efficiency

The discussion of overall energy efficiencies has focused on the gross efficiencies of the appliances themselves, including the heat given off by a typi-

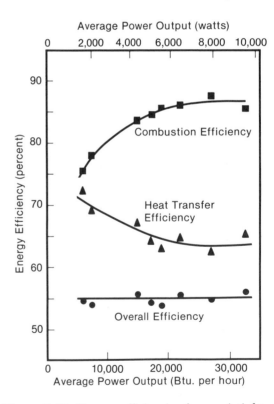

Figure 7-32. Energy efficiencies for an airtight stove over a wide range of power outputs or air inlet settings.

cal 4–6-foot stovepipe connector. The net efficiency of the system can be higher due to heat given off by an interior exposed chimney, or lower if the appliance consumes a significant amount of air relative to the natural house air exchange rate. Usually the house air loss up a chimney is significant only for fireplaces.

Also, for most systems, high overall energy efficiencies are achieved by having high heat transfer. Consequently the flue gases are relatively cool, which means that creosote and draft may be problems.

CREOSOTE POTENTIAL AND AIR POLLUTION

Creosote is the fuel of chimney fires. It can have an insulating effect which decreases the heat transfer efficiency of the system. And creosote is corrosive, particularly when wet. Creosote can form in chimneys, in stovepipe chimney connectors, and in the appliance itself.

Creosote is condensed smoke. Smoke pollutes the air. The best way to minimize both creosote and emissions is to have high combustion efficiency.

STEADINESS AND CONTROL OF HEAT OUTPUT

Stoves are not steady in their average heat output. The room-temperature fluctuations are not usually a serious drawback of wood heating, but most people are more comfortable with smaller fluctuations.

Controlling Combustion

The rate of combustion can be controlled either by controlling the amount of solid fuel in the com-

---

Automobile Analogues

Some of the more important performance characteristics of wood heaters have automobile analogues. Understanding them can help explain some of the limitations and compromises in wood stove performance.

STOVE CHARACTERISTIC	AUTOMOBILE ANALOGUE
Power output range (Btu./hr.): The minimum is achieved with fire just short of suffocation. The maximum is achieved with wide open air inlet.	*Speed range* (mph.): The minimum is achieved with engine rpm. just short of stalling. The maximum is achieved with wide open throttle.
Energy efficiency: Btus. of heat output per Btu. of fuel input.	*Fuel economy:* Miles traveled per gallon of fuel input.
Time between refuelings: Maximized by stuffing the stove full of fuel and then minimizing the consumption rate by starving the fire of air just short of extinction. A consequence is that the stove's heating rate is very low. Nonetheless, many Btus. will be given off before refueling is needed. Energy efficiency does not generally suffer. Stove users should realize that operating a stove this way minimizes the unit's power output, and this may not be convenient. High heating rates and long times between refuelings are incompatable.	*Time between refuelings:* Maximized by filling the tank full and then minimizing the consumption rate by letting the engine run only at idle speed. A consequence is that the car's speed is very low. Nonetheless, many miles will be covered before refueling is needed. Fuel economy does not generally suffer. People do not generally operate their cars this way because time between refuelings is less important than getting there in a reasonable amount of time. High speeds and long times between refuelings are incompatible.
Polluting emissions and creosote potential: Incomplete combustion results in smoke up the chimney, which pollutes the air and can condense in the chimney to become creosote.	*Polluting emissions:* Incomplete combustion of the fuel results in air-polluting emissions from the tail pipe.

---

bustion chamber or by controlling the air supply. In oil and gas furnaces, the combustion rate is fuel-limited—excess air is always available, and the amount of heat generated in the flame is determined by the flow of oil or gas into the burner.

The analogous solid fuel heater is a continuously fed coal, sawdust, wood-chip, or wood-pellet burner. In an automatic wood-chip stoker furnace, a thermostat in the house controls the rate at which chips are fed to the furnace, and this in turn controls the heat output of the furnace. Such a system can have very good combustion efficiencies because an ample air supply and high combustion zone temperatures can be maintained. These systems are discussed later in this chapter.

Controlling the Fuel Supply. Ordinary wood stoves and furnaces also can be operated in a fuel-limited mode to have some of the same advantages of a chip burner. If the air-inlet control is left open all the time and frequent small additions of fuel are made, combustion will be relatively complete. Of course, adding very small amounts of fuel every 5 minutes would be absurd, but it is also unnecessary. For many people adding wood every hour or so is not an onerous chore and usually results in better combustion. Limited tests at SER indicate about the same total amount of fuel is used but the chimneys stay cleaner.

The combustion rate is, of course, not steady if fuel is added every hour or two with the air inlet wide open; it surges and falls with each cycle. In most cases, the resulting room temperature variations are not uncomfortable. But with a little heat storage either in a massive stove, or in masonry surroundings of any stove, the actual heat output into the room can be extremely steady.

Controlling the Air Supply. The other fundamental way to control the combustion rate is to control the air supply. Limiting the air supply can be done manually with air-inlet and/or flue dampers, or automatically with a thermostat or a barometric draft control.

Thermostatically and Manually Controlled Air Inlets. A number of stoves are equipped with thermostatically controlled air inlets. The thermostat works on the same principle as do many wall thermostats for conventional heating systems. Two strips, or coils, of two different kinds of metal are bonded together. Both metals expand (or contract) as they are heated (or cooled), but one expands more than the other. Since they are bonded, the only way one can change its length more than the other is for the shape of the bonded pair to change as illustrated in Figure 7-33 . If one end is held fixed, as the temperature of the bimetallic strip changes, the movement of the other end can be used to open or close an air-inlet damper. In a wall thermostat, the movement opens or closes an electrical switch.

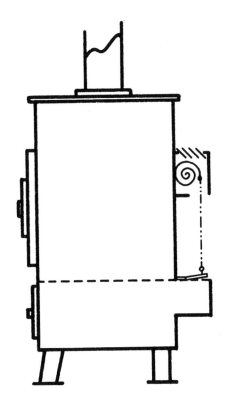

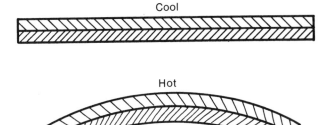

Cool

Hot

Figure 7-33. Bimetallic strips and coils are used for thermostatic controls of air inlets.

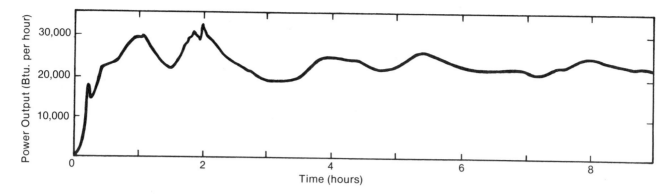

Figure 7–34. The heat output of a thermostatically controlled stove. The graph depicts the burning of a single main load of wood, which was added 20 minutes after lighting the kindling.

In the case of most wood stoves, the coil senses a combination of room temperature and stove surface temperatures. When the room and stove get hot, the movement of the coil closes the air inlet, and combustion slows. Heat stored in the stove walls and in the fuel continues to enter the room, but little new heat is generated inside the stove. The room and stove then slowly cool, which makes the bimetallic strip open the air inlet, allowing combustion to increase again.

A knob on the control can be turned to select the desired temperature. The automatic control on some stoves has a small magnet to pull and hold the air damper shut. When the force of the bimetallic strip exceeds that of the magnet, the damper pops open. As a result, the stove tends to get either a substantial amount of air, or none. The control on most stoves is more continuous— the damper gradually opens and closes and can remain in any position.

Many stoves do not have thermostats to automatically adjust the air-inlet damper. The damper can be set at fixed positions. Most airtight stoves are capable of putting out heat at a reasonably constant rate under these conditions.

Thermostats on solid fuel heaters have a reputation for causing more creosote. This is untested, but plausible. Thermostats tend to decrease the air supply just as the stove is getting hot and pyrolysis is reaching a peak. Decreasing the air supply at this time of the burn cycle usually results in much smoke going up the chimney. In manually controlled stoves, the air supply is more constant and does not decrease at this critical time. More of the smoke is thus burned.

Flue Dampers and Barometric Draft Controls. There are 2 other ways to limit the amount of combustion air entering stoves. Both involve regulating the draft at the air inlets, rather than the size of the openings. A flue damper, usually placed in the stovepipe above the stove, but in some cases, built into the stove, introduces extra flow resistance for the smoke. This uses up some of the chimney's draft so that less air is pulled in through the air inlet.

The other draft-regulating device is a barometric draft control. Whenever the amount of the draft in the chimney exceeds a preset value, the draft regulator opens, admitting air to the chimney. If a combustion surge starts to occur, the flue gases get hotter, which increases the draft; but then the barometric draft control lets more air into the chimney. This decreases the draft (suction) inside the stove by cooling the flue gases and directly relieving some of the pressure difference. Thus, less air is pulled in through the air inlet, and the combustion surge is limited.

Manual stovepipe dampers and barometric draft controls can help achieve long duration burns and improve overall energy efficiency, particularly with nonairtight solid fuel heaters. If the air inlet itself is too leaky, draft regulation is the only way to limit combustion air enough to achieve low power operation. If leaks are located so that their air is not likely to contribute to combustion, decreasing the draft will cut down on this excess air flow, which will improve the efficiency of the stove as well.

In an open stove (or fireplace) or a very leaky closed stove, there is no practical way to control

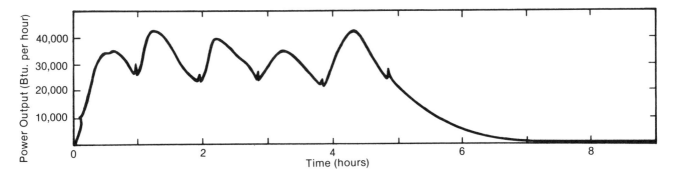

Figure 7–35. The heat output of a Franklin stove with its doors shut. Each surge corresponds to the burning of a load of fuel. Fuel was added 4 times after the initial loading. The small sharp spikes represent the bursts of heat entering the room while the doors were open for refueling.

the heat output by controlling the amount of combustion air. If air is always in excess, limiting the fuel supply is the only alternative. This can be done by controlling the frequency and size of refuelings. Figure 7–35 shows the heat output of a Franklin stove with its doors closed. Each surge corresponds to a refueling. A fairly steady average heating rate can be maintained with frequent refuelings.

Control via Heat Storage. Wood heating rates can be controlled by storing the heat produced in combustion for release later. Heat can be stored in the stove itself, or in the stove surroundings, including the house structure itself, or in an "active" storage area, such as a rock bin.

Heat storage in the stove and its surroundings is usually "passive," in the sense that there is no direct control over when or how much heat is given off or extracted from such storage, but the net effect is improved steadiness of room temperatures.

Mass is always important for heat storage. For instance, if there is a combustion surge in a light, thin-walled stove, there will be an almost immediate surge of heat into the room, which can make the room uncomfortably warm. The same combustion surge in a massive stove will be much less noticeable because much of the heat generated in the surge will be temporarily stored in the stove walls and structure. Most of the heat will still come out, but it will be delayed and spread out over a longer time. Similarly, as the combustion rate declines, the heat output of a heavy stove will fall more slowly than that of a light stove.

Liners, as well as the stove structure itself, can contribute to this heat storage capability.

The amount of heat stored in a 400-pound, mostly metal stove with an average temperature of 400° F. is about 16,000 Btu.[5] Needed heating rates typically range from 10,000 Btu. per hour to 50,000 Btu. per hour. This stored heat would be equivalent to roughly 20 minutes to 1½ hours of heating. The time it takes for a stove to emit or to absorb this amount of heat spans roughly the same range. Thus, a very heavy stove can help moderate room temperatures over time periods up to about an hour. But if a combustion surge lasts longer, the stove will be unable to store the excess heat and the room or house will become too warm; similarly a massive, but otherwise typical, stove cannot store enough heat or let it out slowly enough to help keep a house warm overnight if the fire is allowed to die out.

An interesting possible disadvantage of a massive stove is that more fires may have to be started from scratch. Since its stored heat continues to warm the room for an hour or so after the fire has died down, if you wait until the room feels cool before checking the fire, the fire may already be out, particularly if the fuel rests on a grate.

Massive Masonry Stoves. There are a number of wood heating systems designed to incorporate very large heat storage capabilities. The traditional European masonry heaters are examples—

---

5. (Mass) × (specific heat) × (temperature change) = 400 lbs. × 0.12 Btu./lb. °F.) × (400° F. − 70° F.) − 16,000 Btu.

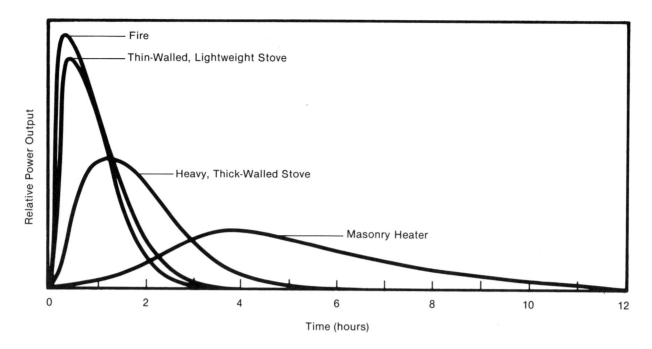

Figure 7–36. The effects of heat storage in stoves on their heat output rates. Each curve represents the heat output rate from a particular type of stove, assuming that the overall energy efficiencies are the same for each stove and that the actual fire is the same in each case. Each curve represents the same total amount of heat; thus, the area under all the curves is the same. The highest curve represents the common combustion rate for all stoves (actually the part of the rate representing heat which ultimately will be used). The effects of heat storage can be seen in the other curves as a lessening of the peak power and a delaying of the heat output—that is, a steadying of the heat output rate. The peak output of the tile stove occurs after the fire in it has died. (These curves are schematic, not the results of actual measurements.)

Swedish kakelugnar, tile stoves, and Russian fireplaces (Figures 7–37, 7–38, and 7–39). Typically, they are built of many tons of masonry and have complex and long internal passageways to transfer heat from the hot gases into the masonry. The fire is started with a bypass damper open so that the flue gases go directly up the chimney. When the fire is burning well, the bypass damper is closed, forcing the hot gases to travel through the passageways in the masonry, warming up the whole mass. When the fire is nearly out, the air-inlet damper is closed to prevent the stored heat from being convected up the chimney. The exterior surfaces of the stove are never too hot to touch; in fact, benches are often part of the stove. But the surface area is so large that the amount of heat given off to the room can be substantial.

With so much mass in the system, the time response is slow. I had an opportunity to use the Swedish kakelugnar shown in Figure 7–38. If the unit starts at room temperature, it can take an hour from starting a brisk fire to sensing the outside of the stove getting warm. It takes a long time for the heat to conduct through the mass. On the other hand, long after the fire has died down, heat stored in the mass keeps coming out into the room. The net effect is a delay as the heat output is spread over time. (Figure 7–36).

One advantage of masonry stoves is that relatively complete combustion can be achieved easily. The existence of the heat-storing buffer allows them to be fired at high rates without overheating the house. High combustion rates result in high temperatures in the combustion regions; if adequate oxygen is available, most of the smoke is burned.

Is the overall energy efficiency of such masonry stoves high? Not necessarily, but it can be. Heat transfer from the flue gases into the masonry is usually very high. But there are 2 ways the heat,

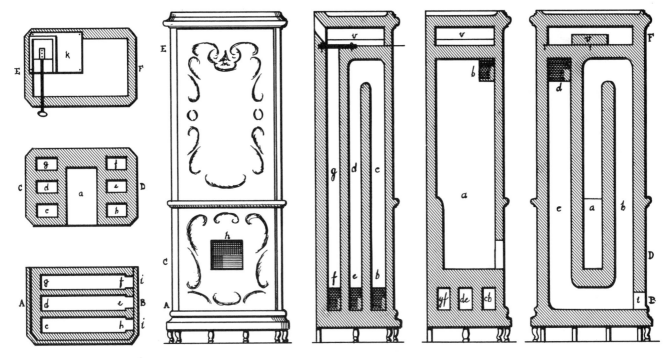

Figure 7-37. A design of a Swedish masonry or tile stove. This particular design was the winner in a competition sponsored by a king who was concerned about the depletion of wood resources in Sweden.

once in the masonry, can get away without warming the house. The heat can be conducted up through the chimney walls to the outdoors. The more massive the chimney is where it leaves the heated space of the house, the greater the effect. Normal 4-inch-thick chimneys are probably not too bad. Changing the masonry chimney over to a factory-built metal chimney before penetrating into the attic or roof will minimize the heat loss by conduction through the chimney structure.

Another potential loss of heat, and this one is probably the more important, is by convection up the chimney. As the fire dies down, the temperature of the air inside the passageways usually will fall below that of the masonry. This is especially true if there is significant air leakage—through the air-inlet damper, the loading door, or through the masonry structure itself. Such air leakage is common. The flow of this air through the passageways will slowly but surely draw heat *out* of the masonry and carry it up the flue to the outdoors.

Minimizing this air flow between fires is important. Making tight doors and air inlets is certainly possible—many ordinary stoves have them. Making the masonry structure itself airtight is more challenging. Masonry has a ten-

Figure 7-38. A Swedish tile stove.

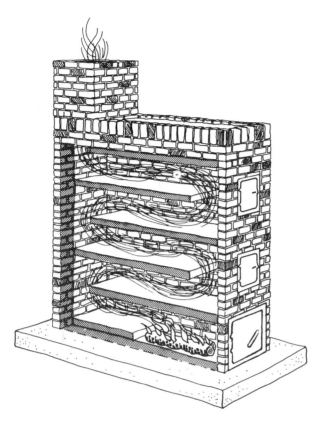

Figure 7–39. One example of the many different designs for Russian fireplaces. This is a schematic only; many important details are not shown.

dency to crack when heated nonuniformly, and nonuniform temperatures are inevitable in any heater. Many experimental masonry stoves and fireplaces built in the U.S. in the late 1970s developed cracks. Part of the problem may have been inadequate curing before using. More research and more careful study of European designs may be useful.

A chimney-top damper or an airtight flue damper high up in the system are potentially effective ways to reduce the air flow through leaks. But such dampers must not be so tight or be shut so soon after each fire has died down that carbon monoxide from the still-glowing coals might get into the house.

It is, of course, essential that these large masonry stoves have good access for cleaning all the interior surfaces. Even if only hot, relatively clean-burning fires are used, it is virtually inevitable that deposits will build up, posing a potential creosote fire hazard and also cutting down on the heat transfer from the hot gases into the masonry.

Heat-storing masonry stoves are not appropriate in all cases despite their potential for high energy efficiency, steadiness of heat output, and relative freedom from heavy creosote and air polluting smoke. The 2 major problems are that when you want heat in a hurry, you cannot get it, and when you want to stop the heat output, you cannot.

I have spent 2 winters living in a house with a steel stove built into massive masonry (adobe) surroundings. Each morning I make a guess about the weather. If I think it will get warm and sunny, I don't light a big fire. Sometimes I guess wrong, and have to start a fire by mid-morning. But the house does not warm up until mid-afternoon. On the other hand, if I incorrectly anticipate that heat will be needed, the house overheats because there is no provision for stopping the heat from coming out of the masonry once it is charged up. A man in New England who built his own *45-ton* Russian fireplace stopped using it after 1 season because of this problem. He claims he had to anticipate the weather by *3 days.*

The lesson in these experiences is that large, passive-storage wood heating systems are most appropriate where the need for heat is *steady.* When the temperatures remain at a fairly constant cold temperature for days at a time, a masonry heat-storage stove is ideal. Where the need for heat can change quickly, such stoves are awkward. If you want a responsive wood heating system—one that can produce heat quickly when you want it—don't use a masonry heater.

Modifications to traditional designs can alleviate some of these problems. Metal fireboxes can be used, and if quick heating is desired, air could be circulated around this firebox and into the room. (Metal combustion chambers also will alleviate the tendency of the masonry structure to develop cracks.) Air passages could run through the masonry structure, and dampers or louvers could control the air flow. An insulating (fireproof) curtain could be drawn around parts of the stove to suppress heat output when desired.

Active Heat-Storage Systems. The possible problems associated with passively storing wood heat are not present in systems with explicit, separate, active storage. In its most extreme form, the heater, which would be a boiler or hot air furnace in this case, delivers no heat directly to the living space. Instead, all of its useful heat output is transferred to a well-insulated rock or gravel bin, or to an insulated water tank (Figure 7–40).

When heat is needed in the house, blowers or pumps are used to transfer it from storage.

In this system, the house thermostat has no effect on combustion. It merely turns on the blower or pump, which circulates heat from storage to the house. If the house starts getting cold, it means the heat storage is exhausted, and it is time for another fire.

The furnace or boiler is fired up whenever needed, or, whenever it is convenient to do so; this might be 3 times a day or once every 2 days, depending on the storage capacity, the heating needs of the house, and how much heat is put into storage at each firing. Large, hot fires can be used, with the attendant advantages of complete combustion and relative freedom from chimney fouling, both of which allow more than the usual amount of heat extraction from the flue gases, and, hence, high overall energy efficiencies.

The mechanics of storage and heat distribution in such systems would closely resemble those of solar heating systems. Roughly 100–1,500 cubic feet of rocks (1,000 cubic feet is the volume of a 10-foot by 10-foot by 10-foot bin) or 1,000–3,000 gallons of water is appropriate. In fact, in many heating sytems of this kind, the building is solar *and* wood heated with common storage. Whenever the amount of heat in storage gets low and the solar collectors cannot replenish it, the wood appliance is fired and its heat is added to storage.

A disadvantage of an active storage system for wood heat is the additional cost of the storage and related heat transfer systems. The increase in overall complexity can lead to decreased reliability due to breakdowns. Also, since electricity usually is required (by pumps or blowers) either to transfer heat from the furnace to storage or from storage to the building, or both, the system will not work during power failures. Finally, using a furnace or storage that is not located in the living space of the building usually results in wasting some of the heat, decreasing the overall energy efficiency of the system.

Quick-Response Stoves. Steadiness of heat output is desirable only if the need for heat is steady.

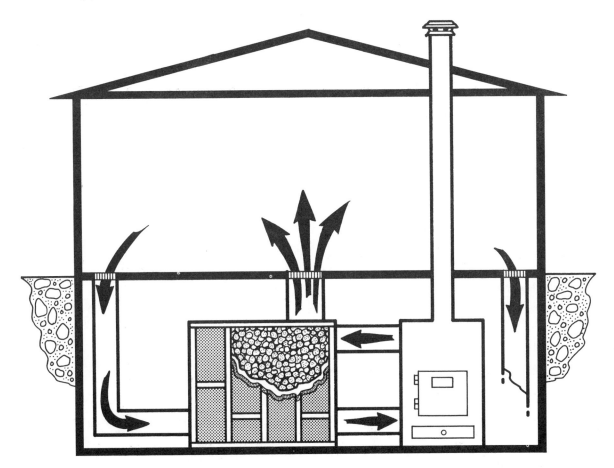

Figure 7-40. Rock-bed heat storage coupled with a wood-fired furnace.

Often this is not the case; heating loads can vary quickly and substantially. Some stoves are used for supplemental or intermittent heat, such as in garage workshops. Here a heater with a quick response time is preferable.

The most important design features for quick heating are low mass, light weight, and thin walls. Of secondary importance is the feature that when the air inlet is wide open, it will let in substantial amounts of air.

EASE OF OPERATION

Very few stoves are really difficult to operate. Yet some people, after using a few stoves for a while, come to have strong preferences for certain design features. Other people may dislike the same features. What some people find easy, others may find bothersome or annoying. There are no hard and fast rules about what makes a stove easy to use—personal opinions vary considerably.

Refueling Frequency

One major factor in ease of operation is how often one must add wood. This is such an important and often misunderstood issue that it was discussed previously in its own section. The important design features for infrequent refueling are a large fuel capacity and airtightness. The important side effects of using these features are creosote and air pollution.

Loading Ease

Stoves vary considerably in wood-loading ease. Fuel is added to most wood stoves through a side door near the bottom of the primary combustion area. With the low location for the loading door, smoke seldom comes out of the stove into the room during refueling. However, it is sometimes difficult to get large pieces of wood in, partly because of weight and leverage problems, and partly because of the danger of burning one's hand. The difficulty is most accentuated with longer and narrower stoves such as some European designs. You do not have to use pieces of wood so big that there is any loading problem, but if the wood is split into smaller pieces and/or cut shorter, then you cannot take advantage of

the burning properties of larger pieces of wood (discussed in chapter 14).

There are stoves that are loaded through a door near the top or even in the top (Figure 7–41), which usually eliminates the possible awkwardness of handling large pieces of wood, for they can just be dropped in (gently). It is also often easier to completely fill the fuel magazine from above.

When the access for loading is placed high on the stove, there is more chance of smoke spilling out of the stove during reloading. Hinged flaps hanging in the top portion of the opening can help minimize the smoking. However, even with top-loading stoves there need not be any smoke leakage, if proper reloading procedures are followed, and if the stove is installed to give reasonably good draft.

To load an airtight stove properly, open the air inlet fully for a few minutes before opening the loading door. If there is not much smoking fuel left, this will help clear the air in the stove. If there is significant smoking fuel, the additional air will usually ignite it, eliminating much of the smoke. Even more important, the resulting combustion surge will warm the fuel gases, increasing the chimney's draft, thus providing larger suction

Figure 7-41. A top-loading stove.

at the loading door. This will tend to keep the smoke from escaping when you open the door.

Large doors, wherever they are located, also make fuel loading easier. Some front-loading stoves have doors that are considerably smaller than the fuel chamber, which makes it a little awkward to fill the stove with wood.

Fire Starting

The ease of starting a fire in a stove depends mostly on the fuel, not on the stove design. The smaller and drier the pieces of wood, the easier the fire will start.

But there is one design feature which *does* have an effect – the air inlet. When a fire is struggling to get going, blowing on it helps, and some air inlets effectively do this by directing the air at the fire with moderately high velocity. The chimney must provide reasonable draft even when cold for this to work most easily, and this essentially requires an interior location for the chimney, and an appropriate flue size.

Fire Restarting

After the fuel load has essentially burned up and the room starts getting cool, it is convenient if the next load of wood starts burning easily and quickly. If there are still hot coals in the stove, this will happen, particularly if the wood is dry.

Stove design can affect the chances of there still being hot coals in the stove when it is opened for reloading. Grates (Figure 7–42) tend to be detrimental in this regard. Grates keep the coals up in the air. Then the coals either burn quickly or cool quickly. In addition, once most of the hot coals have fallen through the grate, new wood placed on the grate will not start as easily because of the distance between the wood and the coals.

In stoves without grates, the hot coals are nestled on or under the ashes. The ashes tend to keep the air away, thus slowing combustion of the coals, and to insulate the coals, keeping them warmer. Coals can stay red hot for as long as 2 or 3 days when buried in ash.

Some stoves tend to have pockets of coals which last for long times. Because of the baffle system in many Scandinavian stoves, not much air circulates to the back of the combustion chamber. With the help of the ash which is naturally there, hot coals are likely to be found there for hours after the main fire has subsided.

Figure 7–42. A traditional Franklin stove. Most Franklin stoves use grates to support the wood.

Ash Removal

The frequency with which ashes need to be removed varies considerably. With heavy use, some stoves require ash removal every day or so; others, every few weeks or even less frequently. A number of stove design features can be important.

The volume of ashes the stove can accumulate without impairing operation is an obvious factor. The volume of an ash drawer or tray under a grate is a limit. Stoves without grates have no clearly defined limit. In many stoves it is normal to take out the ash when its level reaches that of the door openings, but it is possible to operate stoves with considerably more buildup in the back of the stove, unless this blocks the air inlets. (When removing ash, it is wise to leave about an inch on the stove bottom to protect it from overheating.)

Generally, stoves without grates do not need ash removal as frequently. When the wood rests directly on the ashes, the originally flaky and fluffy ash particles get broken and compressed so that more ash can be accumulated in a given volume. In addition, the higher temperatures due to the fire resting directly on the ash and possibly the occasional raking or stirring of the ashes tends to help burn up the particles of charcoal.

An advantage of grates is that most stoves with grates have metal ash drawers for removal of ashes without shoveling. One merely pulls out the drawer to remove the ashes.

In some stoves, a considerable amount of ash is blown out of the stove and up the chimney during normal operation. There can be fairly high air velocities over the ashes and/or through the wood which can cause the ashes to be carried along. Natural dispersal and fallout of ashes which come out a chimney are fairly even and can help to fertilize the surrounding land in regions where the soil is not alkaline.

For most people ash removal is not such a chore that these design feature differences are of critical importance. But on the balance, most people seem to prefer *wood* stoves without grates and ash drawers to avoid the need for frequent ash removal, and because grates substantially reduce the coal retention time.

Cleaning

All stoves benefit from an occasional interior cleaning. Creosote or soot deposits often form in stoves. They can act as thermal insulation and decrease the energy efficiency of stoves. In stoves with extensive baffling or with multiple chambers, ash tends to accumulate in places other than the bottom of the stove, so it is important that all of the inside of stoves be accessible for cleaning. Stoves with insides that cannot be reached from either the door or the flue collar should have removable covers or the equivalent for accessibility. The Swedish kakelugnars have cleanout ports in each of the vertical passageways.

DURABILITY AND STOVE MATERIALS

Most stoves are sufficiently durable that durability need not be a prime consideration in purchasing a stove. But there are exceptions, and there are conditions which are especially wearing on particular types of stoves.

Cast Iron Versus Steel

Most stoves are made of steel plate or cast iron. There are good and bad aspects of both materials. Mechanically, cast iron is harder and stiffer. It is less susceptible to warping because of the shapes and integrity of cast parts. This makes cast iron preferable for doors and door frames where small distortions could result in significant air leakage. But the stiffness of cast iron makes it susceptible to cracking; cast iron cannot "give" very much. If the center of a cast-iron stove side is much hotter than the rest because the fire is against it, the thermal stress can crack it (Figure 7-43). Some cast-iron stove parts warp substantially when hot, but usually return to their original shape when cool.

Steel is relatively soft and malleable. If stressed, it bends, often permanently. Some steel stoves develop distortions in their walls due to thermal stress, which can be displeasing aesthetically, but the functioning of the stove rarely is impaired. Cylindrical and oval shapes are more resistant to distortion than the flat sides of rectangular stoves. Homogeneous steel rarely cracks, but steel stoves can develop cracks at joints and seams.

Both steel and cast iron are susceptible to corrosion. Some oxidation (rusting) of the stove walls from the inside due to the fire is unavoidable. The rate of oxidation at very high temperatures ("red hot") is much higher than at normal stove temperatures. Thin-walled stoves operated at very high temperatures have been known to burn out in 1 season. Even a ¼-inch steel can be oxidized away completely if held at red-hot temperatures for about a day. Although less susceptible than steel, cast iron can also burn out.

Some people are proud of the fact that they can read at night using the glow of their red hot stove

Figure 7-43. A cracked casting on a stove.

for lighting. This is very wearing on the stove and should be avoided. A thin-walled stove, such as a barrel stove, will last for many years if temperatures of the stove wall rarely, if ever, even approach a glowing red condition.

Firebrick or metal liners keep the main body of a stove from getting too hot. This lessens the chances of cast iron cracking, steel plate warping, either material burning out, painted exteriors turning white, and enamel finishes deteriorating.

Not all stoves need liners. This is particularly true of thick-walled steel or cast-iron stoves. Being stronger, thick walls are less likely to crack or distort in the first place; and being thicker, they will not burn out as quickly, even if the oxidation rate is the same. But, in fact, both the thermal stresses and the oxidation rate are less with thick-walled stoves. A thick stove wall cannot have such intense hot spots in it as thin walls can; the thickness makes it much easier for heat to spread out sideways or laterally within the wall, even where a fire or hot coal is against the inside of a thick wall. The increased conductance within the wall also makes the more distant portions (corners) hotter than they would be in a thin-walled stove. The temperature of the stove wall is more uniform, and therefore the thermal stresses are less.

With normal use and care, heat-induced oxidation is the only source of corrosion, but 2 kinds of carelessness can contribute to the premature demise of a stove. Storage in damp conditions has always been a common fate for stoves. But the modern way to ruin a stove, stovepipe, and chimney is to burn trash, including plastics in the appliance. Some plastics contain the elements chlorine and fluorine, which can be converted in a fire to extremely corrosive acids. Stovepipe may last only a few months under such circumstances.

The thermal properties of steel and cast iron are virtually identical (Table 2–1). The reputation cast iron retains for "holding heat better" is almost an historical accident. Until recently, most steel stoves were built of very thin sheet steel. Cast iron got a reputation for both holding heat better and lasting longer principally because it was thicker; it is technically difficult to make stove castings much thinner than ¼ inch. Cast iron and steel stoves of identical design, including wall thickness would be indistinguishable in their thermal performance. These days thick-walled steel plate stoves *are* available, and they also "hold heat."

Overall, both steel and cast iron are suitable materials for stoves. They have some different properties, but neither can be claimed far superior with substantial justification.

Stainless steel is much more durable than steel or cast iron at high temperatures. A few stoves have stainless steel liners, but generally stoves do not need such exotic materials.

Advantages and Disadvantages Of Masonry Materials

Some wood heaters are not made primarily of metal. Examples are the Swedish kakelugnar, European tile stoves, Russian fireplaces, and American soapstone stoves (Figure 7–44), and, of course, traditional fireplaces. In the past, materials such as brick, stone, and tiles were used because they were available.

Today, masonry is a popular material to use as a main part of a wood heater for several reasons.

•Heat storage. Appliances built primarily of masonry usually have relatively thick walls for adequate strength. So the mass of such units is often very large. Mass is sometimes added even beyond that necessary for structural integrity to provide steadier heat output to the house.

Figure 7-44. A soapstone stove (the Hearthstone).

- Aesthetics. Masonry allows more variety of color and shape. European tile stoves are faced with tiles that can be glazed and decorated. The shape of site-built masonry structures can vary widely.

- Pleasant historical connotations. Particularly in the area of wood heating, some people have the feeling that the old traditional ways of doing things are better. These people feel very comfortable and confident using soapstone stoves, a revered New England tradition, European tile stoves, or Russian fireplaces.

Most stove construction materials have potential liabilities, and masonry is no exception. Masonry materials are inherently inferior to iron and steel at heat transfer (Table 2–1). In addition, the thicker walls used in Russian fireplaces and tile stoves further inhibit getting the heat from the fire to the house. However, total heat transfer can still be and often is high—extra surface area is needed to compensate for the lower thermal conductivity. European tile stoves and Russian fireplaces clearly feature large surface areas, both inside for getting the heat from the flue gases into the stove structure, and again on the outside for getting the heat from the masonry structure into the house.

A second potential liability, which can be designed around, is cracking. Masonry materials are not soft or flexible. Like most materials they expand when they get hot. A thick masonry wall in a stove often will be considerably hotter on its inside than its outside surface. Also, the center of a stove side near the fire may be much hotter than the more distant corners of the same side. If there is no provision for the unequal thermal expansion, the material may crack. Recall from chapter 6 that this is one reason the tile liner in a masonry chimney should not be held in place too rigidly.

There are solutions to the problem of cracking. One is to use an appropriate liner to take the brunt of the heat, thus reducing the thermal stress on the rest of the structure. Some European tile stoves have cast-iron fireboxes, partly for this reason. The liners can themselves be masonry—good firebrick is particularly resistant to thermal-stress cracking.

One way to avoid unsightly cracks in soapstone stoves is to use a number of smaller pieces held together in a metal frame for each stove side, rather than one large slab. Then each small piece is at a more uniform temperature, and this means less thermal stress within each piece. Overall thermal expansion can be accommodated at the joints and/or by the frame.

HEAT DISTRIBUTION

Stoves can be classified as either radiant or circulating, depending upon the dominant way they transfer energy into the room. Most stoves are the radiant type, transferring 60–70 percent of their energy output as infrared radiation. A circulating stove is essentially a radiant stove surrounded by an outer jacket with openings at the bottom and top so that air circulates between the stove and the jacket (Figure 7–45). Natural convection can generate a considerable flow of air, but small blowers are sometimes employed. The result is that more of the energy output from a circulating stove is in the form of hot air, although significant amounts of radiation are still emitted by the jacket.

Circulating stoves are preferred by some people. The exposed surfaces are not as hot as they are on a radiant stove; this makes them safer with small children in the house, and allows people, furniture, and walls to be closer to the stove without discomfort or damage. The jackets are often designed to disguise the fact that they house a solid fuel stove.

You can feel the differences between rooms

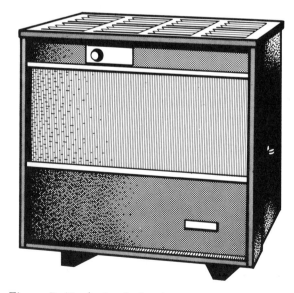

Figure 7–45. A circulating heater.

TABLE 7-4
TOWARDS STEADY HEATING
FROM WOOD

Objective	Design Feature or Operator Action
Fuel-limited combustion	Automatic sawdust, chip, or pellet stoker/burner Frequent, small manual refuelings, with air inlet wide open
Air-limited combustion	Thermostatically controlled heaters Manual air inlets set for restricted air supply Manual flue-pipe damper Automatic barometric draft control
Heat storage with steady release	Heavy stoves (a few hundred pounds) Massive masonry stoves (tons) Mass in stove surroundings or house structure Active storage, as in some solar-heating systems

Design fundamentals and techniques for achieving steady heating.

Figure 7–46. The advantage of a radiant over a circulating stove in a drafty house. Heated air may be swept out of the house without hitting the occupants, but radiant energy will always reach a nearby person, regardless of wind or drafts.

heated mostly by radiation and rooms heated mostly by hot air. Since the radiation from a stove comes from 1 direction only, if you face the stove, your front side will be slightly warmer than your backside. A piece of furniture or a person can block the radiation from reaching parts of the room. But in a room with high levels of infrared radiation, the air temperature can be cooler without causing thermal discomfort, and some people feel better in cooler air.

A radiant stove will provide usable heat more quickly. Even though the room may be cold, as soon as the surfaces of the stove are hot, you can be warmed (on one side) by standing close to the stove. The hot air from a circulating stove rises to the ceiling. You must wait until enough hot air has been produced to warm most of the room before feeling as much warmth. However, since both kinds of stove heat by both radiation and convection, these differences are only a matter of degree.

In an extremely drafty room, such as a leaky cabin, radiant stoves are more effective than circulating types. Air warmed by a stove may be carried out of the room before it can be sensed by the occupants (Figure 7–46). But radiation will contribute warmth whenever it is absorbed, despite drafts. Most of the radiation from a stove travels *through* the air and thus cannot be blown away by drafts.

There appears to be no evidence that jacketing a stove increases the energy efficiency; in fact, preliminary laboratory evidence suggests a slight decrease in efficiency.[6]

Blowers on stoves can improve the temperature distribution in a room. If the blower causes substantially more convection, turbulence, and mixing of room air, it will tend to even out the differences in air temperature between the floor and ceiling. However, natural convection around a stove is strong; the blower must be of significant capacity or used in a careful way to result in better temperature distribution. Blowers or fans located *away from* a stove generally do a better job of distributing heat. One interesting and probably valuable use of blowers is to *counter* natural convection by forcing heated air down to floor level.

---

6. Ibid.

COOKING

Almost any stove designed principally for space heating can be used for cooking food. The stove must have a flat top, and that the top must get adequately hot. The top should have fire or hot smoke underneath it, not air as in circulating stoves. Cooking is much more practical with a cookstove, but in an emergency, most heating stoves can be used for cooking. Real cookstoves, if reasonably airtight, are likely to be very efficient space heaters because of their large surface areas and the long residence time for the smoke.

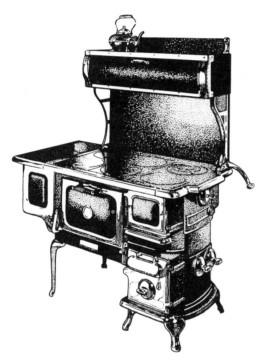

Figure 7-47. Cookstoves have the potential for high heat transfer efficiency because of the large surface area and long smoke residence times. However, air leaks often negate these advantages.

EASE OF BURNING GREEN WOOD

Stove design can affect the ease with which green wood can be burned. Most stoves which claim this ability have a massive refractory lining around the firebox (Figure 7-48), which tends to keep the combustion chamber warmer by retard-

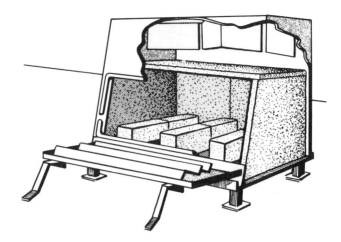

Figure 7-48. A stove with large amounts of refractory material round the combustion chamber (the Greenwood Eagle).

ing the conduction of heat through the walls and by storing heat. Once the stove is warmed and has a bed of coals, it is plausible that this massive and extensive liner makes it easier to burn green wood by providing a warmer environment. Field reports and some informal testing support this contention.

STOVE DESIGN CONFLICTS AND POSSIBLE RESOLUTIONS

Stove technology is evolving. Some day it may be possible to buy a stove that is truly the best in every desirable performance category. But today this is not the case. For a stove to excel in a particular characteristic, some other characteristic must often suffer as a consequence.

Combustion Efficiency Versus Heat Transfer Efficiency

Complete combustion requires high combustion-zone temperatures. One way to achieve this is to insulate the combustion chamber with firebrick or other materials to keep as much heat as possible inside the combustion chamber. However, holding heat in is not conducive to its getting out, and *that* is what is required for high heat transfer efficiency. Thus there is a conflict between 2 desirable but opposing objectives.

In this instance, a possible but not always practical solution exists: design the stove so that com-

Figure 7–49. Two hypothetical stove designs, illustrating the separation of combustion and heat transfer.

bustion and heat transfer occur at different places (Figure 7–49). The combustion chamber can be very well insulated to enhance combustion without significantly hurting heat transfer because heat transfer takes place later in another place in the appliance. In practice, this will usually result in a larger and hence more expensive appliance. But there is no doubt about its technical feasibility. A clear example is the wood-fired boiler based on design work by Professor Richard Hill of the University of Maine (Figure 7–50).

Heat Transfer Efficiency Versus Creosote Accumulation, Corrosion, and Draft

There are many design features that will improve heat transfer efficiency. But regardless of how it is achieved, any improvement in heat transfer efficiency will decrease the amount of heat in the smoke. Energy is conserved. If more heat is taken out, there is less left. Less heat in the smoke almost always means cooler smoke, which in turn means more creosote accumulation.

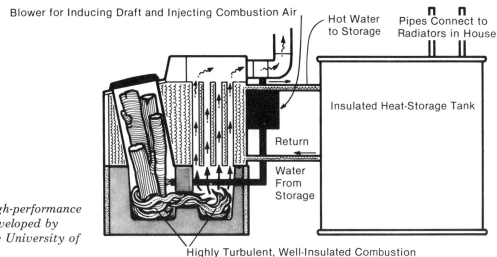

Figure 7–50. A high-performance heating system developed by Richard Hill of the University of Maine at Orono.

Sufficiently low smoke temperatures also result in the condensation of water vapor and decreased draft. The water can create problems of corrosion, odor, and disposal. Here again there is conflict.

Is there a possible resolution to these conflicts? Unfortunately, the answer is essentially no. Combustion in residential wood burning equipment is never perfect; and as long as there is any smoke at all in the chimney, the cooler the smoke, the more it will condense or plate out as creosote on the chimney walls.

It is possible to design and operate systems so that combustion is much better than average. In these cases, the additional chimney deposits resulting from improvements in heat transfer may be so little that in practice the conflict is not very significant.

The remaining conflicts—corrosion and decreased draft—cannot be made to disappear, but the consequences can be dealt with. More expensive and more corrosion-resistant materials can be used. Drips can be contained and collected. If necessary, a fan can be used to induce draft.

High Power Reserve Versus Creosote Potential and Air Pollution

It is my opinion that many people buy oversized stoves. They do this to have some reserve heating power for heating up a cold house more quickly and for more easily handling the very coldest and windiest days of the winter.

Oversized equipment, during most of the year, is underutilized, producing a small fraction of its potential heat output. For most wood heaters, this means more creosote and air pollution. Stoves tend to have their lowest combustion efficiencies when operated at the low end of their power output range (Figure 7–32). It is more difficult to achieve a clean-burning small fire in a large stove than in a small stove.

Are there solutions to the conflict between reserve heating capacity and clean combustion? In a sense, yes, but each has its cost. You can own 2 stoves, a large one and a small one. The small one is used in the fall and spring, and the larger one is installed for use during the coldest months.

Another option is to push a small or medium-size stove to large heat outputs when needed. This will usually involve the inconvenience (or fun if you're a pyromaniac like I am) of more frequent refueling. Using dryer-than-usual fuel, fuel in smaller pieces, and maximum fuel loads will help

to get hotter fires. Chimneys which provide large draft will make it easier to achieve high heat outputs from small stoves.

It is also essential that the stovepipe and chimney be safely installed and be clean, for such hotter-than-usual fires are likely to start chimney fires.

Of course, the problem would be eliminated if a stove could be designed that operates with essentially complete combustion under all circumstances.

The Convenience of Infrequent Refueling Versus Creosote Potential And Air Pollution

If you fill your stove full whenever refueling, the chore will not have to be done so often. Unfortunately, when a stove is filled to the gills, and the air supply is restricted to obtain a long duration burn, the fuel gets hot enough to undergo pyrolysis and smoke, but there is inadequate oxygen and/or temperatures for the smoke to burn. Even at higher air settings where combustion is more complete, larger fuel loads can result in poorer combustion.

Here there *are* some possible solutions. A major cause of the conflict is that each fuel load is burned and stored in the same place. The pieces of wood on the top of the load, which do not really need to start burning for 4 or 5 hours in an 8-hour burn, are in the combustion chamber from the start, getting hot and emitting smoke. With the small amount of air admitted, very little smoke will have a chance to burn.

Is it possible to design a wood heater with separate combustion and fuel storage chambers? The answer is a qualified "yes." An automatic stoker, which burns sawdust chips or wood pellets, feeds the fuel from a storage bin to the burner automatically with, for instance, an auger. The fuel feed rate can be controlled by a temperature sensor. A "pilot flame" is maintained with a slow feed rate to keep the fire from going out. The combustion chamber can be small and hot. In a well-engineered system very little smoke would leave the chamber unburned. Thus, you have a *system* which may need refueling only once a day or once a week depending on the size of the fuel hopper, but which also has the potential for high combustion efficiency.

Obviously such a system is not a simple stove. Most stokers are central heaters—furnaces or

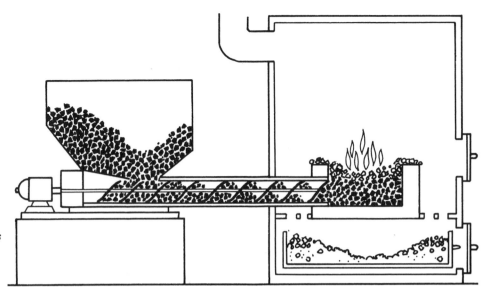

Figure 7-51. A stoker that can burn any solid fuel. In response to the thermostat's call for heat, an electric motor drives the auger and introduces combustion air.

boilers. They require electric power and a large fuel bin, and the owner is dependent on fuel deliveries—it is rarely practical for the owner to make his or her own wood pellets or chips.

These systems are also considerably more expensive than typical stoves. But the potential advantages are considerable: infrequent manual fueling, little creosote and air pollution, and high overall energy efficiency.

Is it possible to design a wood heater with separate combustion and fuel storage areas, but which burns normal fuelwood (sticks, logs, chunks)? Some such units have been built and operated, but careful objective evaluations are not usually available. An interesting design by Jake Lemon (Figure 7-52) has 2 slanted feed tubes which together form a V with its bottom missing. Logs are loaded end to end in each tube and a cover is put on the tube top. Combustion occurs where the 2 logs meet at the bottom of the V. As the logs are consumed, they slip down the tubes by gravity.

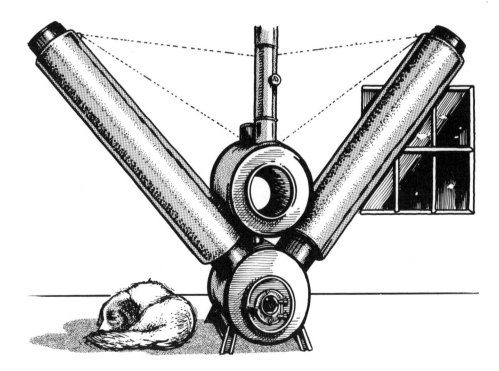

Figure 7-52. A gravity log stoker designed by Jake Lemon.

In an older design by C. E. Jenkins (Figure 7-53), there was only 1 slanted 18-inch-diameter chute, which could hold a number of logs side by side. The logs could be as long as 8 feet, but for convenience of handling, smaller lengths, typically 4 feet, were used. Combustion occurred at the bottom where the logs rested on the ashes. One must be very careful in designs such as this that the fire does not creep up the feed shute.

These designs have the potential for resolving the conflict by keeping the bulk of the fuel load cool until it is time for it to burn. However, their performance is not well documented. The size and shape of the units tend to be large and awkward. But the design ideas are very interesting; further testing and development of these and other ideas might prove valuable.

A second approach to the conflict between the convenience of infrequent refueling and smokeless combustion is use of heat storage. A full load of fuel, containing enough energy to heat a house for 8 hours, *can* burn relatively cleanly if the air inlet is wide open. The problem then is that the

fuel does *not* burn over 8 hours but is consumed in 1-3 hours. The result would be heat outputs over 100,000 Btu./hr. and a substantially overheated house!

Solution? Don't let all that heat into the house at once, but store some of it temporarily for later release to the house. This requires *substantial heat storage capacity.* Masonry heat-storage stoves and fireplaces can be effective. So also are the rock bed or water heat-storage systems used in many solar heated homes. In fact, a number of solar heated buildings used wood back-up heat in just this fashion.

IS COMPLETE COMBUSTION AT LOW BURN RATES POSSIBLE?

The key problem in the design of solid fuel heaters is achieving high combustion efficiency. As has been pointed out throughout this chapter, many design features, although based on valid principles, do not work in practice. Thus, achieving efficient combustion is challenging technically. In contrast, good heat transfer is easy to achieve. And in fact, the amount of heat which can be extracted without encountering creosote problems depends on how good combustion is, rather than on the designer's ingenuity.

Four design approaches help to achieve high combustion efficiency:

- Continuous fuel feed (stokers).
- Heat storage.
- Catalytic combustion.
- Advanced conventional combustion.

Stokers and heat storage have real potential, but the resulting heating systems tend to be expensive, bulky, and more complex than ordinary stoves. Both catalytic combustion and advanced conventional combustion can fit into the space limitations of an ordinary stove.

If somehow combustion in batch-fueled heaters could be complete even at low burning rates, many problems and conflicts would be solved and resolved. What is needed is an afterburner—a device or set of design features that assures combustion of the smoke coming from the fire under almost all conditions, particularly at low firing rates when the smoke's temperature is usually be-

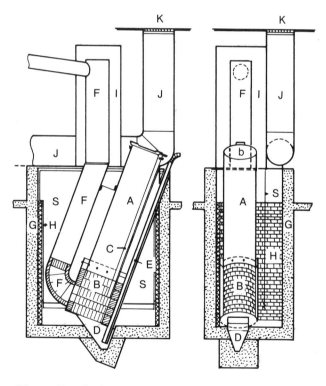

Figure 7-53. A gravity-feed log furnace designed by C. E. Jenkins. Long logs are placed side by side in Chute A. Combustion takes place in Region B, and the smoke rises up the Flue, F.

low that necessary for ordinary ignition (around 1,100° F.).

The only way to absolutely guarantee complete combustion under all circumstances is to supply heat from some outside energy source such as oil, gas, or electricity. Such powered afterburners are sometimes used for control of industrial pollutants. If the pollutant is combustible, it can be eliminated by burning. Burning requires high temperatures and adequate air. A gas or oil burner can be used to assure adequate temperatures, and air is easy to supply. If the pollutant (or smoke)/air mixture has the right proportions so that it is within its limits of flammability (see chapter 5), then the gas or oil burner need only supply enough heat to ignite the mixture. However, if the mixture is too lean to be able to burn by itself, then enough fossil fuel must be supplied to maintain high temperatures of the whole mixture.

There is no doubt this concept could be made to work with wood heaters. Such a system would no longer be just a wood stove; it might be a dual fuel heater, probably a central heater (a furnace or a boiler). The fossil fuel could assure complete combustion when wood was being used as the principal fuel and provide the option of conventional automatic heating. A major unanswered question concerns how much energy output is necessary to assure combustion of the wood smoke. If the entire flow of smoke must be constantly warmed up to about 1,100–1,200° F., then much of the total heat output might be coming from the fossil fuel or electricity. Use of such a system would not dramatically reduce conventional heating costs. If, on the other hand, the fossil fuel is needed primarily to assure ignition of the smoke, then the fossil fuel or electricity consumption might be quite small.

Considerable research is necessary to show whether or not fossil-fueled afterburners are practical. There is the possibility of the fossil fuel flame or electric igniter causing backflashing (small explosions), due to sudden ignition of large quantities of wood smoke.

Catalytic Combustion

Automobile catalytic converters are intended to finish burning whatever fuel the engine did not burn completely, for pollution control. But catalytic converters for wood burning appliances

Gasifiers

Appliances that "gasify" the wood fuel before burning it have received much attention recently. The name suggests that these appliances gasify the fuel and then burn the gas.

Do gasifiers have the potential for improved performance? Basically the answer is no, not because there is anything wrong with the principle, but because there is nothing special about it. All solid fuels are partially gasified before they burn; pyrolysis is another name for the process.

Is there a special type of appliance for which the term "gasifier" is appropriate? I believe there is, but no residential heaters would be included. Wood-powered internal-combustion-engine cars and trucks have true gasifiers which provide the gas which is piped into the engine. Industrial solid fuel gasifiers are used to provide a gaseous fuel for burners for process heat, often as an alternative or retrofit to an existing furnace designed for oil or gas fuel. In both these examples, the gasifier only gasifies; the gas is then transported and burned elsewhere, resulting in a complete seperation of gasification and combustion. (In stoves with secondary combustion chambers, the separation is only partial–some of the gas and all of the charcoal burn in the primary chamber.)

Gasification always results in some waste and/or consumption of fuel; sometimes there is unburned charcoal left over, but some fuel energy is always used to generate the heat necessary to gasify the solid fuel.

could achieve 2 additional objectives–reducing creosote and increasing overall energy efficiency.

A catalyst promotes combustion essentially by lowering ignition temperatures. The catalytic materials themselves are often noble metals, such as platinum and paladium. When a fuel molecule encounters the catalyst, the molecule's bonds are stretched and weakened. The weakened molecule will not need to be hit so hard with an oxygen molecule for the burning reaction to start. This means the burning reaction will start at a lower temperature since temperature and average molecular speed are related.

The normal ignition temperature of wood smoke is probably between 1,000° F. and 1,300° F. With catalytic assistance, the ignition temperature can be brought down at least as far as 500°

F. In a well-designed system, more smoke is burning more of the time, resulting in more heat, less air pollution, and less creosote.

A catalytic system needs to have a large surface area since it is only on the surface of the catalyst that combustion is enhanced. A common geometry is a honeycomb structure (Figure 7-54). The smaller the holes in the honeycomb, the more surface area the structure has, and the more likely it is that the fuel molecules will bump into the catalyst. Many automobile catalytic converters have 400 cells per square inch! In addition, very pronounced microscopic valleys and peaks make the actual microscopic surface area *much* larger than the apparent area.

But large macroscopic surface areas in the flue-gas path can cause problems. If the large area occupies a small volume, such as in the honeycomb geometry, then care must be taken so that the catalyst does not unduly obstruct the flow of the wood smoke. Automobile catalytic converters can offer considerable flow resistance without creating difficulties because the exhaust gases get pushed through by the engine. But most wood heating appliances rely on *natural* draft; only a very small pressure drop across the catalyst can be tolerated.

Some catalyst geometries may be susceptible

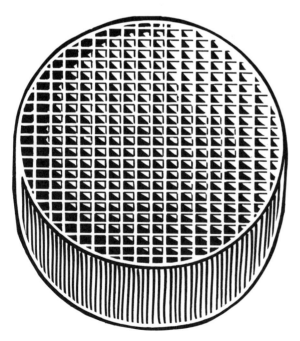

Figure 7-54. A Corning ceramic honeycomb catalytic combustor with a diameter of 5½ inches.

to plugging by creosote, soot, and ash. This not only can keep the smoke from getting to the catalyst surface, but if a bypass is not provided, it effectively can prevent any smoke from getting into the chimney. Thus, designing a catalytic combustion system for wood heaters that is both effective and safe can involve considerable engineering.

Catalysts generally do not remain active forever. Although the precious metals remain intact, they become poisoned or masked by other materials. It will be important to burn only "lead-free" fuel. Wood itself has only very small amounts of lead and other catalyst poisons; but use of painted wood, household trash, fire colorants, tires, and some chemical chimney cleaners may have to be strictly avoided. At extremely high temperatures, the microscopic surface texture can lose some of its surface area, thereby decreasing the activity of the catalyst. For all these reasons, periodic replacement of the catalyst may be necessary.

Catalytic combustion in coal appliances will be much more difficult to achieve because ingredients in coal deactivate most catalysts. By the same token, it is very important not to burn any coal in catalytic wood heaters.

However, the potential for improved performance in wood heaters is substantial. My feeling is that, at the least, catalytic combustion in the real world could reduce creosote and emissions by 50 percent and improve overall efficiency by 10 percent. Testing at SER on a particular and not optimum design (Figure 7-55) showed as much as a 20 percent (10 percentage points) improvement in overall energy efficiency under some firing conditions (Figure 7-56). We have seen overall efficiencies of about 70 percent in some catalytic stoves.

However, do not expect this efficiency in every catalytic unit or under all operating conditions. Some of the earliest catalytic stoves that appeared on the market performed little better than ordinary stoves. Even the best catalytic stoves have limitations; combustion is not perfect. Creosote will still accumulate, but at a slower rate. And there is a burning rate below which a catalyst ceases to be effective.

The durability of stove catalysts is not yet well established. Occasional replacement (perhaps every 1-3 years) is likely to be needed due to declining activity of the catalyst and, in some cases,

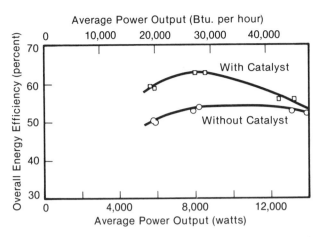

Figure 7-56. Overall energy efficiency of an experimental stove without a catalyst.

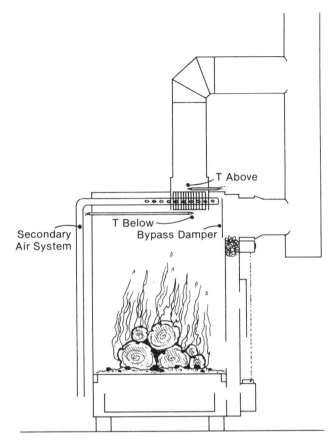

Figure 7-55. A modified Riteway stove used in some experiments at SER to measure the change in energy efficiency due to a catalytic combustor.

due to deterioration of the substrate, the structure on which the catalytic coating is applied.

Noncatalytic stoves featuring advanced conventional combustion systems also offer substantially improved combustion, primarily through secondary combustion systems which really work much of the time. Catalytic combustion is not the only way to improve performance.

These are very exciting times for the solid fuel heater industry. Despite the fact that wood and coal heaters have been with us for centuries, we are now in a decade of real design innovation.

CHAPTER 8

COAL STOVE DESIGN

Coal and wood appliances have more similarities than differences, yet the differences are usually critical. However, designs for good heat transfer and the principles for good combustion are identical. Hence, this chapter is brief; it does not repeat the material in the previous chapter.

Some of the major areas of difference between coal and wood appliances are in the grates, liners, primary and secondary air, and gravity fuel feed.

GRATES

Wood logs burn easily and efficiently either on a grate or resting directly on an ash bed. In contrast, most coals require a grate. There are 3 principal reasons for this difference. The high fixed carbon content of the higher rank coals means more of the coal burns as a solid, as glowing coke, and less as flaming gases, as compared to wood. Coke (and charcoal) burn only to the extent that oxygen is brought right up to their surfaces. With coal resting on a grate, air can pass up through the grate and through the entire fuel bed. Other air flow patterns are also possible, but a grate always facilitates movement of air in and fumes out of a fuel bed.

A second reason why burning coal of any rank may be more efficient with a grate is the tight packing of coal. There tends to be less air space and more tortuous air paths so it is harder for air to penetrate into and through a coal bed compared to a bed of logs.

Grates allow ashes to fall through while holding back the burning coal. Since burning coal results in a considerable accumulation of ash, which must be removed often, grates conveniently allow ashes to be removed while the fire keeps burning. An ash drawer that collects the ashes that fall through the grate is extremely convenient.

Coal grates come in many shapes, sizes, and configurations (Figure 8-1). All are designed around certain principles.

- To let air pass through (usually upwards).
- To prevent coal from falling through.
- To let the coal ash drop through, and usually.
- To help the ash along by a shaking or slicing mechanism.
- To withstand very high temperatures.

The traditional construction material for coal grates is heavy, high-quality cast iron. Even then sagging and burning out are possible if operation is careless. In England, some grates contain as much as 30 percent chromium to reduce the burning out.

To preserve grates, they must be bathed in cooling air (the combustion air) and must not be in direct contact with burning coal. The ash under the grates must not be allowed to build up to where it touches the bottom of the grates and impedes air flow. This is by far the most common cause of grate failure. Also, when you shake your grate or slice the ash above the grate, it does not hurt to leave a little ash on the grates as a buffer between the grates and the burning coal.

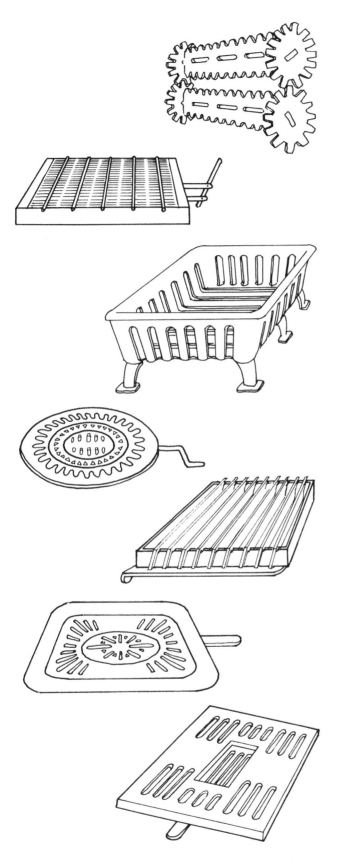

Figure 8-1. Some grate designs for burning coal.

Ease of shaking is a very important design feature of coal stoves. Some grates do not perform the shaking function well under any circumstance. Others shake well but require a great deal of physical exertion. Some work well with one type of coal, but not with others. Specific recommendations are difficult to make. If possible, try operating the stove yourself, fueled with the type of coal you expect to be using, before buying the heater.

With appropriate grates, coal can be burned in open masonry fireplaces.

LINERS

Coal stoves do not have to have liners, but most do. Firebrick is the most common material, but cast-iron liners are used also. Many older coal stoves had no liners but had a very heavy cast-iron construction around the fuel bed.

The reason for extra concern about the combustion chamber environment is the fact that temperatures can be hotter with coal as the fuel, compared to wood. This is because coal is a denser fuel and because coal stoves often have extra air-inlet capacity to help get the fire started.

PRIMARY AND SECONDARY AIR

The distinction between primary air and secondary air is more meaningful for coal stoves than for many wood stoves. In most coal stoves, primary air (or "under-fire" air) enters under the grates and passes up through the coal bed (Figure 8-2). Because coal beds have smaller air spaces, the combustion air is forced to come into close contact with the solid coal and to mix well with the gases. The net effect is that the oxygen tends to get used up.

If the gases have not been burned completely by the available supply of primary air, there is a clear reason to have secondary (or over-fire) air.

However, whether adding secondary air will help is another question. Most of the discussion in chapter 7 on secondary air is applicable here. In order for secondary air to be of maximum benefit, it must be in the right amount (not too little nor too much), it must enter at a location that makes it available to the flames, and it should mix well with the gases. As in the case of wood stoves, it is

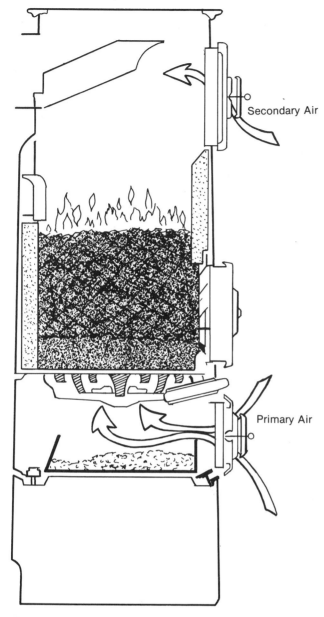

Figure 8-2. A coal stove with primary and secondary (or under-fire and over-fire) air (the Jøtul 507B).

may result in the need for considerable secondary air. Fuel bed thickness can also affect the need, thicker beds usually needing more.

The best you can do as the operator is see how much the amount of secondary air affects the density of the smoke coming out the chimney, and then use the least amount of secondary air that minimizes smoke.

EASE OF REFUELING

As is true for wood stoves, coal stoves are more easily loaded from the top than from the side. In fact, this feature is more important with coal because coal can be poured.

Some coal stoves have a gravity fuel feed. A fuel hopper is part of the appliance, and as coal burns out in the combustion chamber, the fuel automatically flows out of the hopper to take its place (Figure 8-3). Gravity fuel feed extends the time between refuelings without sacrificing combustion efficiency.

Coal appliances without fuel hoppers are said to be "batch-fired" or "surface-fired."

FUEL BED GEOMETRIES

Most coal stoves sold in this country are intended to burn anthracite. Anthracite combustion chambers tend to be relatively narrow and deep. Bituminous coal generally burns best in a shallower and wider bed, more like the combustion chamber in a wood stove. Because of the geometry differences, combination wood and coal stoves are easier to design for use with bituminous coal.

DRAFT ADEQUACY

Coal stoves tend to be more sensitive to draft than wood stoves. If the draft is too little, a coal fire may go out. If the draft is too much, the stove may get dangerously hot. Wood fires are also influenced by draft, but not usually to the same degree. (The term "draft" is used here in its proper technical sense of the suction on, or inside, the appliance—that is, the pressure difference force which pulls air in.)

The minimum draft required by many coal

not sufficient just to put a hole in a stove anywhere above the fuel bed (in conventional updraft coal stoves). Ultimately, laboratory testing is required to determine how effective secondary air is in each particular stove. However, the effectiveness in any particular stove is bound to depend on many details of how the stove is operated.

Regardless of the details, the need for secondary air depends on the fuel. The low volatile content of anthracite diminishes the need, and the high volatile content of some bituminous coals

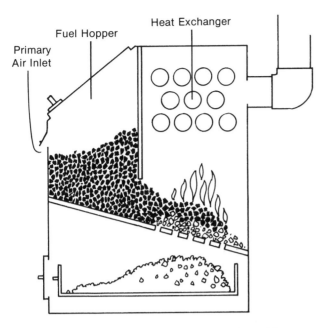

Figure 8–3. A hopper-fed, or gravity-fed, coal heater.

heaters is .04–.06 inches of water; this is a reasonable draft for wood heaters as well. Unfortunately, draft inadequacy usually cannot be ascertained easily in advance; you need to have a similar appliance operating at a typical output on a typical winter day to be able to assess draft adequacy, since draft depends on flue-gas temperatures, among other things. For this purpose a wood stove is a reasonable simulation of a coal stove. If the draft is inadequate, sometimes chimney modifications will improve draft. Increasing the height and decreasing the flue size if it is oversized usually are most helpful. See chapters 6 and 12.

If chimney draft is excessive, there is a simple and effective solution: install a barometric draft control (see chapter 11).

Hopper-fed coal stoves do not need quite as much draft because their fuel beds tend to be thinner.

ENERGY EFFICIENCY

Despite the interest and published claims, there have been very few reliable measurements made of the energy efficiency of coal stoves. A flurry of activity between 1910 and 1925 in England and in the United States produced some interesting results, but the studies did not include types of

appliances in common use today, and the research did not use the best of currently employed methods: room calorimetry.

A careful study in 1940 by Landry and Sherman used room calorimetry and measured overall energy efficiencies of 51–63 percent for a range of designs as part of a smokeless coal stove development program.

There have been some measurements made in Europe on some of the imported coal stoves, but for a number of reasons these numbers are not always obtained in ways which would make them comparable to efficiencies measured in this country.

Thus, at present, there is a dearth of evidence on the efficiency of coal stoves compared to wood. The claims often heard that coal stoves have very high energy efficiencies are generally based either on guesses or inadequate testing.

There are reasons why coal stoves might fare either better or worse than wood stoves in efficiency. Coal's lower hydrogen content and, often, lower water content, means less latent heat loss in the form of water vapor. Also, a good coal stove has the potential for operating with less excess air and hence with higher heat transfer efficiency. But it is likely that more unburned coal

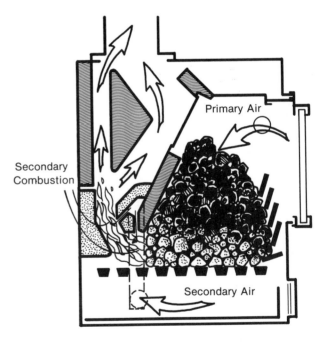

Figure 8–4. A British bituminous coal burner with a thoughtfully designed secondary combustion system and a "back boiler." Backboilers are common in England for both hot water and space heating.

finds its way into the ash drawer than does charcoal from wood fires. Some people who heat with coal screen the unburned coal out of the ashes to lessen this loss.

Overall efficiencies are impossible to predict. It is highly likely that there is a very large range of efficiencies within both coal and wood stove groups. Certainly, it is not safe to assume that any coal stove is more efficient than most wood stoves.

New designs will continue to become available with improved efficiency. The features required for burning coal smoke are virtually the same as those for wood smoke.

It is not yet clear whether coal appliances will be able to take advantage of catalytic combustion. Coal contains many elements which are detrimental to many catalysts. Work on catalysts for coal is proceeding, but no catalytic combustors are yet available for coal stoves.

CHAPTER 9

INSTALLING STOVES: PERFORMANCE FACTORS

Where and how a stove is installed can affect the heat distribution from the stove to the house, the energy efficiency of the system, and the steadiness of the house temperature. Conversely, where and how a stove is to be used should influence the selection of the solid fuel heater.

SIZING STOVES

Choosing the appropriate size solid fuel heater primarily involves matching the heating capability (power output range) of the stove to the heating needs of the building in which it is to be installed.

The literature of many manufacturers is not very useful in this regard. To state that a stove can heat 15,000 cubic feet is an unusual way to rate any heating appliance. Virtually all other heating appliances (furnaces, wall heaters, portable electric heaters, and so on) are rated in terms of power, the rate at which they can produce heat. Fossil fuel heaters in the United States are rated in British thermal units per hour (Btu. per hour). Electrical heaters in America, and all heating systems in the rest of the world, are rated in watts.

The need for heat in a house (or room) is determined not only by its size, but also by how well insulated and sealed it is, how many windows it has, and how cold it is outdoors. An old, leaky house without insulation or storm windows needs much more heat than a new, tight, well-insulated house of the same volume and in the same climate. Likewise, identical houses in Maine and Southern California will not need the same amount of heat because of the difference in climate. Volume by itself is just not enough. Stoves, like any other heaters, should be rated in units of power, such as Btu. per hour, or watts, not in terms of house size.

But it is not easy to give a stove a power rating because, unlike most conventional heaters, stoves do not have clearly and easily defined upper or lower limits to their possible heat-output rates. Chimneys alone can affect maximum power output by a factor of 2. Whatever standard power rating system the industry adopts, power outputs should always be accompanied by the necessary refueling frequency to maintain the heat output. It makes a difference whether the claimed 40,000 Btu. per hour output requires hourly refueling or only 3 refuelings a day.

Determining the heat requirements of a room or house is the other half of the problem, and it also is difficult to do precisely. Standard practices for estimating heat losses from houses are just as applicable to wood-heated houses as to houses with conventional heating systems, particularly if the wood heating is done with a central, wood-fueled furnace. Such estimates are useful despite the difficulty in estimating the contribution of air exchange (or infiltration) to the total heat loss.

If the stove is intended to heat 1 room only, and if the rest of the house will be maintained at a comfortable temperature by some other heating system, then the heating needs of the 1 room can be computed using standard methods by calculating the heat losses through the "cold" surfaces, such as exterior walls (no heat is conducted through a wall, floor, or ceiling where both sides are equally warm).

If the stove is intended to heat as much of the house as possible, and if portions of the rest of the house will be less warm than the room with the stove in it, then computing the needed heat output is more difficult. Relevant factors include how easily heat can move into other rooms, and how large a temperature difference between rooms is acceptable. There are no standardized and readily available guides for computing heating needs under such circumstances.

Because of the lack of reliable stove-performance data and the difficulty of accurately predicting needs, the most practical way to select the appropriate size stove is to seek advice from people with solid fuel experience. Stove retailers should be a good source of information. Friends and neighbors with similarly constructed houses (in the same climate) who already have stoves will have some experience with stove sizing.

My feeling is that in most cases wood heating systems should be sized to handle average, not extreme weather conditions; they should not be oversized. If a stove is undersized, it may be inadequate only on the coldest days. Undersized stoves generally burn more cleanly, contribute less to air pollution, and have less creosote buildup in their chimneys. However, undersized stoves require more frequent refueling.

An oversized stove does have 2 advantages. It has reserve power capability, enabling it to handle a cold snap, to heat the house more quickly in the morning, and to heat up a normally unheated room, such as a shop, quickly. The other advantage of an oversized heater is that long burn durations are possible even at medium heat output rates.

But an oversized stove usually contributes more air pollution and creates more creosote. It is more difficult to obtain low power outputs from a large stove (impossible if the stove is not reasonably airtight), and some purchasers of oversized stoves find their houses are uncomfortably warm much of the time, except when the weather is really cold.

Fortunately, wood stoves are fairly forgiving as heat sources. Most can operate over a wide range of outputs. Coal stoves generally have numerous power output ranges; thus coal stove sizing is more critical than wood stove sizing. Even if your stove is undersized, you can be warm by moving a little closer to the stove. If your stove is oversized and not very airtight, you can move your chair to another room, or open a window. Efficiency and convenience may suffer, but thermal comfort is almost always achievable.

STEADINESS OF HEAT OUTPUT

In houses or rooms heated with stoves, temperatures usually fluctuate more than in conventionally heated spaces due to the irregularity of the heat output from most stoves. The location of a stove affects the size of the temperature fluctuations. The larger the heat capacity ("thermal mass") of the surroundings—furniture, walls, floors, and ceilings—the smaller will be the fluctuations. In most houses, adding heavy heat-storing materials, such as brick or stone, under and behind the stove helps significantly. These heavy materials soak up and store some of the heat output of the stove when it is operating at high power, and release it when the fire subsides, making the air temperature in the house more constant.

A considerable mass is required for the effect to be significant. Suppose a stove is placed in the corner of a room, and both walls behind the stove are lined with a 4-inch-thick layer of bricks from floor to ceiling, extending 5 feet from the corner (see Figure 9–1). Suppose the stove also rests on a 4-inch-thick, 5-foot by 5-foot brick hearth. This set up has a total of 35 cubic feet or a little over 2 tons of masonry. If the stove can raise the average temperature of all this mass from a room temperature of 65° F. to 105° F., the stored heat would be about 32,000 Btu. This is equivalent to between 40 minutes and 3 hours of heating, assuming typical heating loads of 10,000–50,000 Btu. per hour. The actual time it would take for a substantial amount of this heat to enter or escape from the masonry into the room is roughly a few hours.

If the masonry in our example starts at 105° F., and if it cools all the way down to 45° F., the heat that would be released would be 48,000 Btu. This amount of heat might, for instance, prevent water pipes in the house from freezing on a cold night. On the other hand, if you wanted to keep a house

Figure 9-1. Brick facings on the walls and floor around a stove improve the steadiness of heating.

really warm at night with the stored heat, you would need much more than 2 tons of masonry.

The greatest long-term temperature-leveling effect is achieved where the whole house is built of masonry materials. Both the tonnage and surface area of the heat-storing material are then very large. Large amounts of heat can be absorbed, stored, and released. Wood or any kind of heating in such a structure is especially even. Passive direct-gain solar-heated houses have remarkably even temperatures despite the fact that during sunny weather, all the heat needed for a full 24-hour day is delivered during the 5 hours that the sun streams through the south-facing windows. For the other 19 hours, there is no explicit heat source. The key is heat-storing mass. Mass also helps keep a house cool in hot summer weather.

Although in most cases, massive surroundings for wood stoves are beneficial, it should be kept in mind that for quick heating, such mass is detrimental. Mass may not be appropriate in a cabin used only occasionally for short times, or in a workshop that is only heated when used evenings or weekends.

HEAT DISTRIBUTION

Whenever wood stoves are used, it is often desirable that a considerable portion of the heat be distributed to other rooms. The extent of the heat distribution of a wood stove is affected by its location and can be improved by some simple schemes. Placed in a large opening between 2 rooms, a stove can heat both. Placed in 1 room, considerably more heat will get to an adjacent room if there is unobstructed air flow at both the ceiling and the floor. Ceilings, in particular, are often not continuous from one room to the next — warm air at the ceiling often must flow down to get through a doorway, and this impedes heat distribution (see Figure 9-2). The ceiling acts like a lake bottom, and the walls are sides or dams; a doorway that is not open all the way up to the ceiling is like a spillway — the warm air will not flow out until its level is "down" to the lip of the spillway. If the doorway does provide clear passage at ceiling level, warm air at the ceiling will always flow out to the next room. If doorways are not open at ceiling level, large registers at the tops of walls will serve the same purpose. Lateral movement of heat can be assisted with a fan. Small, quiet, room-to-room "computer" fans take up little space and are rather effective.

Vertical heat flow comes very naturally. Stairwells provide an efficient passage for warm air to go up and cooler air to return. The 2 currents do not interfere with each other — the warm flows up along the ceiling, and the cooler air flows down over the stairs. A register installed in the ceiling of a room with a stove in it is very effective for heating the room above, as long as the air in the upstairs room can get out to make way for the incoming air. In very old houses, enough air may leak out through the walls and ceiling. In tight houses, there must be an air return route to the stove room, such as an open doorway, or a register at the bottom of a wall of the upstairs room providing passage to a stairwell.

My suggestion is to not go to the effort and expense of installing registers or fans to aid heat distribution until the need has been demon-

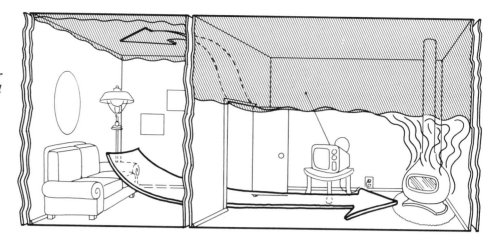

Figure 9-2. Even though the doorway is open, hot air will dam up behind the wall above the doorway, until the hot air lake becomes deep enough to spill through the doorway and up into the ceiling area of the next room. Thus, the heat is concentrated in the room where the stove is located.

Figure 9-3. With a register, or opening, at the ceiling level, hot air can immediately move into the adjacent room.

strated. Use your stove for a season first; help is not always needed. In fact, in some cases people have installed curtains over the bottom of a stairwell to keep more of the heat downstairs. The better insulated and tighter a house is, the more uniform will be the temperature, both from room to room and from floor to ceiling. Added openings not only let heat through, but also sound and odors, and this is not always desirable.

A stove can affect the temperature distribution by its interaction with a conventional, thermostatically controlled heating system. If the stove is located in the same room as a thermostat, the conventional heating system will not come on as long as the stove keeps the room above the thermostat's setting. This, of course, saves conventional energy, but you must be careful that this does not result in frozen pipes in other parts of the house.

If a stove is located far from the thermostats, there is little danger of pipes freezing; the thermostat will control temperatures in most of the house in its normal fashion. But not as much energy will be saved unless the thermostat is set back.

It is not usually efficient to use an existing forced hot air system to help distribute the heat output of a stove to the rest of the house. If the stove can be installed in the same room as a cold-air return register, whenever the furnace is on, the return air can be somewhat preheated by the stove. Of course, if the thermostat for the furnace is in the same room as the stove, the furnace may not come on when the stove is in use. Independent controls for the blower can be installed. A manually operated switch can turn on the blower even when the furnace burner is off; or the switch can be controlled by a thermostat in the room with the stove, so that the blower comes on whenever the temperature in the room exceeds a certain value. However, if the basement or crawl space is cold, a significant amount of the heat

Moving Heat Down

Air that is warmer than the immediately surrounding air rises. If rising warm air hits a flat ceiling, it spreads out. Fan-powered devices, including just plain fans, can mix the air in a room, bringing this warm air down where it is more useful. But hot air at a ceiling is not, in fact, useless for heating. The hot air will heat the ceiling, which in turn can radiate heat downwards. A ceiling surface of 85° F. radiates energy to a floor at 65° F. at a rate of about 18 Btu. per hour per square foot of ceiling area. In a 14-foot by 16-foot room, this means the floor receives 4,000 Btu. per hour of heat from the ceiling.

In a reasonably tight house, this is a significant amount of heat transfer—so significant, in fact, that such a large temperature difference is unlikely to develop. In a drafty and uninsulated house, even larger temperature differences will occur and there is very little you can do about it. Heat will radiate from the ceiling to the floor, but air infiltration and heat loss through the floor will keep taking the heat away.

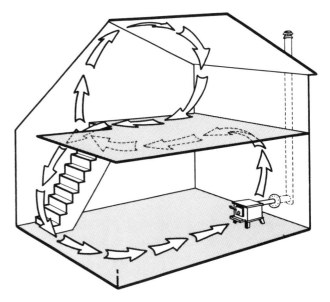

Figure 9-5. Air circulation patterns in a house heated with a stove. At the downstairs ceiling level, hot air moves away from the heat source. At the downstairs floor level, cooler air moves toward the heat source. There is a similar pattern upstairs, with warm air spreading out from the stairwell ceiling and returning along the floor to the stairwell.

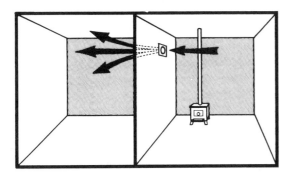

Figure 9-4. Small, very quiet fans can help distribute heat to other rooms unobtrusively. Typical dimensions are 4-⅜ inches square by 1½ inches deep. Power consumption is usually 15–20 watts, and air flow is typically 50–100 cubic feet per minute.

from the stove may be dissipated through the walls of the ducts and plenum. This is often the case.

Of course, the most important factors in choosing the stove's location are where in the house the heat is wanted, and where the chimney is or where it can conveniently be installed. Effects on heat distribution are important, but not always most critical.

ENERGY EFFICIENCY

How and where a stove is installed can affect its net energy efficiency. Since a significant amount of heat is given off by hot stovepipe, increasing the pipe's length increases the energy efficiency. The 4 or 5 feet of stovepipe in typical installations contribute 15–30 percent of the total useful heat. Much longer lengths of stovepipe may result in substantial creosote and serious draft problems. All codes require the shortest possible connector.

A better way to gain some extra heat is to have

an exposed interior chimney. This is discussed in chapter 6. In 2-story and 3-story houses the overall energy efficiency of the system can increase by 10–30 percent (5–15 percentage points).[1] Significant heat will also be given off by factory-built double-wall insulated and triple-wall air-insulated chimneys.

Nickel or chrome-plated stovepipe decreases the efficiency of a solid fuel heating system. Shiny metal surfaces emit much less radiation than do most other surfaces at the same temperature. The net effect for a typical nickel-plated stovepipe installation is probably a 10–20 percent decrease in heat output (a decrease in energy efficiency of 5–10 percentage points). Galvanized pipe also restricts radiation, although not as much as nickel. Flat black and blue-oxide stovepipe are both good radiators.

Most factory-built chimneys have either a galvanized or stainless steel exterior. For interior exposed installations, the chimney will radiate more heat into the house if it is painted any but a metallic color.

Most metal surfaces require an appropriate primer coat first. Do not paint unexposed chimney portions (such as inside floors and ceilings, and inside chimney chases)—just paint those portions which are exposed and can radiate directly to the living spaces of the house.

Locating a heater near an unprotected exterior wall results in a greater heat loss through that wall. For a normally insulated wall, the effect on net efficiency is probably less than a percent under most circumstances. For a glass wall, or other uninsulated wall with no confined air space, the decrease in net efficiency can be a few percentage points. (The glass may also crack.) However, if the wall is covered with a ventilated protector panel, the exterior wall location is no longer a liability. Placing a heater near an interior wall is not detrimental, unless the extra heat conducted into the wall is not fully used because, for instance, it convects up into an unused attic, or because it is conducted to an unheated pantry or laundry room.

Despite these considerations, exterior wall locations are often preferred for the source or entry point for many kinds of heating systems—hot air registers, steam radiators, and baseboard systems are typically placed in or next to cooler exterior walls. The reason is improved comfort. Concentrating the heat there results in more uniform temperatures. Also, the exterior wall location usually bucks the natural convection air flow patterns in a room. This pattern is down the exterior walls, across the floor to interior walls, up the interior walls, and back across the ceiling to the exterior walls. If you place a stove near an interior wall, you encourage this flow pattern, which can result in uncomfortably cool and drafty conditions near the floor. If you place the stove near an exterior wall, the air tends to circulate the other way; the net effect is less air circulation, and more heating of the infiltrated outdoor air before it gets to you.

Of course, in practice, chimney location is more likely to be the deciding factor for stove location.

Installing a stove in or in front of a fireplace need not, but often does, result in a decrease in energy efficiency for 3 reasons.

- There is usually less exposed stovepipe in such installations. (An exposed interior masonry chimney in a 2-story or 3-story house will recover much of this lost heat.
- If the stove is placed inside the fireplace, less of its heat gets into the room, particularly if the fireplace is in an exterior wall.
- Unless installed very carefully, there is more room air leakage into the chimney around the stovepipe and fireplace damper location.

OUTSIDE AIR[2]

All solid fuel burning appliances consume air. Without an outdoor air system, this air is 70° F. indoor house air. Does using a solid fuel heater create an extra heat loss by consuming this heated air? Not necessarily. Does supplying outside air directly to the combustion chamber improve efficiency? Again, not necessarily.

Outdoor Air and Heat Loss

Closed stoves do not use much air compared to typical natural house infiltration rates (see Table

---

1. This estimate is based on data in W. D. Harris and J. R. Martin, "Heat Transmitted to the I-B-R Research Home from the Inside Chimney," *Transactions of ASHRAE* Vol. 59 (1953) pp. 97–112.

2. A portion of the material is adapted, with permission, from Jay Shelton, *Wood Heat Safety*. For a discussion of safety aspects of outside air systems, see *Wood Heat Safety*.

6-5), so the possible extra heat drain due to room air consumption of a closed stove is small to begin with. But, in fact, the effect could be essentially zero – air which would have leaked out through the upper portions of the house may just be rerouted through the stove. If this is the case, using the stove does not increase infiltration and, thus, does not increase the heating load at all.

Large open fireplaces can require 2 or 3 times the amount of air provided by the natural air exchange rate of the house. In this case, using the fireplace certainly increases infiltration substantially, and therefore increases the heating load.

What can be done to prevent throwing away such large quantities of warm air? Not as much as you might think. You can duct outdoor air to a fireplace in a number of ways, but you will not prevent all house air from entering the fireplace; **you can only decrease the flow (Figure 9-7)** – unless you install tight doors over the fireplace opening and supply outside combustion air directly to the combustion chamber. However, if the doors are tight, then the room air consumption is not too serious anyway, and outside air is not needed as much. Thus, with only a little overstatement, the conclusion is that when outside air is needed the most, it cannot displace all the room air loss; and when all room air loss *can* be displaced, its flow is so small that stopping it may not be worth the effort. The principal qualification to this conclusion is that it is not known how much of the room air flow into open fireplaces can

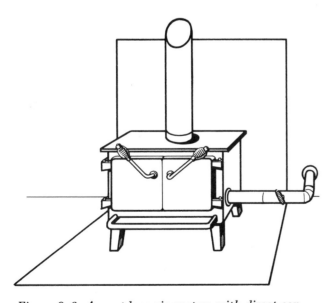

Figure 9-6. An outdoor air system with direct connection to the appliance.

be prevented by introducing outside air. It cannot be 100 percent, but the percentage could be significant. More research is needed.

Ducted Outdoor Air and Energy Efficiency

Supplying outside air to any appliance is most effective if the air is ducted directly into the unit, not just discharged in the general vicinity. This assures nearly complete exclusion of room air, and decreases (but does not eliminate) the heating burden of cold air flow or leakage when the appliance is not in use. In many cases, however, direct connection is not practical because the air inlet is on the loading or ash door.

Will the ducted air save energy? Although it is clearly beneficial to decrease excessive house-air losses, there are other effects of outside air which can result in more oil, gas, electricity, or wood consumption by decreasing combustion and heat transfer efficiencies and by introducing new air leaks.

Combustion efficiencies may be lowered due to lower temperatures in the combustion region.

If the tightness of the house is a limiting factor in the amount of air entering a wood burner, supplying direct outside air may increase the total amount of air entering the appliance. In many cases, particularly fireplaces, this will increase the amount of excess air, which in turn will decrease heat transfer efficiency. I expect this effect is most important for circulating-type open fireplaces.

Using colder outside air decreases heat transfer efficiency, because the fire and all surrounding heat transfer surfaces are cooler. For stoves I estimate that the decrease in heat transfer with cold outside air is roughly the same as the heat saved by not warming the combustion air up to room temperature – a breakeven proposition. Accurate measurements are needed. If it turns out that the heat saved is offset by the decrease in heat output, the overall net effect still is likely to be negative because having a duct for outside air to an appliance in the living space of the house usually results in cold air leaking into the house whether or not the appliance is in use. Even if there is no leakage directly into the house but only into the appliance, the appliance and its chimney will be cooler when not in use and more heat will conduct out of the living space of the house into the appliance and its chimney. This kind of heat loss due to an installation itself and

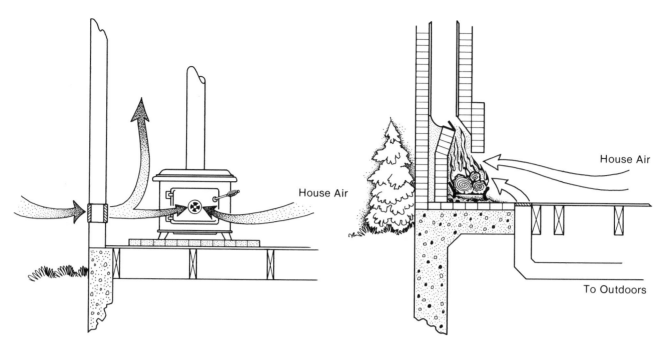

Figure 9–7. An outside air system without a direct connection to the appliance can reduce, but not eliminate, all house air flow into the appliance.

when the appliance is not in use might be termed *standby losses*, and these losses can be larger for systems with direct outside air.

There are some undisputed possible benefits of outside air systems. Fireplaces and fireplace stoves often will not work satisfactorily in a tight house. *Open* wood burners require a certain minimum average velocity (about 0.8 feet per second)[3] of air into the fireplace opening to prevent smoke from coming out the opening into the room. This amounts to between about 100–1,500 cubic feet of air per minute depending mostly on the size of the fireplace opening. If a tight house cannot supply this amount of air, smoke will spill out into the room from the fireplace, and the appliance may be unusable unless a window is opened or an outside air system is installed.

Direct outside air may offset the depressurization of the house caused by using other appliances. In a depressurized house, other vented appliances, such as furnaces, boilers, water heaters, and even other fireplaces or stoves, do not vent properly, resulting in smoke, carbon monoxide, and other lethal gases accumulating in the house.

Another benefit of ducted air is decreased

drafts at floor level in front of a fireplace, which increases human comfort.

Using ducted outside air can also decrease the chances of chimney flow reversal and resulting wood-smoke asphyxiation, although a smoke detector is the best protector. Whether a duct is attached directly into the appliance or there is just a fresh-air register in the vicinity, a pressure-equalizing link to the outdoors on the ground floor or basement of a house will usually prevent flow reversal and ensure that the chimney will be self-starting.

A rough estimate for the appropriate size of an outside air duct system is half the cross-sectional area of the appliance's recommended flue size.

BASEMENT INSTALLATIONS[4]

Installing a stove in a basement will save space upstairs, will confine the mess – bark, sawdust, ashes, and beetles – downstairs, and usually provides easy access to a chimney.

However, there can be some liabilities. Placing

3. *ASHRAE Handbook and Product Directory 1975 Equipment Volume* (New York: American Society of Heating, Refrigerating and Air-Conditioning Engineers, 1975), p. 26.26.

4. Portions of this section are adapted, with permission, from Jay Shelton, *Wood Heat Safety*. For a discussion of safety aspects of basement installations, see *Wood Heat Safety*.

Can Stoves Deplete Oxygen
From House Air?

No one will ever suffer serious injury or death from a lack of oxygen in a house due to a solid fuel appliance.

Fires *do* consume air out of a house, but this air is immediately replaced by fresh outdoor air leaking into the house; it is impossible for a solid fuel appliance to use up the *air* in a house.

Fires *do* selectively use up some of the oxygen *in the air they consume*. But this oxygen-poor air rarely returns to the house—it goes up the chimney. And if it does get back into the house, it is inextricably mixed with the combustion byproducts —carbon monoxide, smoke particles, and all the other ingredients of flue gases. It is these ingredients of smoke, not the slightly lower oxygen concentrations, which are dangerous. If you do not smell and see smoke, the oxygen in the house air is unlikely to be depleted.

(In the rare case that carbon monoxide is the only significant unburned component of smoke, you will not see or smell the carbon monoxide; but, again, the carbon monoxide is much more dangerous than any decrease in oxygen in the house.)

a stove in a basement is not the most efficient way to heat the upstairs. Since 60–80 percent of the heat output of a radiant stove is in the form of radiation, and radiation travels in straight lines until absorbed by a solid surface, much of the heat output of the stove will be absorbed by the basement floor and walls. Only some of this heat will ever find its way into the living spaces of the house.

In addition, not all the heated air at the ceiling will get upstairs; wherever the basement ceiling is warmer than the basement floor, heat will be transferred by radiation from the ceiling to the floor. The effect can be large. For a 500-square-foot ceiling at 75° F. and a floor at 55° F. the radiant heat loss downwards from the ceiling to the floor is about 9,000 Btu. per hour.

You can expect to burn more fuel in a basement stove to get the same amount of heat upstairs compared to the same stove installed upstairs. If your fuel is free, or if you really want a warm floor, or if the stove is for emergency heating only, or if the basement is the only easy place to install the stove, or if heating the basement is a principal objective, basement installations make sense.

If the principal objective is to heat the upstairs of the house, I recommend installing the stove upstairs, or installing a furnace or boiler—appliances designed to transfer heat to other rooms.

CHAPTER 10

INSTALLING STOVES: SAFETY FACTORS

The statistical evidence is clear—stoves are not dangerous, but the way some are installed, operated, and maintained can be dangerous.[1] Chimney safety is discussed in chapter 6, safe stove operation is covered in chapter 14, and the most important maintenance job, chimney cleaning, is covered in chapter 15. This present chapter covers most of the safety issues which are the installer's responsibility—floor protection, clearances and wall protection, and the stovepipe connector.

Much of the material in this chapter is based on National Fire Protection Association (NFPA) standards, particularly the 1980 edition of NFPA 211 (Chimneys, Fireplaces and Heat Producing Appliances).[2] NFPA is the leading national organization writing standards for safe installation of wood heating equipment. The standards have no force of law, unless they are incorporated into local building codes. Most codes are modeled after NFPA standards in areas relating to fire

safety. Thus, the NFPA standards come close to being a national consensus. Nonetheless, where an installation must follow code, it is the prevailing local code that is relevant, not NFPA standards.

No installation is perfectly safe. Even if "perfect" installations were possible, gross negligence in operation and maintenance of the system could still result in house fires. The "perfect installation" itself is essentially unattainable and probably not desirable; if one goes overboard in the direction of safety, the installation can become prohibitively expensive, can have a low energy efficiency, is probably unattractive, and is probably awkward to use. The only perfectly safe installation I have ever heard of was a wood boiler in Massachusetts; due to a code violation, the local building official had the door welded shut.

Reasonable safety *is* achievable. It requires reasonable care in the installation, operation, and maintenance of the system. Installation codes are intended to be reasonable. But opinions differ on how much is enough. In this chapter, I will present my interpretation of the relevant NFPA standards along with my comments and opinions, particularly where the NFPA requirements seem either excessive or inadequate.

Installation standards are different for listed and unlisted stoves. (See p. 151 for the definition of listed.) Underwriters Laboratory (UL) is a well-known organization that lists electrical equipment and solid fuel heaters, among other things. (*Approved* is a term often confused with listed. Approved means acceptable to the authority hav-

---

1. See, for instance, chapter 1 in Jay Shelton's *Wood Heat Safety.*

2. For those who may be familiar with the old (1977) NFPA 211, some of the major changes in the 1980 edition are:

- Asbestos is not included as a protective material because of the possible lung cancer hazard from asbestos fibers.
- Only *air-ventilated* wall and ceiling protectors are allowed for reduced clearances
- Floor protector extents have been increased to 18 inches on all sides.
- Venting a solid fuel appliance into a flue also serving a liquid or gaseous fuel is prohibited unless the solid fuel appliance is listed for such an installation.

ing jurisdiction, such as the local building official.) To satisfy most building codes, listed equipment should be installed in accordance with its instructions, period, even if this is in conflict with other provisions of the code. The reason is that listed equipment has been tested and found reasonably safe installed with the particular clearances and protectors mentioned in the instructions. Also, part of the listing procedure involves close scrutiny of the instructions for their adequacy in most areas of safety.

Unlisted equipment should be installed in accordance with the building code provisions or the instructions, whichever is more conservative. The installation requirements in codes are written to apply to all kinds of stoves; they must be conservative enough to be safe for the worst case – the biggest and hottest stove. Code provisions tend to be excessive for the average stove.

This chapter is mainly concerned with unlisted stoves; however, the principles are applicable to all stoves.

GENERAL PLACEMENT

The previous chapter discussed stove placement for best performance and convenience. Since installations in most locations can be made reasonably safe, performance and convenience should dominate the choice of heater location.

However, 2 safety considerations should be borne in mind.

- The safest stovepipe connector is as short as possible. The heater should be located close to an existing chimney, or the new chimney should be installed as close as possible to the desired stove location.
- Most house doors are made of wood and are therefore combustible. A solid fuel heater should not be located where a hinged door could swing into a position too close to the heater (36 inches for a radiant heater).

FLOOR PROTECTION

Floors under and around stoves may need protection against 2 hazards – radiation from the bottom and sides of the stove and sparks and hot coals that may get out of the stove during refueling, ash removal, or during open-door operation of a fireplace stove.

Any floor containing combustible materials may need protection, even if the surface of the floor is noncombustible. Tile, slate, and brick floors may need protection if the masonry units are placed on wood flooring, unless the masonry floor covering is in full accordance with Table 10-1. Similarly, carpet over a poured concrete floor either needs protection or needs to be cut back.

TABLE 10-1
FLOOR PROTECTION

Appliance Design	Type of Floor Protection
Appliances having legs or pedestals providing more than 6 inches of ventilated open space beneath the combustion chamber.	Masonry units such as bricks, concrete blocks, or stone, providing a thickness of at least 2 inches, placed on top of 24-gauge or thicker sheet steel. The masonry units need not be mortared together, but should be laid tightly together. The sheet steel alternatively may be placed on top of the masonry.
Appliances having legs or pedestals providing 2-6 inches of ventilated open space beneath the combustion chamber.	On top of 24-gauge or thicker sheet metal, 4-inch hollow masonry (hollow concrete blocks, for example) laid to provide air circulation through the masonry layer—that is, with holes parallel and open, even at the edge of the protector. Alternatively, the sheet metal may be placed on top of the masonry.
Appliances having less than 2 inches of ventilated open space beneath the combustion chamber.	None. May only be placed on fully noncombustible floors (with no combustible material against the underside). Such construction must extend at least 18 inches beyond the appliance on all sides.

Extent of floor protection (for all cases):
18 inches beyond the appliance on all sides.

NFPA's required floor protection, as interpreted by the author, for all residential unlisted wood and coal stoves, fireplace stoves, free-standing fireplaces, ranges, and water heaters. Unlisted furnaces and boilers must be installed only on floors of noncombustible construction, as for stoves with less than 2 inches of air space underneath, according to NFPA 211.

The floor protection provisions in Table **10-1** and Figures 10-1 and 10-2 are essentially those in NFPA 211; on the whole I think they are reasonable. The masonry provides the insulating value needed against radiation from the stove, and the sheet metal constitutes a spark-tight layer that is particularly important if the masonry units are not mortared together. Most people prefer to place the sheet metal *under* the masonry for aesthetic reasons. If this is done where the masonry units are not mortared together, some care should be taken to prevent the stove legs from ever slipping down between the masonry units. If the masonry units are small, place oversize masonry units under each leg, or place each leg on a metal or masonry pad on top of the rest of the protector.

There are not very many stoves with less than 2

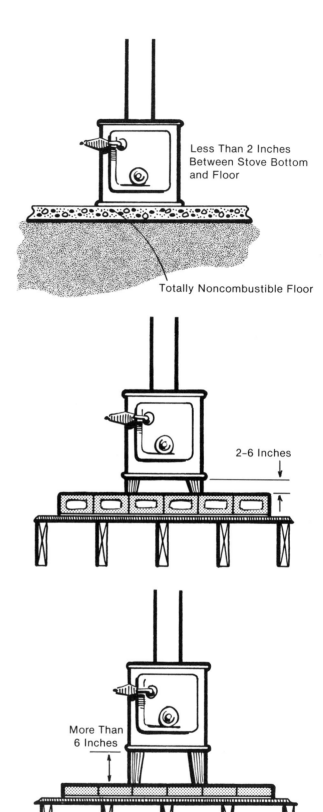

Figure 10-1. Types of floor protection recommended by the NFPA for unlisted stoves.

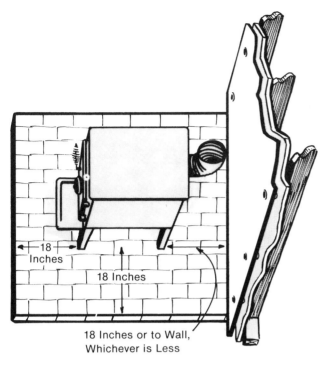

Figure 10-2. Floor protector extent for unlisted stoves, as recommended by the NFPA.

18 Inches

18 Inches

18 Inches or to Wall, Whichever is Less

inches of air space between the combustion chamber and the floor, but they do exist. NFPA 211 prohibits installation of these units on a combustible floor regardless of the protection provided. In my opinion, if there is at least ½ inch of ventilated air space directly under the stove, then 2 crossed layers of 4-inch hollow masonry with 2 pieces of sheet metal (at different levels) is adequate in most cases. If there is no air space at all, I believe a triple-layer version of this ventilated protector system usually is adequate. As always, align the hollow masonry units to allow air flow through the whole protector.

The 18-inch extent of protectors (Figure 10-2) is more than enough in most cases and is more than some codes now require. However, at least this much coverage should be provided in front of fireplace stoves (and free-standing fireplaces and ordinary built-in fireplaces), as protection both against radiation from the exposed fire and against sparks, coals, and errant burning logs. The 18-inch extent is also important on all stove sides with doors. Less than an 18-inch floor protector extent beyond sides without doors can be reasonable for small stoves, or medium stoves with tall legs, or stoves with extensive refractory liners, or stoves with exterior mounted heat shields (floor protector extent can be reduced only on the sides with such shielding). However, none of these features guarantees the safety of reduced extents for floor protectors. Such reduced extents should be treated as experimental. The hand test can be useful (see "How Hot Is Too Hot?").

Manufactured floor protectors ("stove boards") have long been available, but are highly variable in effectiveness. Some have even been manufactured from cardboard and particle board, which, although they may be fire-retardant treated, can still burn. Some manufactured protectors are "fire-rated." This is also basically irrelevant, for this is no guarantee that dangerous amounts of

---

How Hot Is Too Hot?

The most common combustible construction material is wood. Many features of stove installation codes are designed to assure that heat from the stove and its chimney cannot get this wood so hot it smolders and ignites. How hot can wood get and still be safe against ignition? Roughly 200° F. No one has seen wood spontaneously ignite at this temperature in a laboratory, but there is field evidence that at 250–300° F. wood can ignite.

Engineering handbooks list the ignition temperature of wood as around 400–500° F. But it is a fact that prolonged exposure of wood to lesser temperatures gradually decreases the spontaneous ignition temperatures.

You can perform a simple test to gauge the temperature of *wood* near a stove or chimney. The temperature at which it becomes uncomfortable to hold your hand on hot wood is very roughly 200°F. (This test does *not* generally work for materials other than wood.)

The temperature limits in Underwriters Laboratories' standards for wood in the vicinity of a stove or chimney are as follows:

- 90°F. higher than ambient (normal room temperature), or about 160–170°F. for constant conditions for unexposed surfaces, such as beneath a floor protector.
- 117°F. higher than ambient, or about 185–195°F. for constant conditions for exposed surfaces.
- 140°F. higher than ambient, or about 210–220°F. for periods of time of up to about an hour.
- 175°F. higher than ambient, or about 245–255°F. for occasional periods of time of up to about 10 minutes, such as during a short chimney fire.

heat will not conduct through during prolonged exposure to radiation from a hot stove.

Fortunately, standards for manufactured floor protectors are emerging. One can now buy *UL listed* floor (and wall) protectors, and other independent laboratories also have listed some of these products. My advice is to be cautious about buying unlisted manufactured floor protectors.

Two inches of gravel or sand (Figure 10–3) may be used in place of NFPA's 2 inches of masonry. This system is not recognized by NFPA or most codes, but in my opinion it is equally safe. However, there are 4 aspects to be careful about.

- It is important that the 2-inch thickness be maintained.
- The stove legs or base must not rest on the sand or gravel, but must be on solid masonry, such as an 8-inch by 8-inch by 2-inch masonry block. It is not safe to have the legs penetrate through the aggregate to the floor.
- The sheet metal should be placed under the sand or gravel.
- Protect the installation from cats, if sand is used.

The safety of any system is not lessened by adding more noncombustible material either over or under it.

Floors that consist of wood subfloorings covered by at least 2 inches of spark-tight masonry, such as bricks, do not need additional protection for stoves with at least 6 inches of ventilated air space beneath the combustion chamber.

STOVE CLEARANCES AND WALL PROTECTION

Hot stoves can radiate immense amounts of heat. If a stove is too close to a wall, the wall may ignite. There may be no warning of the problem for 2 reasons.

- The process may take years. Prolonged exposure of wood to elevated temperatures can gradually decrease its ignition temperature.
- If the wall has a noncombustible exterior surface or covering, the wall may smolder and ignite on the inside before there is any visible indication of trouble on the outside. An installation that appears to have been safe for a few years may not still be or continue to be safe. This is equally true for floors and walls.

A stove should never be installed touching a wall, regardless of the construction of the wall. Even if there is nothing combustible anywhere in or on the other side of the wall (usually there is), the thermal stress caused by the heat from the stove could crack the wall. In addition, reasonable accessibility to the appliance for operation and maintenance usually requires at least a few inches of clearance. Thus, I recommend that even if a wall is fully noncombustible, solid fuel heating appliances ought to be installed at least 2 inches from the wall; if the appliance is a radiant

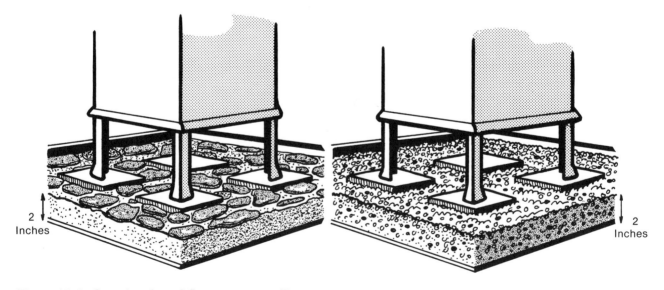

Figure 10–3. Gravel and sand floor protectors. (See precautions in the text.)

type, then I recommend a 4-inch clearance with a ventilated protector in between.

Most wall structures contain wood. Sometimes it is visible as siding or paneling, but many walls contain wood studs hidden by plaster or sheet rock. NFPA considers this kind of wall construction combustible.

If combustible walls have no added protection, NFPA 211 specifies that all radiant stoves be installed at least 36 inches away. (There are a few stoves which do not pass UL 1482 at 36 inches but require more clearance. UL 1482 is the designation of the Underwriters Laboratory's safety standard for stoves. It is my opinion, and apparently that of the NFPA 211 committee in 1980, that 36 inches is still a reasonable clearance for unlisted stoves, for 2 reasons: there are not many stoves that would not pass UL 1482 at 36 inches, and UL 1482 itself has a safety margin built into it.) All circulating stoves should be installed at least 12 inches away from combustible unprotected walls.

Most people would rather not have their radiant stoves so far out into the room. With the appropriate kind of protection, it is safe to install radiant stoves much closer to a combustible wall than the 36 inches required with no protection.

A common, but inadequate and dangerous, protective scheme is to cover the wall with something noncombustible such as sheet metal, asbestos millboard, or cement board, or a thin masonry veneer (such as Z-Brick). The fact that these wall coverings are noncombustible is, by itself, *irrelevant*. The wood in the wall can still burn if it becomes hot enough. The key to effective protection is to keep the wall *cool*, not just covered. A thin layer of metal, asbestos, or masonry veneer mounted in contact with the wall does not do this adequately. The exposed surface of the "protector" gets hot, and because the layer is thin, the heat easily conducts through it and into the wall.

How do you keep a wall from getting too hot? There is only 1 universally effective way—to create a *ventilated air space* behind a noncombustible covering by mounting the covering spaced out from the wall at least an inch and not trimming out the edges in a way which impedes air flow into and out of the gap. It is the moving air that keeps the wall from getting overheated (Figure 10-5). Such protectors allow clearances to be reduced to 12 and 4 inches for radiant and circulating stoves respectively. All reduced clearances are measured to the original wall surface, not the protector.

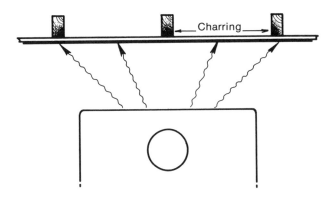

Figure 10-4. A thin layer of any material, no matter how noncombustible, and whether or not it is "fire-rated," is not suitable as a wall protector when placed directly against a wall. Heat can penetrate and char the wood in the wall.

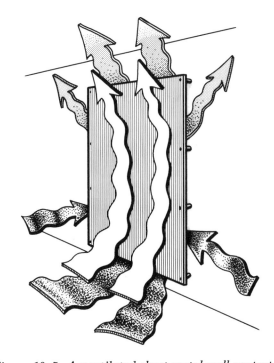

Figure 10-5. A ventilated sheet metal wall protector.

NFPA and most codes require wall protectors to be very large—6-7 feet wide and 5-6 feet tall for some radiant stoves (Figure 10-5). The general rule is that all portions of the wall within the legal minimum unprotected distance (36 inches for radiant stoves) must be protected.

This, in fact, is unnecessarily conservative.[3] In my opinion, the protector size illustrated in Figure 10-8 is more than adequate.

_____

3. See Appendix 2 in Jay Shelton's *Wood Heat Safety*.

The 2 most practical protector materials allowed by NFPA 211 for construction of ventilated protector systems are 28-gauge or thicker sheet metal and bricks laid in their usual orientation, providing a 3½-inch or thicker masonry wall (Figure 10-6). The sheet metal can be mounted with screws or nails and spacers (such as short lengths of thin-wall metal tubing or ceramic electric fence insulators). All spacers and fasteners must be noncombustible. Fasteners should not be placed directly behind the upper half of the stove and the first foot or two of stovepipe if it is closer to the wall than the stove to avoid the possibility of heat from these hottest regions conducting through the fastener and/or spacer into the wall.

Enough spacers should be used to assure that the sheet metal cannot accidentally come in contact with the wall due to thermal or mechanical stress. Spacers installed every 18 inches around the perimeter usually is adequate, but sometimes interior spacers also are needed to prevent the sheet metal from buckling inwards.

Brick ties should be used to help secure the brick wall at least 1 inch from the original wall. NFPA 211 does not detail how much ventilation of the air gap is necessary. It is my opinion that having the sides and top open is adequate; the first course of bricks can be solid and laid directly on the floor. It is also safe to have the bottom and top ventilated and the sides closed. Brick protectors are heavy. Most wall/floor construction can take the weight, but it is wise to check into the

Figure 10-7. A ventilated wall protector made of tiles mounted in a metal frame.

house structure under the wall, particularly for large wall protectors.

An air space of more than 1 inch is just as safe and may be more practical for easy removal of toys, pencils, and dust from behind the protector. Since NFPA's reduced clearances are measured to the original wall surface, increasing the air space does not change where the stove may be installed.

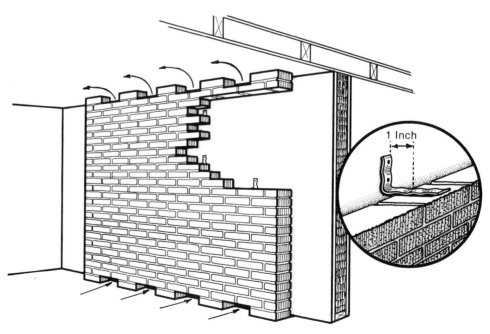

Figure 10-6. A ventilated brick-facing wall protector.

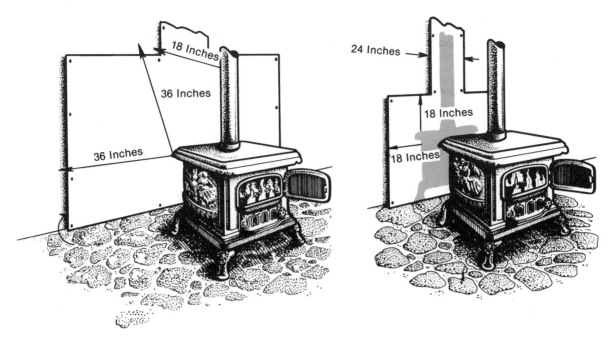

Figure 10-8. Minimum extents for wall protectors according to the NFPA
(left) and the author (right).

NFPA 211 does not give extra credit for doubly protective systems, such as 2 ventilated air spaces or a layer of insulation spaced out from a wall. If it is necessary to place a stove still closer to a combustible wall, such doubly protective systems can make it safe. I feel that 6 inches of clearance for most radiant stoves is reasonably safe with such double protection systems.

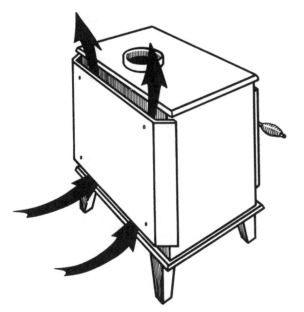

Figure 10-9. A radiant stove with a rear-mounted ventilated heat shield.

Manufactured wall protector panels are available. If listed, they should only be used in accordance with the instructions. Some listed protectors do not require an air-spaced mounting. If unlisted they should be made entirely of totally noncombustible materials and should be used only with a ventilated air space behind them. Manufactured protectors greatly increase the range of aesthetically appealing protectors available.

CHIMNEY CONNECTORS

A chimney connector connects the appliance to the chimney. The connector is not part of the chimney technically. Chimneys are usually either masonry or prefabricated double-wall or triple-wall metal, as discussed in chapter 6. The connector is usually single-wall pipe or "stovepipe," although other kinds of connectors may be used. The connector and chimney together are sometimes called the venting system. Not all solid fuel heaters have connectors; many metal fireplaces are connected directly to a prefabricated chimney, as per the manufacturer's instructions.

Connectors should never be used above a segment of chimney. You may be tempted to use a prefabricated chimney segment to pass through a ceiling or roof, and then switch back to single-wall

Circulating Stoves

NFPA 211, as of the 1980 edition, no longer makes a distinction between radiant and circulating stoves. Apparently, NFPA's reason is that some appliances seem to fall in between these types and are therefore difficult to classify. Three design/performance features can cause difficulties in defining circulating stoves.

- Appliances with ventilated heat shields (Figure 10-8) *are* hybrids. They have the circulating-type stove construction on the shielded sides only. The shields justify some reduced clearance on the shielded sides, but how much is difficult to predict, since the unshielded sides and top of the stove may also radiate to the wall which the shielded side faces.
- Some traditional circulating stoves (Figure 7-45) use metal screens, not solid sheet metal, on their sides. Thus, some primary radiation can penetrate through the jacket.
- Finally, a certain minimum amount of air flow in the space between the jacket and the main stove body is necessary. It can be difficult to describe the design features necessary to assure adequate air flow.

My opinion is that the old NFPA definition of circulating heaters is still basically adequate, and that there are sufficient numbers of unlisted circulating heaters in use that giving reduced clearances for these appliances is useful. This is the old NFPA definition of a circulating room heater:

"A room heater with an outer jacket surrounding the heat exchanger arranged with openings at top and bottom so that air circulates between the heat exchanger and the outer jacket. Room heaters which have openings in an outer jacket to permit some direct radiation from the heat exchanger are classified as radiant types."*

This definition does address stoves with heat shields—they are *not* circulating stoves because not all sides are jacketed. The issue of adequate air flow will very rarely, if ever, be a problem in practice. Manufacturers have incentive to provide adequate air flow for reasons of energy efficiency. Finally, with those circulating stoves using screens on the sides of the jacket, the open area of the screening is generally so small that reduced clearances are usually justified.

For circulating stoves, the old NFPA clearance from the sides or back of the stove to unprotected combustible walls was 12 inches. In the vast majority of cases this clearance is adequate. However, be watchful; if the nearby wall is ever too hot to hold your hand on, or if the wall ever shows any discoloration, either the clearance should be increased or the wall should be protected.

A circulating stove may be 4 inches from a combustible wall with a ventilated protector.

---

* NFPA 97M, "Glossary of Terms Relating to Heating Producing Appliances, 1972" (Boston: National Fire Protection Association, 1972).

---

pipe, either to save money or to recover more heat. This violates NFPA 211 and most building codes.

Materials and Types

Stovepipe connectors are probably the weakest link in most installations. They are subject to burning and rusting out, they can be shaken apart during chimney fires if not adequately fastened, and they can get extremely hot. Proper clearances are critical.

NFPA recommends using 24-gauge (0.023-inch thick) steel stovepipe connectors for pipe diameters of 6–10 inches, and 26-gauge (0.029-inch) for pipe diameters less than 6 inches.

I recommend against using galvanized pipe. There is zinc in the galvanic coating. When the pipe gets hot, some zinc vapor may be released. Zinc vapor is not healthy to breathe. It is not clear if the vapor released is enough to be a serious hazard. But when there is a possible hazard and alternative materials are available, the alternatives should be used.

NFPA 211 suggests, but does not require, galvanized pipe because of its resistance to corrosion. The biggest corrosion threat in solid fuel heating systems is from the inside out—the fumes and condensate inside the pipe. It is not clear how much additional protection, if any, is afforded by galvanization under this usage. In any case, galvanized pipe is not required for reasonable longevity.

Steel pipe is commonly available with a blue-steel finish and/or painted black. Other color paints, and baked enamel finishes, are sometimes available. All are suitable – if the gauge is adequate.

Nickel-plated pipe is available in a few stores. It is more a decorative than a functional feature. The nickel plate substantially reduces the radiation output from the pipe so its use is not desirable usually.

Stainless steel stovepipe is increasingly available. Its emissivity for radiation also is somewhat low; its use will decrease the energy efficiency of the system compared to use of ordinary blue-steel or black pipe. NFPA does not specify minimum thicknesses for stainless connector pipe.

An alternative connector type where clearances will be less than the standard 18 inches (see below) is the "Type L vent." It is a double-wall prefabricated chimney designed for oil heating systems. It should not be used as a chimney for solid fuel heaters; but as a connector, it reduces the safe no-added-protector clearance to combustibles from 18 inches to 9 inches.

Mechanical Security

Connectors should be mechanically secure. Joints that are only a press fit and have no other mechanical security may come apart, particularly during chimney fires. For standard stovepipe, 3 sheet metal screws distributed around each joint is adequate. The connector must also be secured to the appliance and to the chimney. Screws or

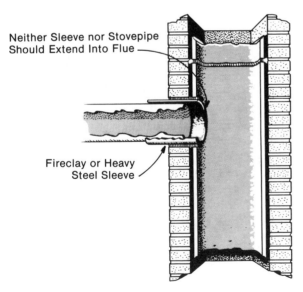

Neither Sleeve nor Stovepipe Should Extend Into Flue

Fireclay or Heavy Steel Sleeve

Figure 10–10. Stovepipe connection to a masonry chimney.

bolts and nuts should be used at the flue collar on the appliance; sheet metal screws or an equivalent method should be used in joining stovepipe to prefabricated chimney. Masonry chimney connection is illustrated in Figure 10–10.

Length of Connector

NFPA 211 specifies that total connector length should be as short as possible.[4] This is good advice usually, but I do not recommend taking it too literally. Excessively long connectors are not advisable – they usually are less reliable mechanically, and they exacerbate creosote and draft problems. However, excessively short connectors may deny the installation some useful heat transfer into the house. Also, in a few rare cases, having too short a connector can stress the chimney by getting it too hot because flames from the appliance reach up into the chimney. The connector not only radiates heat to the room but also serves to cool the flue gases a little. In most cases, connectors turn out to be between 2 feet and 8 feet long, and this is a desirable range. I would not intentionally try for still shorter connectors. However, if connectors naturally come out shorter than 2 feet, I would not worry unless all 3 of the following conditions are present.

- The chimney is of questionable safety.
- The appliance tends to have high flue-gas temperatures because of poor heat transfer despite having normal amounts of excess air.
- The operator of the appliance tends to push it hard much of the time.

Slope of Connector

NFPA 211 specifies that "the horizontal run of an uninsulated connector to a natural draft chimney ... serving a single appliance shall be not more than 75 percent of the height of vertical portion of the chimney ... above the connector."[5] This is a good general rule to follow. Horizontal runs decrease draft, whereas vertical portions of the venting system generally increase draft. In practice, most horizontal runs of connectors will be shorter than NFPA's specification, and this is desirable, especially in the case of open appli-

4. NFPA 211 1980, paragraph 5–3.
5. NFPA 211 1980, paragraph 5–3.1.

ances where draft is more critical (to avoid smoke spillage into the house).

A minimum upward slope for nearly horizontal runs of connector of ¼ inch rise per foot of length is required by NFPA 211. This is usually not critical. The problems of draft, creosote dripping, and accumulation of creosote and fly ash will all be indistinguishable between strictly horizontal pipe and minimum upward sloping pipe. Connectors should not slope down. Aiming for horizontal slope and allowing for normal construction tolerances should result in a reasonably effective and safe installation, assuming the horizontal run is no longer than about 6 feet.

FITTINGS

Elbows, tees, breachings, and abrupt changes in diameter all increase the resistance to flow for the smoke. However, this added resistance is only important where the capacity of the venting system is marginal to begin with. This is rarely the case with small closed appliances, such as stoves. Thus, the number and sharpness of elbows and so on usually is not critical for stoves. For open appliances, such as fireplace stoves and fireplaces, having too many elbows or tees is likely to result in smoke spilling out of the appliance when the combustion chamber is open to the room.

There is no hard and fast rule for what type and how many elbows, tees, and so on can be tolerated, because total capacity depends on the chimney itself. However, more than two 90-degree turns is not advisable. Two 30-degree or 45-degree fittings are usually tolerable. But the best chance for smokefree operation comes with a straight-up, no-turns installation.

ACCESSIBILITY

NFPA 211 specifies that "the entire length of a connector shall be readily accessible for inspection, cleaning, and replacement."[6] This is important! You must be able to inspect for creosote inside connectors and clean as necessary. Also, the

_____
6. NFPA 211 1980, paragraph 5-7.12.

connector pipe itself is susceptible to corrosion and burn out. Unlike chimneys, connectors are _expected_ to deteriorate. It is vital that such deterioration be readily apparent. Thus, connectors should be used only in locations fully exposed to normal daily traffic in a house, and not in closets, chases, underneath stairs, nor through walls or ceilings.

CREOSOTE DRIPPING POTENTIAL

Creosote dripping out of stovepipe connector joints is a sign that something may be wrong. Most likely, the appliance needs to be run hotter. But since liquid condensate in stovepipes is not uncommon, it is wise to install the pipe in such a manner as to minimize leakage.

Creosote will leak out of vertical stovepipe very rarely, if the pipe is assembled so that the smaller diameter end (usually crimped) is pointing downwards. This orientation is recommended and, contrary to some beliefs, will not have any significant impact on draft and will not cause smoke to leak out of the joints.

Horizontal pipe, elbows, and tees cannot be made perfectly leakproof. Furnace cement in the joints can help, but is no guarantee. Longitudinal pipe seams should face upwards.

STOVEPIPE CLEARANCES AND ADDED PROTECTION

Stovepipe can get hot—red hot during a chimney fire and even sometimes under abusively hard "normal" use. Clearances from combustible walls and ceilings are important.

Vertical stovepipe should be at least 18 inches away from unprotected combustible walls, and horizontal stovepipe should be at least 18 inches away from unprotected combustible ceilings or floors. Clearances where pipes penetrate ceilings or walls are discussed in the next section.

Reduced clearances are safe with appropriate protection. Only ventilated protectors should be used on walls and ceilings, the same kinds of protectors discussed for reducing stove clearances earlier in this chapter. (But don't try a ventilated

TABLE 10–2
REDUCED CLEARANCES*

Principal Application Type of Protection Minimal Dimensions	Radiant Stoves		Stovepipe		Circulating Stoves	
	Recommended Clearance With No Protection					
	36 Inches		18 Inches		12 Inches	
	As Wall Protector	As Ceiling Protector	As Wall Protector	As Ceiling Protector	As Wall Protector	As Ceiling Protector
4-inch thick (nominal) brick or stove facing without a ventilated airspace	24 (–)	NA	12 (–)	NA	9 (–)	NA
1-inch ventilated air space behind a 4-inch thick brick facing	12 (12)	NA	6 (6)†	NA	6 (–)	NA
1-inch ventilated air space	12 (12)	18 (18)	6 (6)†	9 (9)	4 (–)	6 (–)
Two 1-inch air spaces	6 (–)	12 (–)	3 (–)	6 (–)	3 (–)	3 (–)
Sheet metal on 1-inch of insulation mounted with a 1-inch ventilated air space between the insulation and the wall or ceiling	6 (12)	9 (18)	3 (3)	4 (4)‡	3 (–)	3 (–)

Clearances are measured from the outer surfaces of the appliance or stovepipe to the original wall or ceiling surface. In no case should the appliance or pipe touch the protector; there should be at least a 1-inch air space. All materials, including fasteners, should be noncombustible, be adequately strong, and retain their shape, strength, and insulating properties at high temperatures.

*Author's recommendations, with NFPA recommendations in parentheses. A dash (–) indicates NFPA does not have recommendations covering this case. NA means not applicable.
† For stovepipe connectors, NFPA allows only a 9-inch clearance.
‡ For stovepipe connectors, NFPA permits a 3-inch clearance.

brick protector on your ceiling!) With such ventilated protection, clearances of stovepipe from walls can be reduced from 18 inches to 6 inches (NFPA requires 9 inches), and clearances from ceilings can be reduced to 9 inches.

According to NFPA and most codes, these protectors should be sufficiently wide that the distance from the pipe to an unprotected wall or ceiling is 18 inches or more. To achieve this, the protector is usually about 3 feet wide. In my opinion, 2 feet is more than adequate.[7]

7. See Jay Shelton, *Wood Heat Safety.*

An alternative method for reducing stovepipe clearances is to use a half-round (or full) shield mounted on the stovepipe. The shield should be spaced about an inch (or more) away from the pipe, and air must be allowed to freely circulate in the space between the shield and the pipe.

Stovepipe itself is a convenient material from which to make such a shield. For convenience and neatness, a full-round shield is sometimes preferred. The shield should be made from stovepipe with a diameter 2 inches larger than the pipe being shielded, giving a 1-inch air space all around. Mount the shield with long sheet-metal screws.

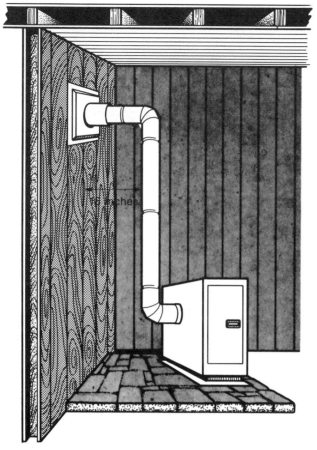

Figure 10–11. Stovepipe should be at least 18 inches from an unprotected combustible wall. Orienting the stovepipe sections with the smaller-diameter ends pointing back toward the appliance will lessen the chances of creosote dripping.

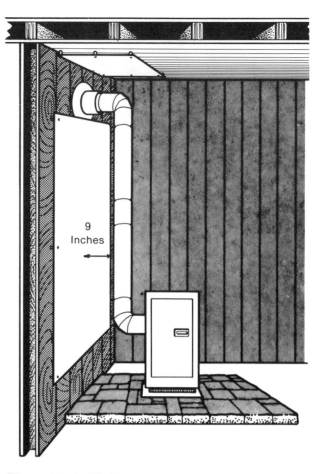

Figure 10–12. With a ventilated protector, stovepipe can be 9 inches from the original wall surface, according to the NFPA.

Spacers (not made of aluminum) can be used, although they are not always necessary.

This shielded stovepipe connector is not explicitly recognized by NFPA or building codes. However, primarily because of its similarity to the Type L Vent, which is recognized, I expect that shielded stovepipe can be installed 9 inches from a parallel wall or ceiling with reasonable safety. However, because portions of the connector are hidden from view by the shield, extra effort is necessary to inspect the covered stovepipe to be sure it has not corroded or burned out.

Of course, using Type L double-wall pipe is an alternative. It has the advantages that it is recognized by most codes, and that its inner pipe is made of stainless steel and is much less likely to deteriorate.

If a stovepipe connecting at the rear or side of an appliance is closer than 18 inches to the floor

and the floor is combustible, the normal floor protection for the appliance should be extended, if necessary, as far as the connector extends, and preferably a few inches beyond the elbow or tee if there is one. If the connector is less than a foot from the floor, which is very rare, additional protection in the form of a ventilated floor protector may be necessary.

WALL PASS-THROUGHS

The best advice on how to safely install stovepipe through a combustible wall or ceiling is, "Don't do it!" If at all possible, the solid fuel appliance should be located in the same room as the chimney so that the problem never arises.

If you do require a wall pass-through, remember that any venting component located inside a

Figure 10–13. Ventilated stovepipe shielding is not recognized by most building codes, but makes a reasonably safe 9-inch clearance.

wall and not visible for inspection should be of the highest quality. A single-wall steel stovepipe connector is *not* sufficient.

NFPA and most building codes prohibit, in all cases, passing a chimney connector through a *ceiling* and/or *floor*. This is partly because the required 18 inches of clearance is less likely to be maintained near floor level in an upstairs room—furniture, rugs, and toys may inadvertently get too close to the pipe. Another reason given for the distinction between ceiling and wall pass-throughs is the higher rate at which fire and/or smoke spreads through a house where a fire starts at a ceiling/floor pass-through.

NFPA and codes permit connectors to pass through combustible *walls*, with particular kinds of protection. It is my opinion that NFPA will soon revise its recommendations primarily because most of the given methods have liabilities

of one sort or another (some unrelated to safety). Below are discussed each of NFPA's currently recognized methods, along with my comments, and, in some cases, suggested modifications.

Using Prefabricated Chimneys

A section of listed prefabricated all-fuel or Class A chimney, installed with appropriate wall spacers, makes a reasonably safe wall pass-through (Figure 10–14). Some manufacturers of prefabricated chimney have explicit instructions and components applicable to wall pass-throughs. In these cases, follow their instructions. Usually, the wall pass-through is intended to be the bottom of a prefabricated chimney. But most manufacturers do not intend their chimneys to be used in this context. In this case, the points below should be observed.

It is vital that the clearances normally required for the prefabricated chimney be observed. Clearances must generally be 2 inches to combustible parts of the wall. It is also recommended that the chimney sections extend well beyond both sides of the wall—8 inches or more. There should be no chimney joint inside the wall—a single section should be used for the pass-through.

Using double-wall mass-insulated chimneys is generally preferable in an exterior wall pass-through application because triple-wall chimney sections usually allow room air to pass out.

This general approach is recognized by the NFPA and most codes, *if* the pass-through is "made of factory-built chimney material and *installed in accordance with the conditions of the listing and the manufacturer's instructions.*"[8]

However, few if any of the manufacturers intended their products to be used in this application (unless the pass-through connects directly to the same brand chimney, in which case the pass-through is technically part of the chimney). Since using prefabricated chimney as a simple pass-through is not generally a "listed" use of the product, such pass-throughs generally do not satisfy NFPA or most building codes. However, if my recommendations are followed, I believe it is reasonably safe.

Masonry Patch

Stovepipe itself may pass through an interior combustible wall if enough of the wall near the

8. NFPA 211 1980, paragraph 5–7.4.

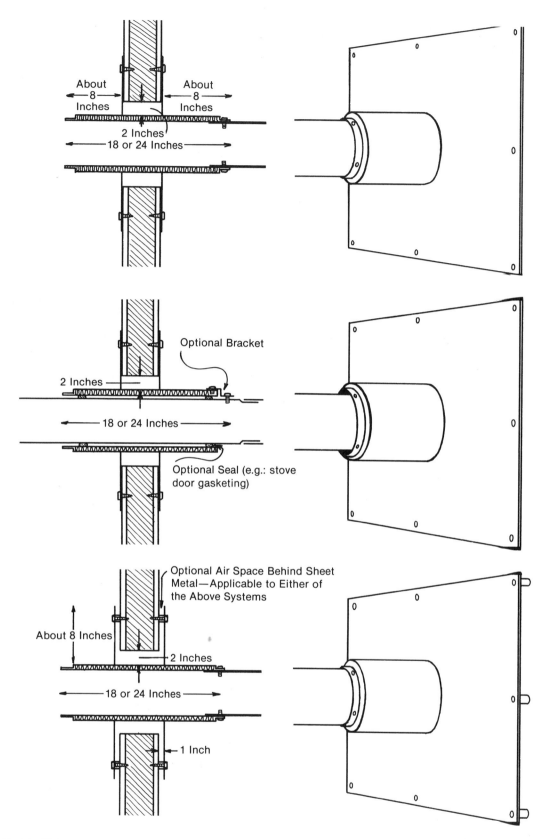

Figure 10–14. Using a section of prefabricated chimney to vent gases through a combustible wall.

pipe is made noncombustible, and if the stovepipe is easily serviceable and at least partly visible. Specifically, there should be at least 8 inches of mortared brick around an oversized sleeve or thimble (Figure 10-15). The sleeve of heavy steel, or stainless steel, or fireclay should be permanently fixed in the masonry and should extend 8 inches beyond both sides of the wall. A metal sleeve is recommended because of the relative ease of mounting spacers and fasteners, if needed. An air space, open to the room, of about an inch between the stovepipe and the sleeve should be left – both for visibility of the pipe and to keep the masonry from getting too hot.

This general approach to wall pass-throughs is recognized by NFPA and many codes, but without the ventilated air space between the stovepipe and the sleeve. I believe the ventilated air space is an important addition. The stovepipe can be kept centered in the sleeve with screws or bolts. If animals are likely to build nests in the air space, screening can be used.

If the installation is in an interior wall, both ends of the air space can be left open. As a pass-through in an exterior wall, the increased ventilation of the house is not desirable. Testing is needed for verification, but I suspect the system is reasonably safe with one end of the air space sealed, as long as the other end is open. If the sleeve is not horizontal (it should be), the higher end should be the open one. In exterior walls, the indoor end should be open to provide better accessibility and increase the likelihood that you will inspect the pipe. The air space can be sealed with a narrow ring of ceramic-wool insulation or with sheet metal.

This system may be especially applicable where a masonry chimney lies just on the other side of the wall. In some cases the sleeve can then pass right on through into the masonry chimney. However, this application should be treated as experimental until more experience is gained.

Cut-back and Shield

An alternative is to cut out the wall 9 inches all around the stovepipe and install ventilated protection over all combustible material directly facing the stovepipe, such as the framing in the wall (Figure 10-16). This method is not explicitly listed in codes or mentioned in NFPA 211, but is consistent with other provisions of NFPA 211.

As shown in Figure 10-16, this system has hidden stovepipe. As long as there is adequate mechanical support of the stovepipe, one or both of the sheet metal side covers may be deleted to make the pipe visible. This system (without side covers) is especially suitable for passing through a wall and immediately into a masonry chimney large enough to cover the hole.

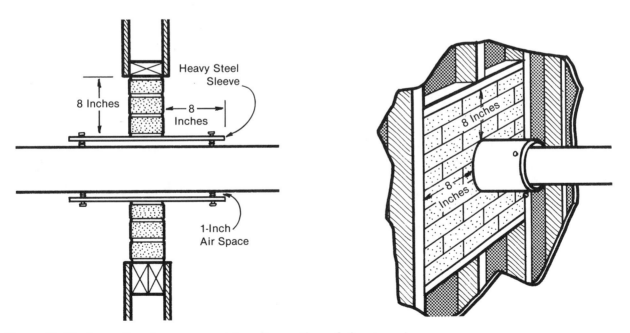

Figure 10-15. A ventilated masonry-patch wall pass-through for stovepipe.

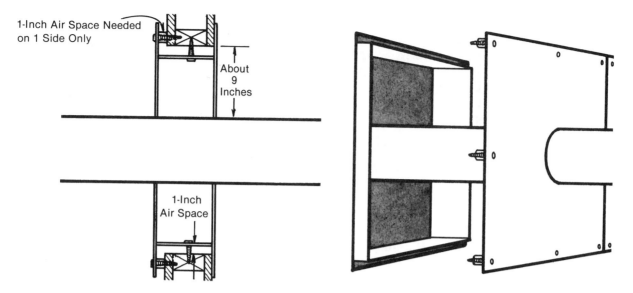

1-Inch Air Space Needed on 1 Side Only

About 9 Inches

1-Inch Air Space

Figure 10–16. Using a 9-inch cutback with ventilated shielding as a wall pass-through for a stovepipe. All or part of 1 side of the sheet metal may be removed to improve the visibility of the stovepipe inside the wall.

Ventilated Thimbles

NFPA and most codes recognize ventilated metal thimbles (Figure 10–17) as a safe way to pass stovepipe through a combustible wall. However, the specified thimble is not readily available—it must be custom-made. NFPA requires that the thimble provide 6 inches of ventilated air space all around the stovepipe. Thus, the thimble for a 6-inch pipe would be about 18 inches in diameter! Manufactured thimbles are available, but they have only a 1-inch or 2-inch air space; they are not designed for solid fuel heaters, but rather for gas appliances.

One liability of metal ventilated thimbles is the lack of visibility of the stovepipe. Another is the increased air exchange rate within the house in the case of exterior wall installations. It is not known if adequate cooling occurs if the ventilating holes on 1 end of the thimble are sealed.

I am not enthusiastic about traditional ventilated metal thimbles. More effective designs are possible which would be more compact, would keep the combustible wall cooler, and would not let in outdoor air. But until such improved thim-

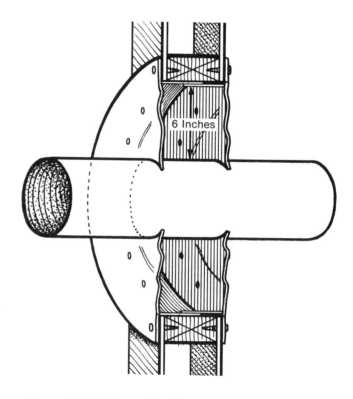

6 Inches

Figure 10–17. A ventilated metal thimble for passing stovepipe through a combustible wall.

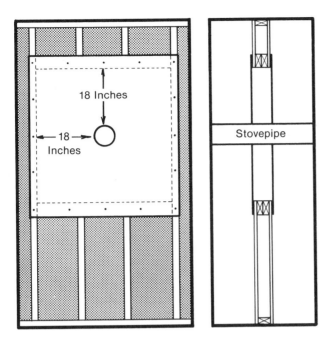

Figure 10–18. A full 18-inch-cutback wall pass-through for stovepipe. The hole in the wall can be round instead of square; but in practice, a square hole is easier to reinforce.

bles are available and tested, I recommend using 1 of the other methods discussed above.

SMOKE AND GAS DETECTORS

Smoke detectors in working condition should be part of every solid fuel heating system. As a minimum, I recommend 2–1 in the room with the heater to warn of possible problems there before the situation is beyond easy control, and 1 near bedrooms.

There are some choices in selecting smoke detectors: battery-operated versus plug-in and ionization versus photoelectric principles of operation. I have heard no compelling arguments favoring any of these choices. The plug-in units are susceptible to being unplugged and will not work during a power failure, whether fire-related or not. They also will not work if the fire damages the circuit they are using, but such occurrences are relatively rare. On the other hand, the battery units will not perform if you forget to replace the battery, required about once a year. The photoelectric type has a bulb that needs periodic replacement.

The photoelectric type is more sensitive to smoke and gives earlier warning of fires starting with smoldering conditions. The ionization type is more sensitive to invisible products of flaming combustion and may give earlier warning of fires starting with flames.

Overall, the choice of which smoke detector to use is much less critical than the decision to use smoke detectors.[9] All types work satisfactorily and provide substantially reduced risks of fire damage, injuries, and deaths. The units should be listed by a major testing laboratory, such as Underwriters Laboratories, and should be installed according to the manufacturer's instructions.

With the increased use of coal as a fuel, there has been more interest in carbon monoxide detectors (also called gas detectors). Both wood and coal fires can generate considerable carbon monoxide. But it may be more likely with coal that under some circumstances, there is not much odor-producing gas generated along with the carbon monoxide. Since carbon monoxide itself is odorless and invisible, you may not notice it leaking into the house.

Are gas detectors needed in addition to smoke detectors? Under most circumstances, it appears that significant carbon monoxide will be accompanied by other products of combustion that would set off an *ionization-type smoke detector.* If you have such a smoke detector, a gas detector usually will not be necessary.

---

9. Consumers Union has tested a number of smoke detectors. See *Consumer Reports*, August, 1981.

CHAPTER 11

STOVE ACCESSORIES

Accessories for stoves and other solid fuel heating appliances are intended to serve one or more of the following purposes.

- Increase energy efficiency.
- Improve convenience.
- Extend the capabilities of the system.
- Decrease creosote.
- Enhance safety.
- Improve aesthetics.

This chapter covers the first 3 categories. Accessories dealing directly with creosote are discussed in chapter 15, and most safety-related accessories (other than those dealing only with creosote) are covered in chapter 10. Check the index to locate specific accessories not covered here.

HEAT EXCHANGERS

Heat exchangers (or heat extractors) are intended to improve overall energy efficiency by extracting additional heat from the system. Most are installed on or in place of a section of stovepipe. Others are installed on the stove itself. Some are passive, operating on radiation and natural convection; others are active, employing a blower or fan to help transfer the heat by forced convection.

The most common type of heat extractor is the tubular device illustrated in Figure 11-1. In this device, the flue gases enter a closed box with pipes running through it. The gases flow around the outsides of the set of pipes, while room air is blown through the pipes. Heat is conducted from the flue gases through the pipes and into the moving air, warming it.

Variations on this design include reversing the functions of the box and the tubes so that the flue gases are directed through a set of vertically oriented tubes while air is blown around the outside of the tubes. The blower or fan on most heat extractors is thermostatically controlled to be on only when the flue temperature is above a given value.

It is impossible to predict precisely the net effect the heat exchanger has on heat output of a stove. Careful laboratory measurements are difficult and are lacking for many such devices. Where laboratory measurements are available, they require careful interpretation since details of particular installations and how the stove is used have important effects. My guess is that tubular stovepipe heat extractors are relatively effective, and that the strap-on fin type is not. The convective cylindrical type has some analogies with circulating stoves — convective heat transfer may be enhanced, but radiant is decreased; thus the net effect is difficult to predict. Stove heat exchangers are similar in this respect; it is possible that the enhanced convective heat transfer is largely offset by the reduced radiant transfer.

The tubular stovepipe heat extractors probably increase overall energy efficiencies by 3-15 percentage points, which can result in a 6-30 percent reduction in wood consumed to produce the same amount of total useful heat. The amount of elec-

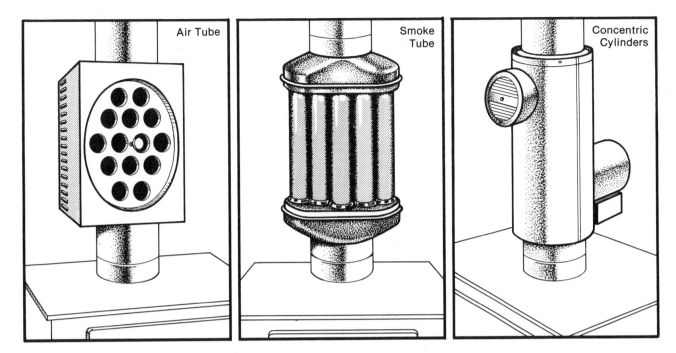

Figure 11-1. Stovepipe heat extractors.

tricity consumed to run the blowers is not significant usually. The range is very large due to differences in design of the extractors, and because the effectiveness of an added heat extractor depends very much upon the heat transfer efficiency of the stove.

Stovepipe heat exchangers generally must have substantially more total heat-transfer surface area than the stovepipe they replace or are mounted onto in order to be effective. A doubling of the area is substantial. Most stovepipe tubular heat exchangers meet this requirement, both by virtue of the larger outside surface area of the box (relative to stovepipe) and, of course, the surface area of the tubes themselves.

Any simple up-draft or diagonal-draft stove operated at high power will benefit from an effective heat extractor—in such cases flue-gas temperatures within a foot or so of the stove can be as high as 800–1200°F., indicating large amounts of available heat. But stoves with a high heat transfer efficiency have so much surface area relative to their combustion rates that flue gases are rarely above 500°F., even with a large hot fire. At medium and low firing rates, typical temperatures are under 300°F. Since the stove extracts so much more of the available heat in these cases, an additional heat extractor device is relatively superfluous.

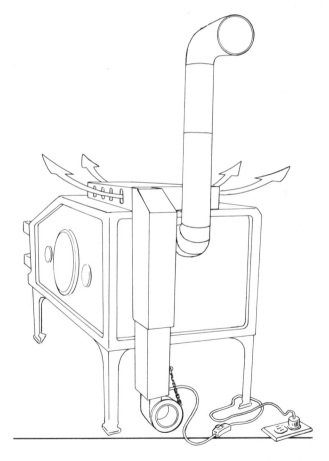

Figure 11-2. Heat exchanger accessories, such as the type illustrated here, cover up part of the radiating surface of the stove, which is a definite disadvantage. The net effect is difficult to predict.

Regardless of stove design and firing rate, if a stove is vented into a 2-story interior exposed masonry chimney, a heat exchanger accessory will not be of much benefit because the chimney itself is an excellent heat exchanger.

Overall, the more sensible heat that is wasted by the whole system, the more beneficial will be a good heat exchanger accessory.

Active (electrically powered) heat extractors generally are more effective than the passive ones and can significantly improve the heat distribution because of the room air mixing and circulation the blower induces. In some cases, the unit can be oriented to induce a general air circulation pattern around a room or even a small house.

Importance of Easy Cleaning

A major consideration for all types of heat extractors is the ease with which they can be cleaned on the inside. The heat extractors get more heat from flue gases by creating cooler surface areas inside the flue. Chimney deposits from smoke build up preferentially on cooler surfaces. These deposits are generally good thermal insulators; so the buildup decreases the amount of heat extracted by the device. The buildup of deposits is inevitable. To maintain their effectiveness, heat extractors require frequent cleaning, typically about once a week, but perhaps as often as daily, or as seldom as monthly.

Some heat extractors are not easy to clean, which is a serious flaw. It is inconvenient and often messy to have to partially disassemble a unit or sections of stovepipe, and carry the dirty parts to an appropriate place to brush them clean.

Most tubular heat extractors have easy cleaning built in with a movable plate with holes in it, which fits snugly around the tubes. The plate can be moved back and forth, scraping the tubes clean, merely by pulling and pushing a rod on the outside of the unit. The deposits either fall back into the stove or are carried up and out of the chimney. The whole operation takes seconds and creates no mess.

All effective heat extractor accessories reduce draft by cooling the flue gases. Some also may reduce draft by presenting flow resistance to the flue gases. These accessories should not be considered for installations with inherent (calm weather) draft problems.

Heat exchanger accessories tend to increase creosote accumulation in the system. Heat exchange accessories are not advisable in systems where creosote accumulation is already a problem.

Another factor to keep in mind is noise. Not all fans and blowers are noisy, but they are definitely audible. For a quiet heating system, you may prefer a nonelectric accessory, or none at all.

In deciding about heat exchanger accessories, keep in mind the other ways that can help to extract a little more heat from the system—a stovepipe damper, an extra elbow or two in the stovepipe connector, a few extra feet of length in the connector, and using an interior exposed masonry chimney in a multistory house. The pros and cons of each of these are discussed elsewhere in this book.

Measuring the Performance Of Heat Extractor Accessories

Measuring the net effect of a heat extractor accessory can involve some subtleties. In addition to the heat output of the device itself, 3 other factors should be included in an assessment of the *net* effect of the device.

First, the more heat the device extracts, the cooler will be the stovepipe that comes after the device; hence use of the device decreases the amount of heat given off from the downstream stovepipe and chimney. There is still a net gain, but it is smaller than the heat output of the extractor itself.

Second, if the extractor is not present, the ordinary stovepipe that would be in the same place transfers considerable heat to the room by itself. Some of the decrease in temperature of the flue gases passing through the extractor would have occurred without the device.

Third, all the electrical energy used by the active heat extractors is ultimately converted to heat. Consequently a small amount (50–100 watts, or 170–340 Btu. per hour) of electric heating is a side effect of active devices, but should not be included as part of the extracted solid fuel heat when computing efficiencies.

Thus, the net heat-extracting effect of a heat extractor is its total heat output, minus the output of the length of stovepipe the extractor replaced, minus the decrease in heat output from the stovepipe (and chimney) beyond the extractor, minus the electrical power consumption of the device.

FANS

A medium-sized (4–12-inch blade) household fan placed a few feet from a radiant stove and blowing at the side or rear of the stove can increase the system's overall energy efficiency slightly—by just a few percents. A ceiling fan may improve heat transfer a little, but generally it does not create as high a velocity at the stove as a floor fan. Ceiling fans are probably more effective at breaking up temperature inversions in a room. Both types of fans can assist in moving the heat from a stove into other rooms.

CATALYTIC ADD-ONS AND RETROFITS

Most catalytic combustors are incorporated into appliances themselves. But catalytic afterburner add-ons, which can be installed on top of many existing noncatalytic stoves, are available.

A catalytic combustor incorporated into an accessory is no different in form or function than a catalytic combustor built into a stove. Both are designed to improve the combustion of gases and smoke from the wood. The result can be increased overall energy efficiency, decreased creosote, and decreased emissions. Add-on catalytic units have the slight disadvantage, compared to catalytic stoves, because the flue gases entering the combustor will tend to be cooler, so performance likely will be somewhat less.

Since catalytic afterburners are relatively new, there are not a lot of test results yet available. A test conducted by Shelton Energy Research (SER) in cooperation with *Mother Earth News* demonstrated that substantial creosote reduction is possible; a 45 percent reduction was observed in this particular test. Actual reductions in the particular installations will vary substantially. There is no doubt that heat output was increased and total emissions reduced, but these effects were not quantified in this test.

A few stove manufacturers offer retrofit kits to transform their wood heaters into catalytic stoves. The catalytic combustor is fitted into the stove itself, which eliminates the problem of cooler flue gases.

Does it Pay To Use an Electrically Powered Heat Extractor Accessory?

Some heat extractor accessories use electricity for a blower or fan. The electricity required usually is not significant in terms of its cost relative to the value of the extra heat extracted by the accessory.

A 100-watt fan motor running 8 hours a day will cost roughly 5–20 cents a day in electric energy (for 6–25 cents per-kilowatt-hour electricity). If the average power output of the stove is 25,000 Btu. per hour, and if the accessory results in 10 percent extra heat output, then 20,000 extra Btu. will be extracted by the accessory during the 8 hours.

The value of this extra heat ranges from roughly 15 cents to 75 cents depending on what kind of heat it displaces (wood, gas, oil, electricity, and so on). In most cases, the value of the extra heat produced exceeds the cost of the electricity used to extract it. However, this may not be the case for relatively inefficient electrically powered accessories that displace relatively cheap heat, such as wood heat for those who cut their own wood. The overall economics also must take into consideration the original cost of the device. If it is expensive and/or the stove is not used very much, then it may not pay to buy the accessory.

Most, if not all, catalysts on the market are designed for wood fuel only, not coal.

Virtually all catalytic combustors are somewhat restrictive to the smoke flow. Thus, if your system has a marginal draft, a catalytic accessory may not be appropriate.

Figure 11–3. A catalytic add-on accessory (the Smoke Dragon).

When a catalytic combustor is working well, a substantial amount of heat is added to the flue gases. To reap the benefit of this increased energy efficiency, you must have this heat transferred to the house. Catalytic add-ons have a built-in extra heat exchanger for this purpose; retrofits do not.

But in either case, lack of explicit extra heat transfer capability is not as serious as many people think. Stovepipe is an effective heat exchanger. A few feet of stovepipe after a catalytic combustor will recover most of the extra heat. And even in the case of a catalytic retrofit with little exposed stovepipe, much of the heat generated in the catalyst will make the stove surface hotter, and this in turn will greatly increase the radiant heat transfer from the stove (see Table 2-2).

STOVEPIPE DAMPERS

In the old days, the main purpose served by manual stovepipe dampers was control of combustion. In a nonairtight stove, even shutting the air-inlet damper may not slow down the fire as much as desired. By closing the stovepipe damper, the draft in the stove is reduced, reducing the air flow. Or, put another way, imposing additional resistance to flow anywhere in the system will decrease the flow everywhere in the system.

In an airtight stove, this function of manual stovepipe dampers is not needed—just closing the air inlet will suffocate the fire to extinction.

However, there may be a second purpose served by manual stovepipe dampers. Experiments at SER suggest that in the case of at least 1 airtight stove, using a stovepipe damper may increase the overall energy efficiency of the system. This is due in part to the turbulence induced in the smoke in the stovepipe, but it also may be related to much more subtle and complex changes in air, flame, and smoke flow patterns in the stove itself. Since a closed stovepipe damper imposes additional flow resistance, to compensate the air inlet on the stove must be opened somewhat to achieve roughly the same heat output from the system. Then roughly the same amount of air enters the stove through a wider inlet opening. Thus, the air velocity is less, and this could change the flow patterns in the stove, altering the energy efficiency. Further experimentation is needed to verify the effect and to determine if it generally is true with all stoves. In the meantime, there is not any significant detrimental effect of manual stovepipe dampers; so it certainly does no harm to install them. My guess for the best location is roughly halfway along the longest straight segment of stovepipe.

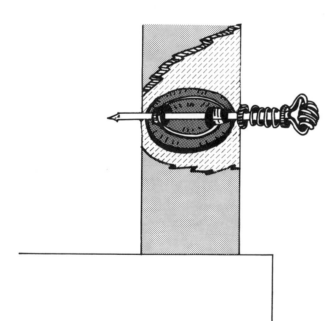

Figure 11-4. A stovepipe damper.

BAROMETRIC DRAFT CONTROLS

A barometric draft control (Figure 11-5) works on a very different principle. It contains a hinged and weighted flap that is normally closed when the appliance is not in use. During use, when the draft in the chimney exceeds a preselected value, adjusted by moving a small weight attached to the flap, the draft pulls the flap open. This lets room air into the chimney.

Barometric draft controls are commonly used with oil furnaces and boilers to improve energy efficiency. The controller maintains the right or optimum amount of air. Without the draft controller, when the draft exceeds the optimum value, more air than necessary is pulled into the combustion chamber, and excess air always decreases heat transfer efficiency.

In the case of wood-fueled heating appliances, there are both benefits and possible liabilities of using barometric draft controllers.

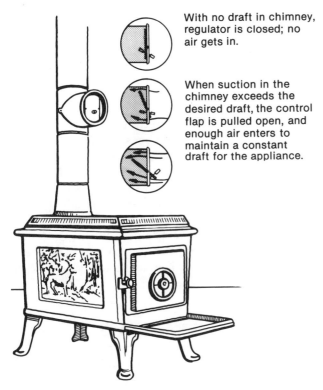

With no draft in chimney, regulator is closed; no air gets in.

When suction in the chimney exceeds the desired draft, the control flap is pulled open, and enough air enters to maintain a constant draft for the appliance.

Figure 11-5. A barometric draft control installed in a stovepipe connector.

Benefits of Barometric Draft Controls

A significant benefit is a steadier burn and therefore steadier heating. In solid fuel heaters, combustion rates are usually limited by the air supply. (In oil appliances, the combustion rate is limited by the fuel supply.) Thus, by limiting the draft, barometric draft controls shave off the peaks and extend the duration of the burn (Figure 11-6).

Energy efficiencies do not seem to be affected dramatically. Based on careful measurements at SER of 1 typical radiant stove, it seems the main effect of a barometric draft control on energy efficiency is due to the cooler stack temperatures resulting from the air entering through the control. If the control is placed immediately after the appliance, the stovepipe will be cooler and give off less heat. However, if the control is placed at the upper end of the stovepipe just before the chimney, then only the chimney itself will be cooler; there will be no effect on heat transfer unless the chimney is interior and exposed.

Even though a barometric draft control cools the chimney, *less* creosote will accumulate, because dilution air in a chimney decreases creosote

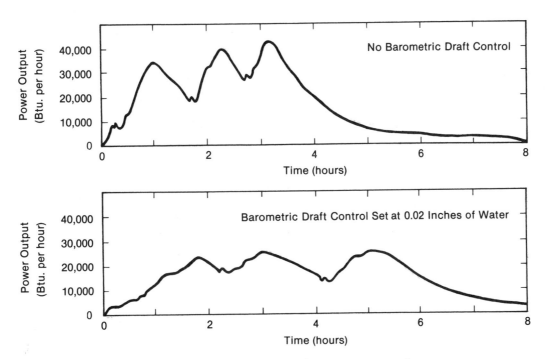

Figure 11-6. The effect of barometric draft control on the power output of a typical stove not controlled by a thermostat. The air inlet of the stove was wide open in both cases. Each hump in the graphs corresponds to the burning of a fuel load; 3 loads were burned.

accumulation. Creosote reductions of up to 75 percent have been observed in experiments at SER.

This reduction in stack temperatures has another benefit. There is less thermal stress on the chimney and a decrease in possible fire hazards to surrounding combustibles. This can be of significant value, particularly in marginal chimneys, such as unlined masonry or masonry without adequate clearances. The benefit also is particularly relevant with large capacity heating systems, such as furnaces and boilers, where flue temperatures can be high.

A barometric draft control, by limiting draft, prevents dangerously hot fires in the appliance.

Liabilities of Barometric Draft Controls

There are 2 possible liabilities of barometric draft controls. Using a barometric draft control results in additional house air lost up the chimney. The net heat loss this represents depends on the excess draft in the chimney, the outdoor temperature, and on how much of the air flow represents extra house exfiltration caused by the damper (versus a rerouting of air through the damper—air which would have left by another route). Tests to determine how much new air infiltration is caused by using a barometric draft control are difficult to perform, since they must be done in actual homes. Since the amount of air flow into the control is typically small compared to house air exchange rates, it is likely that most of the air flow into the control is merely a rerouting of the air, not new infiltration, and, therefore, it does not represent a new heat loss.

In the worst case of all—the air flow being a new net loss—the net effect on the energy efficiency of the system is roughly a few percentage points. Thus, the actual effect cannot be large, but could be significant.

Do not duct outside air to a barometric draft control to eliminate the house air loss. This will defeat the primary draft-controlling function of the device, particularly during windy weather. Also, admitting *cold* air into the flue might not have the same effect on creosote as room temperature air.

The other possible significant liability of a barometric draft control is that during a chimney fire the control will be wide open, admitting considerable air. This could make the chimney fire more intense. But it is also possible that the extra air might have a neutral or even positive effect in

terms of safety. Chimney fires are sufficiently complex that only testing can determine the net effect. It seems most likely that the effect is not good, but it may not be very large. However, note that if the chimney is kept clean, this possible problem literally disappears.

NFPA standards (NFPA 90–B) require a barometric draft control on hand-fired thermostatically controlled wood furnaces. My feeling is that with large appliances, such as furnaces, or with any appliance hooked up to a chimney with excessive draft, a barometric draft control is of benefit, but only if the chimney is kept sufficiently clean that a large chimney fire is impossible.

Keep in mind that a barometric draft control is of no use in systems with a marginal draft since the control works by decreasing draft.

THERMOSTATS

Some stoves come equipped with thermostatically controlled combustion air. For those that do not, accessory thermostats are available to fit some stoves, including most stoves that have spin air caps.

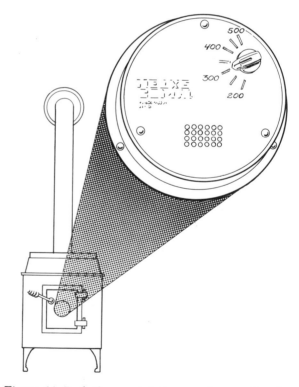

Figure 11-7. A thermostat that can be retrofitted to many stoves. (This thermostat is manufactured by the Condar Co.)

There is little doubt that thermostats usually serve their primary purpose of regulating the heat output of the stove. Some manufacturers also claim that thermostats improve energy efficiency in 2 ways: an inherent improvement in the efficiency of the appliance itself and an indirect fuel savings, due to the fact that with an effective thermostat the house rarely becomes overheated. This latter effect on energy efficiency is not easily measured in laboratory testing but rather requires testing in actual homes. Each of these claimed efficiency-improving mechanisms is plausible, but neither has been verified by tests conducted by an independent laboratory.

A thermostat is more effective at controlling heat output in a relatively tight appliance than it is in a relatively leaky appliance.

WATER HEATERS

Heating tap water with solid fuel is a logical extension of space heating with wood or coal for those who want to use an absolute minimum of gas, oil, or electric energy. Water heating typically accounts for 10–15 percent of all household energy use.

Heating tap water is a natural and common part of boiler central heater. There are also very small solid fuel appliances designed exclusively to heat water. Small add-on boilers can be used as tap water heaters. But the most common tap water heating systems are accessories or adjuncts to solid fuel stoves.

Heating tap water with solid fuels is dangerous! The principal hazard is steam explosions.

Heating water with solid fuels can be done safely. I strongly recommend taking all of the following steps before putting in a tap water heating system.

- Consult with a competent professional plumber.
- Read the manufacturer's instructions carefully (if you use a purchased heat exchanger).
- Read pages 110–120 in Jay Shelton's *Wood Heat Safety*.

I have had many wood-fired water heating systems in my homes, and I recommend the concept. But there are many subtle and some not-so-subtle precautions that must be taken to assure the safety of such systems.

One kind of installation is shown in Figure 11–8. It is typical to use a conventional water heater as an automatic back-up to the solid fuel system.

Here are some of the requirements of this kind of water-heating system.

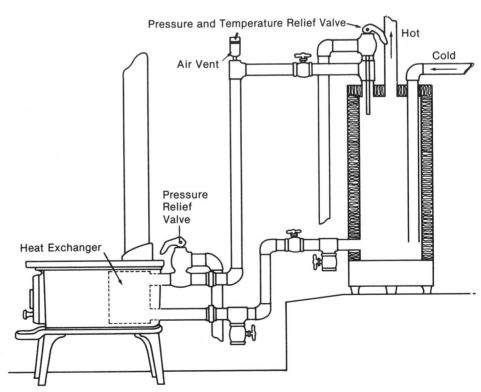

Figure 11–8. An installed water-heating heat exchanger in a stove. The size of the plumbing is exaggerated.

- Use strong and durable materials which can take high pressures and which will last. Heat exchangers can be homemade, but the better manufactured units may be worth the price for the assurance of the required quality.

- In "closed" systems, there must be provision for the expansion of the water as it is heated.

- All water heating systems require an automatic pressure relief system to let water and/or steam out before the pressure becomes dangerously high. Closed systems use pressure relief valves. Open systems use the fact that they are open for emergency relief.

- Some systems need special features to assure that water cannot be sent into a hot but dry (empty) heat exchanger.

- Heat exchangers either should be located in a hot part of the stove or entirely outside the combustion chamber and flue to minimize creosote build up problems.

- Heat storage is a necessity, both to provide enough hot water when needed and to keep the water in the exchanger from boiling when hot water is not being used.

- The water must be circulating through the heat exchanger whenever the fire is hot. This can be done either by a pump or by natural thermosyphon (or gravity) flow. There are critical geometrical constraints on thermosyphon systems. All piping must slope upwards so that there is no place an air pocket could form that would block the flow.

This is only a brief outline of some of the important aspects of water-heating systems. Such systems have important subtleties; the details could fill a small book. Please take the precautions noted at the beginning of this section before putting in a water-heating system.

THERMOMETERS

Thermometers are a popular accessory item. Surface thermometers are used on stoves and on stovepipe connectors, and stem thermometers are used to measure flue-gas temperature in the stovepipe. Most surface thermometers and 1 type of stem thermometer use a bimetallic coil as the sensor, and another kind of stem thermometer uses a thermocouple as the sensor.

Thermometers indicate at a glance the intensity of the fire, and thus can warn of unusually hot conditions or indicate that reloading is necessary.

The most common use for thermometers is for creosote control. Since creosote condensation is so related to temperature, there is an opportunity to monitor flue-gas temperatures with the purpose of keeping it above a critical temperature to prevent creosote accumulation. The critical temperature will vary with the installation.

Typically, the only convenient place to measure the smoke temperature is in the stovepipe connector to the chimney. But the temperature in the stovepipe is not the same as the temperature in the chimney. The smoke cools as it moves along. The cooling can be very significant in uninsulated chimneys, in very tall chimneys, in exterior chimneys, and in oversized chimneys.

Regardless of the temperature of the smoke throughout the chimney, it is the temperature of the interior wall surface of the chimney that is most relevant, for that is where the creosote actually condenses.

Finally, there is probably a wide, critical-temperature *range*, and the range itself probably depends on smoke composition and smoke density.

Because there is no 1 magical temperature ap-

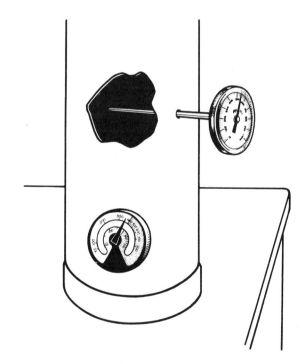

Figure 11–9. Stem and surface-mounted thermometers. Most surface-mounted thermometers are attached magnetically; wire is sometimes used as a backup should the magnet fail.

plicable to all installations, experience with your particular system must be your guide for the correlation between indicated temperature and creosote accumulation.

A stem thermometer with a heat-sensitive element that is inserted into the pipe yields a more accurate measure of flue-gas temperature than a surface thermometer and responds more quickly to temperature changes. However, a change in the surface temperature under the same firing conditions can indicate creosote buildup. Both types of thermometers have their merits.

STOVEPIPE OVENS

An accessory which has been on the market for decades is the stovepipe or "granny" oven. Its use as an oven requires careful operation of the stove to maintain proper temperatures inside the oven, and such operation may or may not keep the house at a comfortable temperature. In many installations, it is difficult to achieve high enough temperatures for baking.

Stovepipe ovens act as heat exchangers, whether or not they are also used for cooking, due primarily to their large exterior surface area.

Stovepipe ovens have limited durability. They are usually made of thin-gauge sheet steel.

Because of the extra weight of a stovepipe oven and the fact that additional weight and jarring will result from its use, it is particularly important that such ovens be installed securely into a secure connector system. Be sure to use 3 or more sheet metal screws or the equivalent in *all* joints, and be sure the stovepipe is securely connected to both the stove and the chimney. For side and rear exit stoves, the lower elbow or tee (if applicable) may require extra support beneath it to help hold up the weight.

For serious cooking, a real cookstove is a must.

FIRE EXTINGUISHERS

A small fire extinguisher in the home can easily and quickly suppress many small fires, preventing them from becoming major problems. I rec-

ommend some form of fire extinguishing capability in *every* home.

Most fires associated with solid fuel heating systems are Class A, involving combustion of ordinary materials such as wood, paper, and fabrics. For these fires, Class A extinguishers are appropriate. Water is also appropriate. A bucket of water and a cup or pan kept near a stove can quickly and neatly handle small fires.

Many household extinguishers are designed to handle all 3 fire types—A, B, and C. Class B fires involve burning liquids, and are not likely with solid fuel heaters. Type C fires are any fires involving, but not necessarily caused by, live electrical equipment.

Extinguishers are rated in terms of their capabilities for these 3 kinds of fires. For fires related to solid fuel heating systems, a good rating for Class A fires is most important. Most household extinguishers are rated for all 3 types of fires. These offer the greatest flexibility in controlling fires—they can safely and effectively be used on any type of fire.

GLOVES

A pair of gloves kept near the stove can be useful. Some stoves have door handles and air-inlet controls that are too hot to touch with a bare hand. Gloves are especially important for handling burning wood in an emergency. Should a burning log roll out of a fireplace stove with its doors open, or a hot coal get beyond the floor protector during ash removal, or should the last log in a new fuel load have to be removed because it would not go in far enough to permit closing the door, or should the hot cooking-hole cover be dislodged from its hole by a backpuff, or. . . . A good pair of gloves could save the day.

Heavy leather gloves are suitable, but keep in mind that leather can burn, or at least smolder. Do not grasp a burning log or coal any longer than necessary, and then do not carelessly toss the gloves aside.

Asbestos gloves should not be kept around the house because of the possible lung cancer hazard from asbestos fibers. Safe substitutes, such as ceramic fiber gloves, are available.

CHAPTER 12

FIREPLACES

Traditionally, a fireplace was a simple masonry structure with a fire chamber permanently open to the room. Today the term "fireplace" can encompass masonry and metal units, site-built and factory-built units, free-standing and built-in units, units with and without doors (and both see-through and metal doors), and systems with heat distribution by radiation, hot air, and/or hot water. In fact, the traditional distinctions among fireplaces, stoves, and furnaces are now blurred. There is so much variety and diversity in wood heating equipment today that simple classification systems are bound to fail.

TRADITIONAL MASONRY FIREPLACES

Traditional masonry fireplaces (Figure 12–1) differ from stoves in 2 important performance aspects. Fireplaces are 1-sided, while stoves have 6 sides. Significant heat transfer occurs through 1 side only—the front of a traditional masonry fireplace. Also, open fireplaces consume very large quantities of air. This large air flow must be taken into account when discussing the energy efficiencies of fireplaces—it can constitute a substantial heat loss.

Because of this large air consumption, it is important to make a distinction between the gross heat output (or gross energy efficiency) and the net heat output (or net energy efficiency) (see Table 12–1). The net heat output of any appliance can be different from the gross heat output due to 3 factors.

- Heat loss during fireplace operation due to the flow of heated room air up the chimney (Figures 12–2 and 12–3).
- Heat gain from an exposed interior chimney, if applicable.
- "Standby losses." Extra heat is lost from the house just due to the presence of the fireplace when it is not being used (Figure 12–4).

Room-Air Heat Loss

How much room air goes up a fireplace chimney? A very rough rule of thumb is that an operating open fireplace consumes about 50 cubic feet per minute of air for every square foot of the fireplace opening. A fireplace with a 3-foot by 3-foot face would consume 450 cubic feet per minute. Typical fireplaces need this amount of air flow into the opening to prevent smoke from spilling out of the opening into the house.[1]

The air originates outdoors. Assuming there is no direct outdoor air feed system, the air gets in through cracks around doors and windows, through clothes drier vents, and through leaks between the house and its foundation. By whatever means the air infiltrates the house, usually it is warmed up to room temperature before it reaches

1. *ASHRAE Handbook and Product Directory, 1979 Equipment Volume* (New York: American Society of Heating, Refrigerating and Air-Conditioning Engineers, 1979), pp. 26–27.

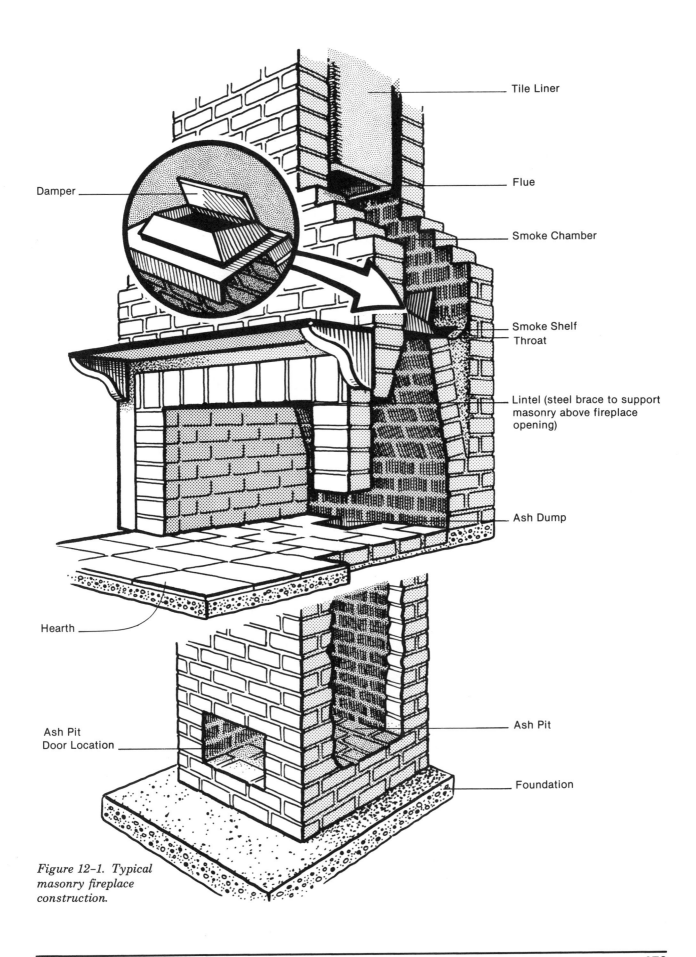

Damper

Tile Liner

Flue

Smoke Chamber

Smoke Shelf
Throat

Lintel (steel brace to support masonry above fireplace opening)

Ash Dump

Hearth

Ash Pit Door Location

Ash Pit

Foundation

Figure 12–1. Typical masonry fireplace construction.

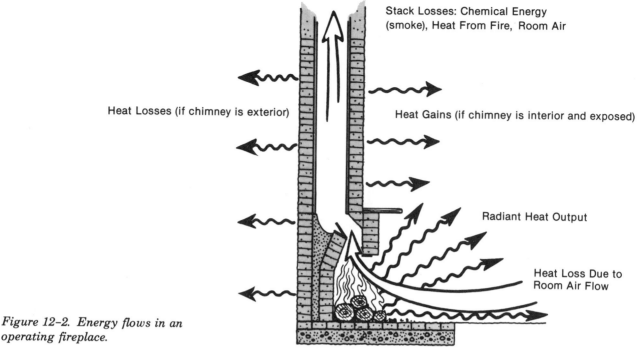

Figure 12-2. Energy flows in an operating fireplace.

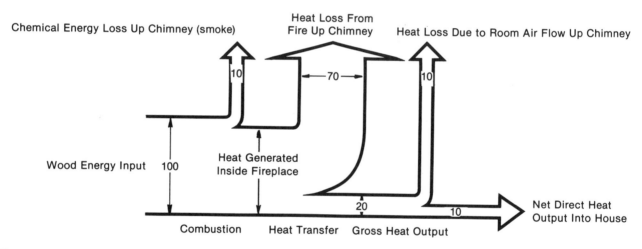

Figure 12-3. Energy flow diagram for an operating fireplace. The numbers are rough estimates only; in particular cases, the amounts may be quite different.

the fireplace. Throwing this warm air up the chimney is a waste of heat. And it really makes no difference whether the air was warmed by heat from the central heating system or from the fireplace—the amount of heat wasted is the same. Using a fireplace increases the heating needs of the house by increasing the air infiltration rate.

Not all of the air flow into an operating fireplace is extra infiltration caused by the fireplace. Some is air which would have entered the house anyway but would have left by some other route—through the attic or through cracks around upstairs windows. The fireplace should be penalized only for the extra infiltration it causes.

Typically, the air consumption of operating fireplaces is roughly comparable to the normal air exchange rates of houses; so it is not obvious what fraction of fireplace air consumption is extra infiltration. Theoretical and experimental works by Madera and Sondregger[2] indicate that

2. M. P. Madera and R. C. Sondregger, "Determination of In-Situ Performance of Fireplaces," Lawrence Berkeley Laboratory, University of California, Berkeley, CA (1980).

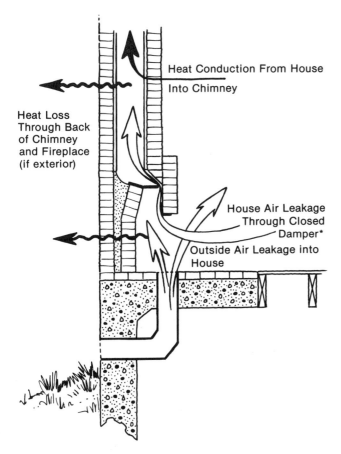

Figure 12-4. Possible standby heat losses from a masonry fireplace (heat losses when the fireplace is not in use).

___

* In a multistory house with an exterior chimney, outdoor air may come down the chimney, rather than house air go up. See chapter 6.

the extra infiltration can vary from a small fraction for a small fireplace in a loose house to a large fraction for a large fireplace in a tight house. Since the trend these days is towards tighter houses, and for the sake of simplicity, most of the following discussion assumes that all the air flow up the chimney represents extra infiltration. This is an overestimate in some cases.

How important is this heat loss up the chimney? It depends very much on how cold it is outdoors, since this affects how much heat was invested in the air to get it up to room temperature before it entered the fireplace. For a 3-foot by 3-foot fireplace opening, the loss is about 8,700 Btu. per hour when it is 40° F. outside, and about 20,000 Btu. per hour at 0° F. (Figure 12-5)! The 20,000 Btu. per hour figure represents a substantial fraction of the total heat needed to keep a

typical house warm when it is 0° F. outside (30,000–80,000 Btu. per hour). So, yes, the heat loss due to room air going up the chimney of an operating open fireplace *is* substantial.

Heat Gain From Chimneys

If a masonry chimney runs up the inside of a 2-story or 3-story (or more) house and has its sides exposed to the living spaces of the house, there can be a considerable heat gain from the chimney. There is also some (but less) heat gain from factory-built chimneys of the double-wall insulated type or the triple-wall air-insulated type.

Quantitative estimates are difficult because of the variability of fireplace design and operation, and because there is not much measured data. Measurements with masonry chimneys are more difficult because of the time delay between the fire in the appliance and the heat gain from the chimney. I can only speculate that the effect may increase the energy efficiency of the system by

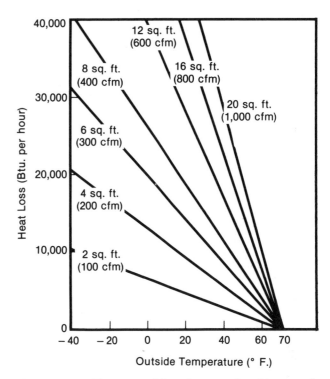

Figure 12-5. Estimate of heat loss up the chimney of an operating open fireplace due to the loss of heated room air. The area of the fireplace opening is expressed in square feet. Air flow up the chimney is expressed in cubic feet per minute. The principal assumptions are that the air flow is 50 cubic feet per minute (cfm), and that all of the air flow is extra infiltration into the house caused by the fireplace operation.

5–15 percentage points. The effect is certainly enhanced if there is no air flow through the chimney after the fire has gone out; flowing air would tend to pull the heat back out of the masonry into the flue and up the chimney. The dampers should be closed when the fire has gone out.

Standby Losses

Even when a fireplace is not in use, its presence results in a little extra heat loss from the house (Figure 12-4). Even with the damper closed, a little air usually leaks up the chimney. House heat may conduct through the chimney walls and then be convected up the chimney. This effect is probably lessened by using a chimney top damper.

Heat loss through the back of the masonry fireplaces built into exterior walls is significant also. Brick is not a good insulator. A brick wall would have to be many feet thick to have an insulating value equal to that of an ordinary wall with 3½ inches of fiberglass insulation.

All these standby losses taken together may result in a heat loss rate of 500–3,000 Btu. per hour on cool days when the fireplace is not in use.

Overall Energy Efficiency

So when all these factors are taken into account, what is the overall energy efficiency of a traditional masonry fireplace? (By "traditional," I mean with no glass doors, no tubular grates, no air circulation behind a metal shell—just a plain masonry fireplace.) You can say that a stove has an overall efficiency of 45–65 percent and have accurately described the vast majority of stoves regardless of how installed or operated. However, with traditional fireplaces, how they are built and operated makes a significant difference.

First, there is the 5–15 percentage point advantage to a multistory interior exposed masonry chimney. But, this is usually not the case and will be presumed not to be the case in the rest of this chapter.

Second, when a traditional fireplace is in steady use, its gross efficiency depends on the size of the fire (Figure 12-6); the efficiency increases with fire size. Of course, larger fires put out more heat, but to say they are more efficient means that there is more heat *per log burned* in larger fires.

There are 2 reasons for this effect. Typically, combustion is more complete in larger (hotter) fires. Also, radiant heat transfer is more efficient

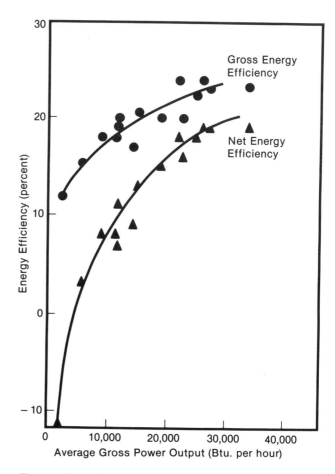

Figure 12-6. Measured energy efficiencies of a (simulated) traditional open masonry fireplace. Net energy efficiencies in this graph differ from gross energy efficiences as they include the heat loss of warm house air going up the chimney. Possible heat gains from exposed chimney walls and possible standby heat losses have not been considered. Negative net efficiency means more heat is lost up the chimney than the fireplace contributes to the room.

from hotter fires (see Table 2-4). The higher temperature results not only from the larger fire size but also from the fact that *relatively* less "excess" air enters the fireplace with a big fire in it. This results in relatively less convective cooling of the fire chamber walls. Thus, both the fire itself and the surrounding bricks radiate more efficiently when the fire is large.

The effect of fire size on *net* energy efficiency is even more dramatic because the air consumption of a fireplace does not increase in proportion to fire size; it increases more slowly. Quantifying the effect requires knowing how much air is going up the chimney at each firing rate and how cold it is outdoors, so that the heat loss from air going up

TABLE 12-1
GROSS AND NET ENERGY EFFICIENCIES

Gross overall energy efficiency	=	gross heat output [heat output for the appliance itself plus a standard amount (4 feet) of exposed connector, if applicable]
		fuel energy input
Net overall energy efficiency	=	gross heat output plus heat gain from chimney minus heat loss due to consumption of room air minus standby heat losses
		fuel energy input

the chimney can be computed. Regardless of the details, the trend shows clearly in Figure 12-6.

The smaller the fire, the larger is the correction for converting from gross to net energy efficiencies. The energy penalty for air flow up the chimney becomes a larger fraction of the heat output with a small fire. For sufficiently small fires, the room-air heat loss up the chimney exceeds the heat output from the fire; the net efficiency then becomes negative! For no fire at all, but an open damper, there is no output but still an energy loss.

Thus, the net energy efficiency of a fireplace depends on its use—what size fires for how long, and such things as whether or not the damper is shut between fires. Typical fireplace use is illustrated in Figure 12-7; there is an evening fire, and the damper is shut the next morning. When the fire is well established, the net energy efficiency may be 10-20 percent, but when the fire is starting and dying down, the efficiency is much less. After midnight, there is a significant net heat loss until the damper is shut the next morning. In the example, the areas on the graph of net loss must be subtracted from the areas of net gain before computing the net energy efficiency of the full 24-hour period. The result is a net efficiency close to zero percent.

Taking into account other use patterns than the one outlined above, my guess is that most traditional masonry fireplaces operate with average net overall energy efficiencies between −10 and +10 percent. Most of the above discussion is based on tests at SER.[3] Although very few good measurements have been made of masonry fireplace efficiencies, the available information is in general agreement.

A direct experimental comparison between a closed stove and a fireplace was conducted by Tatnall Kollock in 1911.[4] Equal amounts of wood were burned, either in a fireplace or a stove. Three different rooms and two fireplaces were involved.

3. J. W. Shelton, "Measured Performance of Fireplaces and Fireplace Accessories," (Santa Fe, NM: Shelton Energy Research, 1978).
4. T. Kollock, "Efficiency of Wood in Stove and Open Fireplaces," *Forest, Fish and Game*, 3 (1911), pp. 95–97.

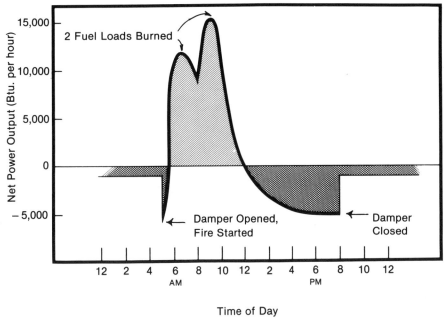

Figure 12-7. *Estimate of net power output from an open fireplace over 24 hours. Loss of warm room air up the chimney results in a negative net heating effect when the damper is open and there is no fire. Even with the damper closed, slight air leakage through the damper and heat conduction through chimney walls and fireplace backs in exterior walls results in standby heat losses. In this hypothetical example, overall net energy efficiency over this 24-hour period is close to zero.*

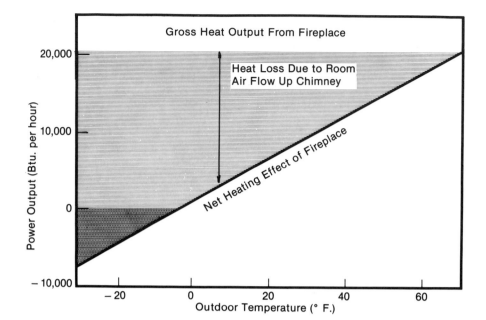

Figure 12–8. Estimate of the effect of outdoor temperatures on net heat output rate from a traditional open masonry fireplace. In this particular case, using the fireplace has a net cooling effect on the house when the outdoor temperature is below −3° F.

The stove was built of sheet metal and was called "Hot Stuff." Room temperatures were measured constantly, and the appliances were compared on the basis of how long each could maintain the room temperature above 65° F. Although not all precautions were taken or corrections made to ensure a fair comparison, the results clearly indicated the superiority of the stove, which kept the room warm 4–10 times longer than either fireplace. Since stoves are known to have efficiencies in the range from 40 percent to 65 percent, the above comparison suggests fireplace efficiencies are about 5–15 percent, which is consistent with other measurements.

One of the most careful and accurate studies was done by the University of Illinois Engineering Experiment Station, using a highly instrumented test house.[5] The investigators measured the change in gas consumption of the gas-fired forced hot air furnace as a function of the use of a fireplace. Coal, not wood, was burned, but the results with wood should be roughly similar. Careful monitoring of outdoor conditions made it possible to correct for varying weather. The fireplace damper was opened, and a fire lighted at 4 P.M. was kept burning briskly until about 7:30 P.M. By 10:30, the fire was reduced to glowing coke. Only a few glowing embers were left at 11 P.M. The fire-

place damper was closed at 7 A.M. This sequence was then repeated.

The results were very interesting. With the furnace thermostat located in a different room than the fireplace, the net effect on temperature was that the room with the fireplace was a few degrees warmer than the rest of the house. Gas consumption was slightly less than without the fire during the evening, but slightly more during the rest of the night until the damper was closed. The net effect was a slight (3–4 percent) increase in gas consumption – a negative net efficiency for the fireplace. Using a tight-fitting metal cover over the fireplace opening from 10:30 P.M. until 4 P.M. the next day decreased the nighttime air loss enough so that the overall gas consumption was

---

5. S. Konzo and W. S. Harris, *Fuel Savings Resulting From Closing of Rooms and From Use of a Fireplace*, Bulletin 41, No. 13 (November 16, 1943), University of Illinois Engineering Experiment Station Bulletin Series No. 348.

Increasing the Net Energy Efficiency Of Masonry Fireplaces

Here are 5 ways to increase the net energy efficiency of traditional open masonry fireplaces.

1. Build full-size fires.
2. Have long-lasting fires.
3. If the fireplace damper is adjustable, use it at the most restrictive setting that does not result in smoke spilling into the house.
4. Keep the damper shut when the fireplace is not in use.
5. Do not have fires on very cold days.

---

about the same as without using the fireplace (implying zero net efficiency).

With the furnace thermostat located in the same room as the fireplace, all the rest of the house dropped below the temperature setting when the fireplace was in use. (The heat from the fire kept the thermostat from turning on the furnace.) Less gas was consumed, but probably about the same amount less as if the thermostat had been set back to the cooler temperature of the other rooms.

There is no doubt a fireplace can at least keep 1 room in a house quite warm. But if this is the only heat source, the rest of the house will be cold. One can save on conventional heating costs and still be warm as long as living is concentrated in the room with the fireplace. However, much less fuel would be consumed doing the same heating job with a wood stove. In fact, the net energy efficiency of an ordinary open fireplace is so low that, in many cases, it would cost less to heat 1 room with a portable electric heater than to burn purchased wood in a fireplace.

Am I against traditional fireplaces? No. I enjoy one in my home. But I do not use it for heating – I have a stove for that job. I use it for the enjoyment of watching and feeling the radiation from an open fire.

FIREPLACE THEORY AND DESIGN

Fireplace designs that are intended to optimize heating effectiveness must take into account the 3 main ways a fireplace influences the energy balance in a house.

- Radiant energy from the fireplace heats the room it is in.
- Heat in the hot flue gases may conduct through the fireplace and chimney walls into the house.
- Warm house air is pulled into the fireplace and up the chimney.

The ideal energy efficient fireplace should maximize the amount of radiation it emits, minimize the amount of air escaping up the chimney, and have an exposed interior chimney (and fireplace back and sides). The room air loss is not only a direct loss of heat from the house, but its diluting effect decreases the amount of heat which will conduct through the chimney walls.

Maximizing Radiant Energy

It is useful to think of 2 sources of radiant energy coming out of a fireplace. Primary radiation comes directly from the coals and flames of the fire into the room, and secondary radiation is reflected and emitted from the fire chamber walls. Direct radiation is maximum for fireplaces which are especially shallow, tall, and wide (Figure 12-9), since with this shape less radiation is intercepted by the fireplace walls. Taken to the extreme, this geometry becomes a fire built against a flat wall, with a smoke-gathering hood far above it. The idea is to bring the fire into the room as much as possible (Figure 12-10).

The secondary radiation is mostly emitted radiation, not reflected primary radiation. Bricks are highly absorbing of infrared radiation, just as black objects absorb visible radiation. Most of the direct radiation that is intercepted by the fireplace walls is absorbed by them. Parts of the walls also are heated by direct contact with flames and hot gases from the fire. The hot bricks then radiate some of their energy; the rest of their energy is either conducted away from the fireplace through the masonry or is conducted into the air and/or gases that are moving through the fireplace.

The radiation from a surface travels in all directions and its amount increases with the temperature of the surface. To maximize the secondary radiation for heating a room without diminishing the primary radiation, 2 things can be done. The upper part of the back of a fireplace should slope forwards, and the sides should not be perpendicular to the front face, but angle inwards toward the back. This allows a larger fraction of the secondary radiation to get out of the fireplace and increases the intensity of the radiation because these slanted surfaces are closer to the fire and thus hotter.

In addition, the temperature of all the surfaces will be still higher if highly insulating materials are used in the construction of the fireplace. This makes the surfaces hotter by decreasing the heat conduction back into the structure of the fireplace. Since the kind of brick that is better at insulating is also physically weaker, a compromise is in order. Insulating brick used next to the logs should be covered with a layer of more durable firebrick. Loose, high-temperature-insulating material can also be used behind firebrick (Figure 12-12).

Figure 12-9. A larger portion of the direct radiation from a fire gets into the room from a shallow fireplace than from a deep, cavernous fireplace.

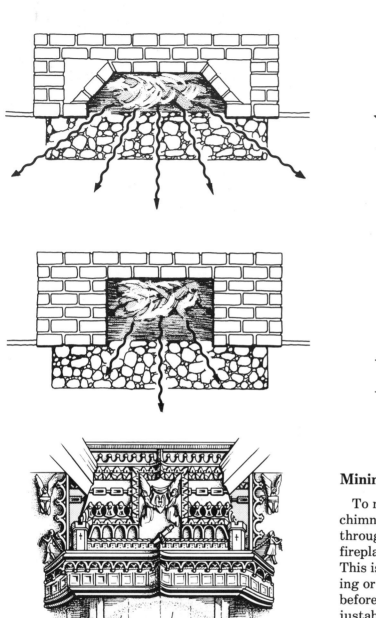

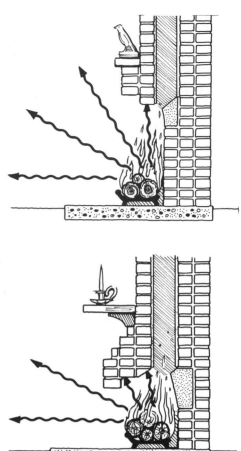

Figure 12-10. Many old fireplaces were very shallow. This 15th century fireplace was built against a wall with only slight protrusions for the sides and a hood overhead.

Minimizing Heated Room Air Losses

To minimize the loss of heated room air up the chimney and to maximize the heat conducted out through the chimney walls, the air flow into the fireplace should be reduced as much as possible. This is usually done, in part, by having a narrowing or a constriction in the throat of the fireplace before it enters the chimney. A damper with adjustable settings is ideal to meet the varying conditions of fireplace use.

The more challenging problem in fireplace design is to shape the fireplace so the constriction can be as substantial as possible without making smoke come into the room. The turbulence and eddies of the flames in a fireplace tend to eject smoke unless countered by an inward flow of air through the fireplace opening. The required average velocity of this air is about 0.8 ft./sec. This implies an air flow, for each square foot of fire-

Figure 12-11. The importance of angled sides for efficient indirect radiation. For clarity, the indirect radiation is shown coming from only 1 point; in reality it comes from all the surfaces of the fireplace. The angled sides not only radiate and reflect to a larger part of the room, but, being closer, they are hotter and emit more intense radiation.

place opening, of about 50 cubic feet per minute. The most obvious way to decrease the amount of air needed to prevent smoking is to reduce the frontal area of the fireplace. This requires making the fireplace narrower or less tall, which in turn inhibits the radiant energy transfer to the room.

Thus, the central compromise in fireplace design is between very open and shallow fireplaces, which are the best radiators, but may require relatively large total air flow to prevent smoking, and a more closed boxlike design with less frontal area for the same size fire, which decreases total air requirements, but intercepts much radiation. It is also probable that the necessary inward frontal air velocity to prevent smoking is not the same for all fireplace shapes. As is the case in stove design, the number of design variations and interactions is so large that thought and prediction alone are inadequate—experimental testing is essential.

To my knowledge, no careful testing has been done on these issues of fireplace design. Various

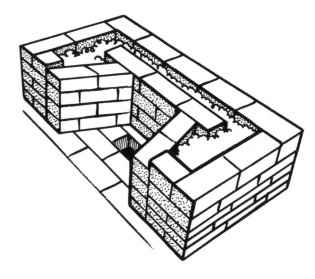

Figure 12-12. Adding insulation behind the back and sides of the firebox will make the fireplace more energy efficient.

people, most notably Count Rumford,[6] have given supposedly ideal shapes and dimensions of fireplaces (Figure 12-13). Since Rumford was, among other things, a scientist, it is remarkable that he apparently never published a paper on his scientific *evidence* that his fireplace designs were more efficient.

Most fireplaces are built with a "smoke shelf." It is my belief that under most circumstances the shelf serves no critical function, but is an incidental side effect of fireplace designs with forward sloping backs. Smoke shelves often are thought to be important for preventing downdrafts in chimneys.[7] However, smoke shelves are not used

6. Count Rumford, "Of Chimney Fireplaces," *The Complete Works of Count Rumford* (Boston: The American Academy of Arts and Sciences, 1875), pp. 484-557.

7. Paul R. Achenbach, "Physics of Chimneys," *Physics Today* (December 1949), pp. 18-23; Vrest Orton, *The Forgotten Art of Building a Good Fireplace*, Yankee, Inc. (1909), p. 44; David Havens, *The Wood Burner's Handbook* (New York: Harpswell Press, 1973), p. 93.

in the venting systems of any other kind of fuel-burning appliances (free-standing fireplaces, stoves, furnaces, water heaters, and so on). Many prefabricated fireplaces have no smoke shelf.

Smoke shelves *may* be useful in oversized chimneys, or when a fireplace has only a very small fire in it, particularly if the chimney has a side exposed to the outdoors. The upward velocity of the flue gases then may be quite small. This permits cool outdoor air to enter the top of the chimney. The air may or may not survive as a distinct entity all the way down to the bottom of the chimney, for as it descends it will mix with the rising warm flue gases, and it will be warmed by the chimney walls. If it does reach the chimney shelf, the shelf may help keep it, along with some smoke, from coming out into the house. Exterior masonry chimneys are especially susceptible to downdrafts; the relative coolness of their exposed side(s) can assist or even induce downdrafts against the cold flue surfaces.

Oversized chimneys are likely to cause smoking

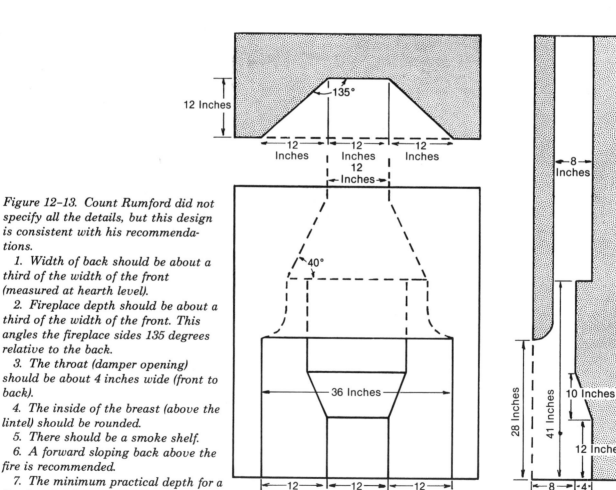

Figure 12-13. Count Rumford did not specify all the details, but this design is consistent with his recommendations.

1. Width of back should be about a third of the width of the front (measured at hearth level).

2. Fireplace depth should be about a third of the width of the front. This angles the fireplace sides 135 degrees relative to the back.

3. The throat (damper opening) should be about 4 inches wide (front to back).

4. The inside of the breast (above the lintel) should be rounded.

5. There should be a smoke shelf.

6. A forward sloping back above the fire is recommended.

7. The minimum practical depth for a fireplace is about 12 inches.

whether or not there is a smoke shelf. The low velocity of the flue gases results in more cooling than is usual, which decreases the draft. The low velocity also makes a chimney more vulnerable to wind-induced draft problems. If there are down currents in the chimney (whether or not they reach the smoke shelf), they decrease the draft by their own cooling effect, and they physically impede the upward flow of the flue gases. If the only problem a smoke shelf might help prevent occurs with an oversized chimney, then the only general solution is a good chimney, not a smoke shelf. The best chimneys are appropriately sized, are not on an exterior wall of the house, and have caps.

Traditional fireplaces are not dead. They will continue to be built and used for a long time. It would be worthwhile to do some basic research on the energy efficiencies of various fireplace designs. If some designs are much more efficient than others, there is the potential for considerable savings of wood and conventional fuels.

FIREPLACE ACCESSORIES

What can be done with a traditional masonry fireplace to make it more efficient? The choice of products is wide, and so is the range of prices and effectiveness. Glass doors, heat exchangers, radiation-enhancing grates, inserts, and, finally,

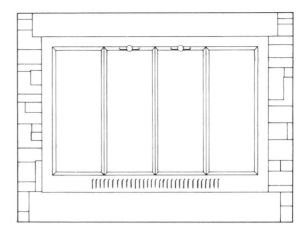

Figure 12-14. Glass doors on a masonry fireplace.

stoves installed in fireplaces will be discussed in this section.

Glass Doors

Glass doors (Figure 12-14) are the most misunderstood fireplace accessory. The usual rationale for glass doors is that they will allow a view of the fire while reducing the flow of room air up the chimney. What is often not realized is that with the types of glass normally used, the doors block much of the infrared radiation from the fire (Figure 12-15). With the competing effects of reduced heat radiation (bad) and reduced room air loss

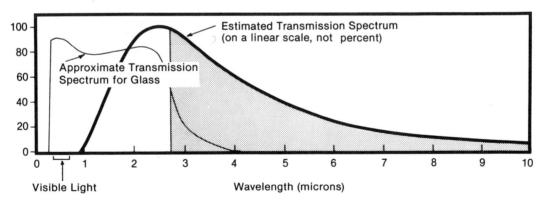

Figure 12-15. Approximate transmission spectrum for glass and rough estimate of the emission spectrum from a fire. The shaded area is approximately the portion of the radiation from the fire that does not get through the glass. Glass absorbs almost all radiation at wavelengths greater than 3 microns. The glass gets hot, but its own radiation into the room only partially compensates for the radiation it blocks. (The emission spectrum in the figure is that of a "black body" at 1,650° F.). This is not an accurate representation of a fire; the purpose of the figure is to illustrate qualitatively *the effect of glass on heat radiation.*

(good), the only way to know the net effect is to measure it.

SER conducted 2 all-day tests with glass doors closed over a simulated masonry fireplace, using typical fueling rates for the test fireplace of 10–15 pounds of wood per hour. There was a 40–50 percent reduction in total gross heat output compared to the same size fire in the same fireplace but with the glass doors open. The converse effect on heat output was also observed when the doors were opened (or closed) for 5–10 minutes during tests during which they normally were closed (or open) (Figure 12–16). The gross energy efficiency with the glass doors closed was about 10 percent, compared to about 20 percent with the doors open.

The reduction in air flow up the chimney was not as large as might be expected—it was only about 20–25 percent. (During these tests the air inlets under the doors were closed.) Whether this small amount of reduction in air flow is typical is

not certain, but I suspect it is. The total length of cracks or joints around the doors is typically large—10 feet or more is common in folding-type doors and the fit is rarely tight. When the main frame is mounted over a fireplace opening, an airtight seal against rough masonry can be difficult to achieve. A fiberglass strip helps, but fiberglass was never intended to be used as an airtight gasket, and it is not. The glass itself often is not sealed tightly in its frame. The draft (suction) tending to pull air in through the fireplace opening, or cracks in the doors, becomes quite large when the doors are shut.

In the SER experiment, correcting the measured gross efficiencies of 10–11 percent for the air flow yielded net energy efficiencies of about 4 percent, less than half of the net energy efficiency with no glass doors and at the same fueling rate. Even if our doors were unusually leaky, it seems unlikely, based on our results, that using any glass doors on any operating, plain masonry fireplace would result in an increase in net energy efficiency. The leakiness of most glass doors is attested to by the fact that in most cases with the doors and the air inlet dampers closed, a fire can continue to burn brightly. This indicates an ample supply of oxygen.

There are some clearly beneficial effects of glass doors. They are effective as spark screens for fire safety and for curing a smoking fireplace. They are also probably effective at reducing air flow up the chimney with a dying fire (or no fire) in the fireplace. As the fire dies down in the evening, the throat damper cannot be shut without risk of asphyxiation; but closing glass doors achieves some of the same objectives. But the effect of glass doors on net energy efficiency, with a normal fire in the fireplace, is almost always negative.

The best way to use glass doors is to leave them open when there is a fire and efficient heating is desired, and to close them over a dying fire or if a spark screen is needed. (According to preliminary tests at SER, a typical metal wire spark screen lets through most of the heat from the fire, and in this sense is better than glass screens.)

Glass doors have a tendency to get dirty with creosote when used. The film of creosote is usually sufficiently thin and/or irregular to still permit a clouded or partial view of the fire. For a clear view, frequent cleaning is necessary.

Different kinds of glass are used as fireplace (and stove) doors. The most common kinds are

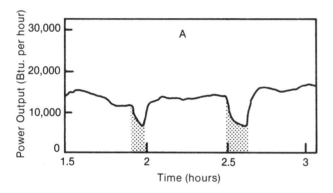

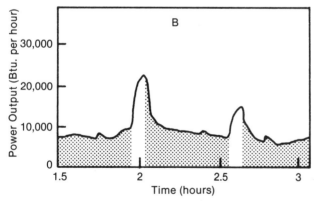

Figure 12–16. The measured effect on power output of momentarily closing (A) or opening (B) glass doors on a fireplace. The doors were closed in the shaded portions of the graphs. Having the doors closed resulted in about half as much heat output (gross and net).

tempered Pyrex, Vicor, and Pyroceram. (These are trade names used by the manufacturer who supplies most of the U.S. market—Corning Glass Works. Other companies, such as Nippon and Schott, often can supply approximately equivalent glasses under other names.) Tempering adds strength to Pyrex glass, but the tempering is lost at high temperatures. If the glass breaks, the tempering causes it to break into many small pieces. However, in a properly designed and used system, tempered Pyrex is suitable. Vicor is more expensive and is not a clear glass—it lets through light but does not give a clear image of the fire. But Vicor is much more durable at high temperatures than either of the other materials. Pyroceram is less expensive than Vicor and has a smooth, image-preserving texture. It also has a slight amber color.

FIREPLACE HEAT EXCHANGERS

In this book, a *fireplace heat exchanger* is a device installed in a fireplace which in intended to extract additional heat from the fireplace and which does not block the front of the fireplace with doors.

Tubular Grates

Tubular grates (Figure 12–17) were among the first devices on the market in the 70s designed to extract additional heat from an open fireplace. Room air enters at the bottom of each tube, is heated, and emerges from the top end where it is supposed to reenter the room. Blowers are sometimes used to circulate the air at a higher rate. Extension tubes (6–8 inches long) for the upper ends are available for some models to help insure that the heated air does not get swept back into the fireplace by the general air movement.

SER has tested a tubular grate in a simulated masonry fireplace using fueling rates around 15 pounds per hour. The tubular grate itself did not perform significantly differently than normal grates or no grate; but when used with the blower accessory or the tube extenders, gross energy efficiencies were significantly higher. Gross energy efficiencies were 27–30 percent, and net efficiencies were about 21–24 percent for the tubular grate with either the blower or the extenders. These represent increases of about 20–30 percent (5–8 percentage points) over a normal nontubular grate. Apparently without the extenders or the blower, much of the hot air output from the tubes *is* swept back into some fireplaces.

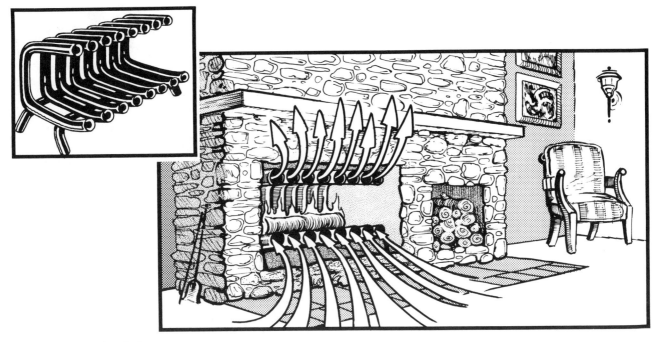

Figure 12–17. A tubular grate.

Tubular grates may not always need extenders or blowers to be effective. The grates should be installed with the tube ends flush with the fireplace opening or face, not recessed into the fireplace, and the tops of the tubes should be no more than 2 inches below the top of the fireplace opening. A fireplace with excessive draft may inhibit the hot air from entering the room. In general, my recommendation is to get either tube extenders or a blower with a tubular grate.

The quality of tubular grates and other fireplace heat exchangers is important for safety. These devices can burn out. If they do, smoke, carbon monoxide, and sparks can get into the tubes and be blown into the house. There is obviously a potential asphyxiation and fire hazard. But this is only a *potential* hazard. It can be avoided. Some heat exchangers are made of stainless steel, a material which is much less susceptible to burning out. Heavy-gauge ordinary steel is better than light-gauge steel. Periodic inspection can detect weaknesses before they become hazards. Also, smoke detectors are a good investment in all homes whether wood heated or not.

Other Fireplace Heat Exchangers

There is a wide variety of fireplace heat exchangers available (Figure 12–18). It is difficult to generalize about their performance because of the variety. But there are 2 general principles which you can use as an approximate guide, if you are considering buying one.

• The more total surface area of the device is exposed to the flames and hot coals, the better. For the amount of extra heat to be significant, the exposed surface area should be approximately equal to (or larger than) the area of the fireplace face.

• A reasonable amount of air flow is required through the device. This does not necessarily require a blower; but in some units where the air inlet and outlet are at the same height, blowers are necessary to get any significant air flow at all. But devices such as tubular grates work well on natural convection alone.

Blowers can serve 2 useful purposes. One is to better distribute and mix the hot air with the room air. In leaky houses, this is needed more than in tight, well-insulated houses. The other (and primary) purpose is to extract more heat from the fireplace heat exchanger. For this objective, it is difficult to judge informally the effectiveness of a blower because the temperature alone does not indicate the amount of heat. One must also know the air flow rate, and this is much more difficult to measure. A blower must increase the air flow substantially to be effective.

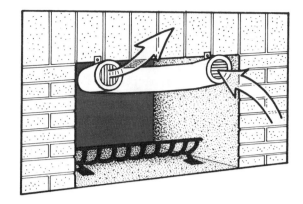

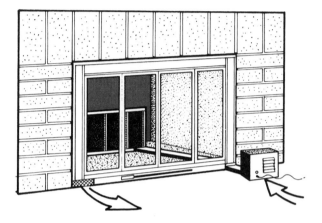

Figure 12–18. Fireplace heat exchangers. Many can be used with or without glass doors.

Fireplace heat exchangers which heat water instead of air are also available. Just as with air systems, surface area is important. The biggest difference is the potential in any water-heating system of a steam explosion. Safety and performance aspects of water-heating systems are discussed in chapter 11.

RADIATION-ENHANCING GRATES

There are a number of special grates on the market, intended to direct a larger fraction of the radiation from a fire, especially from the hot coals, out into the room. Most are 2-level or 2-layer grates that permit a more vertical array of wood, or a C-shaped array of wood. The idea is to expose more of the glowing wood directly to the room and to prevent so much of the radiation of the fire from hitting the back of the fireplace.

One unit tested at SER is designed to keep the fire constrained to a cavity formed by the wood (Figure 12–19). The wood itself then blocks radiation from going upwards, backwards, or downwards, focusing it out into the room. The unit was operated according to the manufacturer's instructions, attempting to keep the grate loaded with fuel. Because of the small size of the grate, this required quite frequent refueling.

The measured energy efficiencies of the grate were not significantly higher than for normal grates when fired to give the same power output. The unit performed at the upper end of the power output range of the fireplace because of the large amount of well-aerated fuel. Large fires seemed to be a necessary consequence of using the grate as instructed. Although in general, larger fires result in more energy efficient heating, using this special grate resulted in no higher energy efficiencies than using normal grates or no grates with the same fueling rate.

SER's tests of this radiation-enhancing grate were of relatively long duration, involving a number of refuelings, about 8 hours of firing at a normal rate, followed by a cooldown period. It was difficult to constrain the fire to the cavity region. Logs in the upper layer must be quite straight so that there are no gaps between them when they lie beside each other. Otherwise, the flames penetrate through and the fire becomes "normal."

Even with very straight logs, the fire penetrates through the upper layer at some point. Then the fire is again normal until the upper logs have burned away enough to permit reloading. But by then the whole fireplace structure has warmed up and a hot coal layer on the hearth is established. The charcoal layer is outside the cavity, so its radiation is not "focused." Also, the heat from the coals and the fireplace structure cause the whole next fuel load to ignite more quickly both inside and outside the cavity, again giving an unconstrained normal fire. The cavity fire geometry lasts the longest with the first fuel load. If the tests had been shorter (one load only), there might have been an effect.

The SER tests involved only 1 brand of special radiation-enhancing grate, and only the smallest model of this brand. Larger sizes and different brands might behave differently. But all such grates are limited in their potential effect since they only attempt to manipulate the radiant heat output of a fire, while a significant amount of heat is also in the form of hot gases rising above a fire. Also, they do nothing to limit the room air loss up the chimney.

FIREBACKS

Firebacks are (usually) cast iron plates installed on the back of a masonry fireplace to increase the heat output of a fireplace (Figure 12–20). I am aware of no scientific testing of firebacks. I would expect a small increase in energy efficiency.

Figure 12–19. Some designs for radiation-enhancing grates. The room is to the right in all cases.

Figure 12-20. A fireback installed in a masonry fireplace.

FIREPLACE INSERTS

The fireplace accessory that has the potential for improving a fireplace's energy efficiency the most is a fireplace insert – a device which both extracts heat and blocks off some of the room air flow up the chimney. But the range of products is large, and they do not all perform the same (Figure 12-21).

Principles of Insert Design

The heat exchanger surface area in the hot part of the fire should be relatively large – at least twice as large as the fireplace face. With shell-type inserts, if air is allowed to circulate around the entire shell, the entire combustion chamber shell is a heat exchanger. Many tubular grate units also have large heat exchanger areas. Inserts with heat exchangers restricted either to the grate area or to the area above the flames will not perform as well generally. A few square feet of heat exchange surface in an average size fireplace is not enough to be very effective.

Heat loss from an insert into the surrounding masonry should be minimized. A layer of insulation as part of the insert, or applied to the fireplace surfaces before installing an insert, is effective.

Inserts should be relatively airtight. Airtightness actually has 3 potential benefits.

- Controls the fire and, hence, the heat output.
- Limits the amount of warm house air that is lost up the chimney.
- Limits the amount of excess air in the combustion chamber, which in turn improves the heat transfer efficiency in the fireplace.

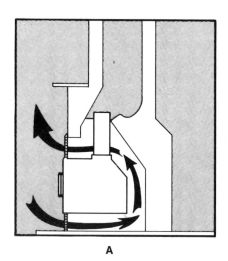

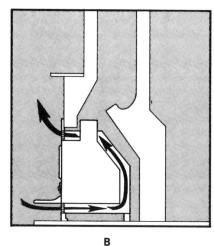

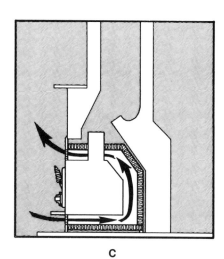

A   B   C

Figure 12-21. Three common insert designs. C allows the least leakage of heat into the surrounding masonry; A has the most leakage. The performance of single-wall inserts (A) is improved if the fireplace is lined with a high-temperature insulation (such as ceramic-fiber insulating boards). If the fireplace and chimney are interior and exposed, then neither the insulation nor the design type is critical since much of the heat absorbed by the masonry will ultimately heat the house.

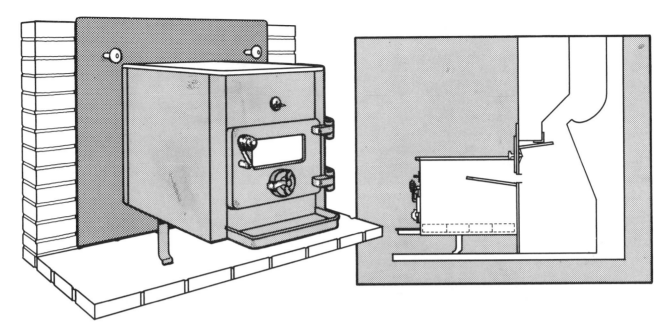

Figure 12–22. An "outsert," or hearth stove, designed to be installed just outside, rather than inside, a masonry fireplace.

The glass doors on many inserts are not very tight, particularly if they are mounted in light-weight frames, as is true for most glass door accessories alone. However, inserts built more like heavy steel or cast-iron stoves can be very tight. There are of course exceptions to both the above generalizations.

A relatively tight throat or flue-collar damper can achieve some of the objectives of airtightness in an insert (or any other wood heater), by reducing the burn rate and reducing the excess air enough to improve energy efficiency. SER tested a zero-clearance fireplace to determine which of 3 operator controls was most important for reducing air flow up the chimney: closing the glass doors, closing the air inlets at the bottom of the doors, or closing the throat damper.

In the "closed" position, there is roughly a ⅛-inch crack around the damper plate. As is true for most glass doors, closing the doors did not have a big impact on the amount of air entering the fireplace. Because of the leakage around the doors and possibly elsewhere, closing or opening the air inlet had little effect. But closing the throat damper substantially reduced the air flow.

The throat damper could not be closed without prodigious amounts of smoke coming into the room, unless the glass doors were closed as well. The net effect on this particular appliance was a substantial increase in net overall energy efficiency when operated with both the doors and the throat damper closed.

Tests at Auburn University indicate that most prefabricated fireplaces and most inserts have lower efficiencies when their doors are closed, despite the provision for convective heat transfer from the surfaces inside the fireplace. The key to turning door-closing into a net gain is limiting the excess air to the fire, as well as having adequate surface area and air flow for heat transfer.

The third design fundamental for inserts is that the air flow through the heat exchanger must be adequate. This may, but does not necessarily, require a blower. Natural convection can be adequate if all the air flow passages are relatively large and open. But in inserts, because they are made to fit into existing fireplaces, there is not always room for large air passages. Thus, blowers usually help.

Measurements of insert performance indicate that the best ones substantially improve the energy efficiency of a traditional masonry fireplace. Energy efficiencies can be as high as 40–45 percent. Higher efficiencies than this may be possible, but it is difficult to approach the efficiencies of average stoves: 50–60 percent. The radiant-plus-convective heat transfer from a radiant stove tends to be more efficient than the mostly convective heat transfer in inserts (and circulating stoves), and inserts do not have the exposed stovepipe that typically contributes 10–30 percent of the total heat output of stoves.

INSTALLING INSERTS AND STOVES INTO EXISTING FIREPLACES

Many people have found out too late that installing an insert or a stove into an existing fireplace is not as simple or effective or safe as they at first thought. The job can be done right, but it takes more effort and costs more than you may expect.

Potential Problems

• Fireplace chimneys are usually oversized for inserts or stoves. Fireplace flues are designed for large, open burners. Most inserts and stoves do not use as much air, particularly when operated with their doors shut, and so do not need such large flues. Oversized flues result in decreased draft and increased creosote accumulation. The large surface area of the chimney and the low velocity of the smoke cause the smoke to cool off excessively.

• Most fireplaces are built into exterior walls of houses. In winter, the cold outdoor temperatures can worsen the draft and creosote problems. The draft problems can be so severe that the whole system can go into reverse; air flows down the chimney, and the smoke comes into the house through the air inlets. Multistory houses with exterior chimneys are most susceptible. In less severe cases, it is just not possible to operate the stove or insert with its doors open without smoke spilling out into the house.

• A related problem with fireplaces in exterior walls is that the inside surfaces of the fireplace can be cold in the winter because brick is not a very good thermal insulator. In installations where smoke from the insert or stove discharges into the fireplace chamber itself, chimney sweeps sometimes find creosote deposits as thick as 3 inches on the back of the fireplace. This creosote is often the most difficult kind to clean—a hard thick glaze of tar.

• Most inserts and stoves are installed in fireplaces in a way that makes it difficult to inspect the system for creosote accumulation. One must either look down the chimney from the roof (which doesn't allow inspection of the fireplace itself) or remove the insert or stove. If inspection for creosote is awkward, it is less likely to be done; this indirectly increases the probability of a serious chimney fire. The *only* reliable way to know when a chimney needs cleaning is to look inside it.

• Inserts often are hard to remove. This increases the cost of every professional cleaning job, because a thorough cleaning usually involves removing the insert and then reinstalling it. A chimney sweep may need a full day to do a complete job. This same problem occurs with stoves that are vented into fireplaces and not connected directly to the fireplace chimney.

• Many old masonry chimneys are in poor condition and are not accessible enough for a thorough inspection.

• Some masonry chimneys, which are in good condition and are safe when used with an open fireplace, are not safe with an insert or stove installed in the fireplace. The brick chimney of an open fireplace does not get very hot, because considerable room air dilutes the flue gases, and because open fireplaces generally are used only occasionally for short periods of time. The same chimney may get much hotter when used to vent an insert or stove because there usually is substantially less excess air and because such appliances are likely to be used more. With an air-limited appliance, more creosote is likely; hence, chimney fires are more likely.

• Finally, in some insert installations, the combustion gases are vented directly behind the fireplace lintel. This area is normally kept relatively cool by the flow of room air in an open fireplace. Wood structural members are sometimes built into the masonry near this area. This wood will ignite if it gets too hot.

• Installations into prefabricated metal fireplaces involve some unique risks. Where house air flows around the combustion chamber and/or up nonflue passageways in the chimney for cooling, installing an insert or stove may block this needed air flow. Since thermosyphon chimneys are used with many prefabricated fireplaces, and since smoke density tends to be higher from inserts and stoves than from fireplaces, creosote accumulation can be substantial.

In some designs, the creosote may drip or fall into parts of the fireplace where it blocks needed cooling air in the fireplace and where, if ignited, it causes dangerously high exterior surface temperatures on the fireplace. Finally, regardless of the details, installing an insert or a stove in a listed factory-built metal fireplace usually violates the terms both of the listing and the manufacturer's warranty and is done at the owner's own risk.

Problem Solving

If you have a non-self-starting chimney where you must push a lighted newspaper up into the chimney to get the draft started in calm, cold weather, you have a chimney that is susceptible to flow reversal. Flow reversal is more likely when an insert or stove is installed in a fireplace because the lower flue-gas velocity means more cooling before the flue gas gets out of the chimney. It is also more dangerous because more smoke is likely to enter the house whenever reversal occurs. The principal danger is asphyxiation. Even if flow reversal does not occur, the appliance may smoke whenever the doors are opened.

There are 2 possible solutions.

- Install an outside air feed directly to the appliance.
- Install a draft-inducing fan in the chimney.

Adding a liner to the chimney will sometimes help, but it is not a guaranteed solution.

Outdoor air generally is under higher pressure than indoor air; so a duct feeding outdoor air to the fire usually helps prevent flow reversal. Outside air systems are most effective if the air is ducted directly to the combustion air inlet of the appliance. But many inserts and stoves have their air inlets in their doors, making such direct connections impractical.

A draft-inducing fan is an electric fan that forces or pulls the smoke up the chimney. This is a fairly drastic and expensive solution; but it is guaranteed to work if done right.

All things considered, in most cases I recommend just not trying to use a non-self-starting chimney for an insert or a stove.

For most of the other potential problems there are reliable solutions.

Relining a masonry chimney has substantial benefits. It is, of course, necessary if the chimney does not have a good liner already; but a new liner is advisable even where the chimney is in good shape. Relining typically improves draft, reduces creosote, and increases the general safety of any chimney. The many ways to reline masonry chimneys are described in chapter 6.

Keep in mind that most relining systems reduce the flue size. The reduction may be so much that the original fireplace will no longer work without smoking. If the option of returning the fireplace to its original use is important, choose a relining system that is easily removed—such as stainless steel stovepipe without added insulation.

Often it is wise to connect an insert or stove directly to the flue rather than venting it into the fireplace (Figure 12–23). This keeps the smoke from touching the fireplace and throat walls, eliminating creosote in those areas. It also eliminates the possible hazard of hot combustion gases impinging on the upper forward portions of the fireplace.

Making a direct connection can be difficult. Fireplace geometries vary enormously. But increasingly, products are becoming available to

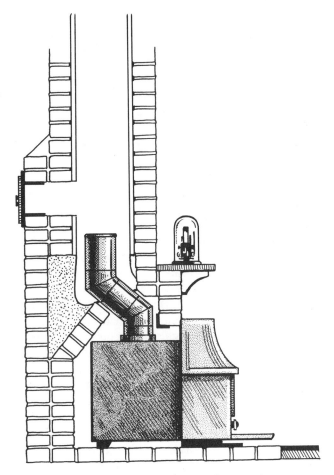

Figure 12–23. A direct connection from an insert up into the lined portion of the fireplace flue.

help. Some manufacturers make various oval, off-set, and T-shaped stainless steel components expressly for this purpose. The pieces do not fit every fireplace, but they do fit most. Standard stovepipe fittings in stainless steel are becoming widely available also. Flexible stainless tubing is used by some installers. Some insert manufacturers supply components for a direct connection.

To prevent creosote flakes falling from the chimney down onto the insert and accumulating around the insert in the fireplace, a seal is necessary around the direct connection. A sheet metal or cast-iron plate at the original fireplace damper location is common. When the chimney is cleaned, it will be necessary to clean the debris off the upper side of this plate. To maximize the convenience of cleaning, without having to remove the insert, this plate should be installed as high as possible so that its topside is accessible when cleaning from above (see discussion below).

In systems with a direct connection from the appliance to the tile liner, but without an additional liner as well, creosote debris will accumulate both on the sealing plate and on the original fireplace smoke shelf. This happens during normal operation and during cleaning. In most installations the insert or stove must be removed to gain access to these areas, unless a cleanout door is installed on the back wall of the chimney, above the smoke shelf. The door provides access to the shelf and to the lower part of the flue and upper part of the fireplace throat.

If the flue is relined in addition to providing a direct tight connection between the appliance and the new flue, a new cleanout door usually is not necessary. Debris cannot fall onto the smoke shelf—instead, it falls through the liner and into the connector or appliance, bypassing the shelf. The liner usually can be cleaned by reaching down into it with brushes from the roof. It should not be necessary to remove your insert or stove. Remember, however, to clean out the stovepipe connecting the appliance to the flue. Special brushes are becoming available to clean out the oval fittings used in many direct connections.

Stove Installation Options

You have more options when it comes to installing stoves compared to inserts. Three common ones are shown in Figure 12-24. The best is shown as "A." The more common installations are

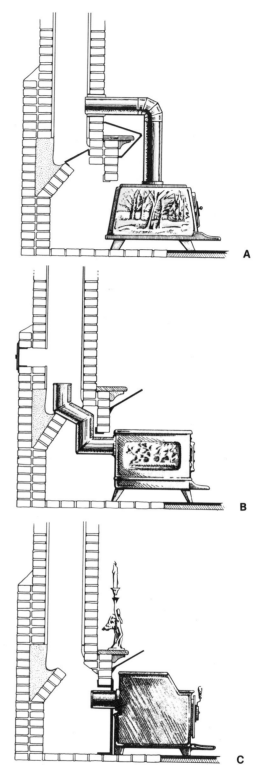

Figure 12-24. Three common ways to install stoves into masonry fireplace flues. A is best. B is acceptable; the area around the pipe where it passes through the damper location should be sealed. C is not generally recommended.

shown as "B" and "C." Installation "A" usually puts the stove further out into the room and requires that a hole be made through the masonry chimney, but the advantages of this installation are substantial.

- Better draft, due to the vertical segment of appropriately sized, hot stovepipe.
- Better energy efficiency due to more exposed stovepipe.
- Better accessibility for inspecting and cleaning the chimney.

Installation "C" is the least satisfactory and is particularly inadvisable in an *exterior* fireplace/chimney. Because the back of the fireplace can be cold, creosote condensation can be a substantial problem. In some cases, liquid creosote has reportedly flowed out of the fireplace onto the floor. Fireplace "outserts" are designed for installation "C" only.

A safer installation still would be "A" with the addition of an added liner from the breaching to the top of the chimney. This addition would be particularly beneficial with an exterior chimney. With an interior chimney in good condition, which is kept clean of large creosote deposits, this extra liner is not as important.

Only installation "B" is advisable in the case of a factory-built fireplace. Here again, adding a stainless steel liner all the way to the top of the chimney cannot hurt. However, it is best *not* to install stoves or inserts into factory-built fireplaces.

General Recommendations

My overall recommendation for installing inserts or stoves in an existing masonry fireplace is to provide a new, durable, continuous flue passage all the way from the appliance to the top of the chimney (Figure 12-25). This is not cheap, but it solves many safety problems. There will be cases where this is not necessary—interior chimneys in good condition with adequate clearances, and an operator who does not overfire the appliance and who always cleans the creosote before it reaches dangerous levels. Safety is a matter of degree. But since operating and maintenance habits are difficult to predict, since most fireplace chimneys are exterior, and since complete inspections of most masonry chimneys are not possible, a new continuous flue passage often is advisable.

Figure 12-25. A direct connection and new liner (modeled after the Dura-Vent system).

MASONRY FIREPLACES WITH STEEL SHELLS

In the last few decades a number of masonry fireplaces have been (and are still being) built with a built-in heat-extracting steel shell (Figure 12-26). Room air circulates behind the shell, is heated, and comes back into the room. This is not an accessory that can be added to an existing fireplace.

These shells function like many heat-exchanger and insert devices. If provided with adequate size registers (and passageways) for the air to get in

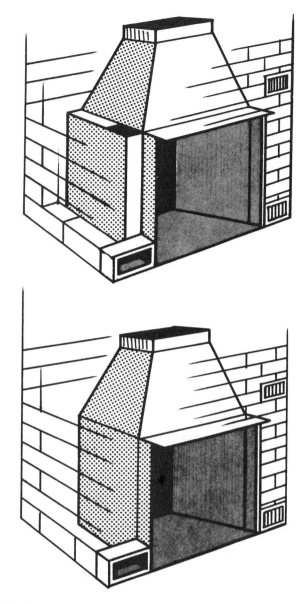

Figure 12-26. Steel fireplace forms. Some have a single-wall construction; others have 2 walls with built-in passageways for air circulation.

and out, there is little to be gained by having a blower. There are often 4 registers, each measuring 6 inches by 12 inches.

Although I am aware of no tests, my guess is that these steel shells increase the energy efficiency of fireplaces a little, perhaps by 5 or 10 percentage points, which is roughly comparable to effective fireplace heat exchanger accessories. Their surface area is large, which is good. But considerable heat is lost through the masonry structure if it is exposed to the outdoors, and most are; thus, not all the heat which conducts

through the steel shell contributes to heat the house.

Accessories and inserts should change the efficiency of steel shells about as much as they would in ordinary masonry fireplaces. The effect of glass doors usually will be negative, that of radiation-enhancing grates will be about neutral, and using heat exchangers, inserts, and stoves will usually improve the efficiency.

FACTORY-BUILT FIREPLACES

Factory-built prefabricated fireplaces can be divided into 3 basic categories: freestanding, built-in and of average design, and fireplace furnaces. Most are built predominantly of sheet metal; insulation, refractory liners, and glass doors are common components.

Freestanding Fireplaces

Freestanding fireplaces traditionally have been thought of as decorative, not as heaters. The most common shape is conical, either open all around or with a covered opening on 1 side (Figure 12-27). Many have no doors or other means of

Figure 12-27. Freestanding fireplaces.

sealing the opening. I am not aware of any energy efficiency measurements on such units. I suspect many of them are more energy efficient than traditional masonry fireplaces because of the exposed sides of the combustion chamber and the exposed chimney, but this depends on the structure of the fireplace and its chimney. To preserve the appearance of the colored finishes often used, the units and their exposed flues are sometimes of multiwall construction. This keeps the exterior cool, which is good for preventing discoloration of the finishes, but it also keeps heat from getting out into the room.

One interesting unit is enclosed in glass on all sides (Figure 12–28). Substantial air enters between the panes, but in such a way that most of the glass is kept relatively clean by being bathed in air, and a helical flow pattern develops which is interesting to watch. I am aware of no energy efficiency measurements on this design. I would not expect efficiencies to be much different from other conical freestanding fireplaces.

Built-In Prefabricated Fireplaces

In addition to freestanding fireplaces, there are built-in prefabricated units. Most of these are *zero clearance* fireplaces (Figure 12–29)—metal fireplaces with enough insulation and/or spaces and cooling air flow through these spaces that the back and, in some cases, the sides can be placed safely directly against a wood wall, with zero clearance between the unit and the wall. You can then build around them, using any normal construction materials, and end up with what can look like a traditional fireplace. Prefabricated chimneys are used. It is important to use the manufacturer's specified chimney brand.

Prefabricated fireplaces are increasingly common in new construction, because the installed cost is less than that of masonry fireplaces. They are also lightweight enough that they can be installed on any normal floor; this is especially important in multistory houses, apartments, and condominiums.

Prefabricated fireplaces are not difficult to install, even in existing houses; but it is critical to follow the instructions. In a few parts of the country, zero clearance fireplaces have wrongly acquired a reputation for being dangerous. As is true with all wood heating systems, poor installations cause most house fires, and inadequate maintenance and negligent operation cause most of the rest.

Figure 12-28. A freestanding fireplace with glass on all sides.

Figure 12-29. A zero clearance fireplace.

Prefabricated fireplaces have the potential for high energy efficiency—better than 50 percent in my estimation. Unlike inserts, which are constrained to fit into a tight space, built-in metal fireplaces can use the space above the combustion chamber for additional heat extraction, somewhat like the stovepipe in most stove installations.

However, most built-in prefabricated fireplaces do not have high energy efficiencies because it was not considered important in the past. Better designs are now available. These are the features to look for.

- Large heat-transfer surface area.
- Reasonable heat-transfer air flow either by natural convection or via a blower.
- Reasonable airtightness, achieved with tight doors or with a relatively tight throat or flue-collar damper.

Some prefabricated fireplaces have provisions for ducting some of the heat output to other rooms. If the fireplace is efficient and is to be used very much, this can be useful; it avoids overheating the fireplace room and allows the fireplace to be more fully utilized. In fact, in some designs, house air is pulled in at the base, circulated around the combustion chamber, and then this heated air is thrown away up through part of the chimney! This is really a double heat loss—room temperature house air is lost, and heat from the back and sides of the fireplace is wasted. The designers only had in mind keeping the outside of the fireplace from getting dangerously hot.

Some built-in prefabricated fireplaces have provision for outside air for combustion and for heat transfer (Figure 12–30). Outside *combustion* air is unlikely to improve net energy efficiency very much; but outside *heat transfer* air is another matter. It certainly should reduce floor-level drafts in front of the fireplace by eliminating most of the flow of house air towards the fireplace. In addition, other rooms in the house will not get as cool. Bringing outside air (warmed in the fireplace structure) and forcing it into the house tends to pressurize the house slightly. This reduces cold air infiltration.

However, the effects on net energy balance in the house are not so clear. The cold air passing through the fireplace heat exchanger picks up more heat than room temperature air would be-

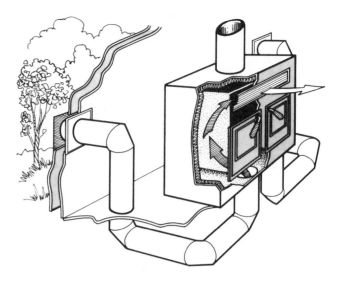

Figure 12–30. A prefabricated fireplace with outside heat transfer air.

cause the temperature difference between the hot steel and the cold air is larger. But the key issue is whether or not introducing this outside air into the house increases the total flow of outside air into the house. If the total flow of outside air into the house has increased, then so also has the flow of house air out of the house; this extra out-flow is an extra heat loss or cost which should be subtracted from the energy output of the fireplace to get the net effect.

Fireplace Furnaces

Fireplace furnaces (not a precisely definable category) are the natural next step as designers strive for better energy efficiency and better heat distribution (Figure 12–31). They differ from ordinary prefabricated built-in fireplaces only in degree—more design effort is put into heat transfer and distribution.

Fireplace furnaces tend to be taller, often extending all the way to the ceiling. The capacity for ducting heat to other rooms is more of a standard feature than an option. Some units incorporate heat storage to make the heating rate steadier. Most use blowers.

Fireplace furnaces often differ from ordinary wood furnaces by having smaller fuel capacities and by lacking some of the automatic controls. If used for serious heating, the fuel consumption

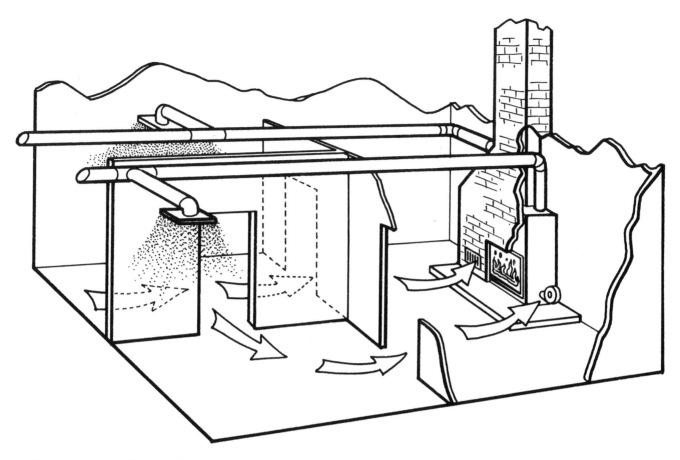

Figure 12-31. A fireplace furnace.

will be large – and the fuel storage and the errant bark, sawdust, and beetles can be a problem.

RUSSIAN FIREPLACES

Russian fireplaces are site-built, massive, heat-storing wood heaters. They are functionally equivalent to the European tile stoves, but usually with these differences.

- The fire is tended from the room in which the fireplace is primarily built.
- There is the option of directly viewing the fire.
- The exterior finish is typically ordinary rough masonry, whereas tile stoves have glazed tile exteriors.

Russian fireplaces and European tile stoves are discussed in chapter 7.

CHAPTER 13

CENTRAL HEATERS

A *central heater* is an appliance designed to heat the whole house uniformly. Its heat output generally is distributed by hot air in ducts or by hot water or steam in pipes. A hot air central heater is called a *furnace*. A hot water or steam control heater is called a *boiler*.

CENTRAL HEATERS VERSUS ROOM HEATERS

A central heating system can heat a whole house to a relatively uniform and steady temperature. Several stoves (or any other kind of room heater) can heat a whole house, but not as uniformly and not as easily – tending several small fires is more work than tending 1 larger one. Central heaters are especially appropriate in large, poorly insulated houses, where a large number of stoves would be required to heat the whole house. Also, prime living space is rarely used by a central heater; most often they are installed in basements or garages. Thus ashes, bark, and bugs are kept out of the main part of the house.

However, central heaters also have disadvantages compared to stoves.

• Their response time is slower; it takes a longer time after starting or revitalizing a fire on a cold morning to create warmth in the house.
• There is no extra-warm place in the house. With a radiant stove, no matter what your temperature is and no matter what the average temperature of the house is, you can always satisfy your desire for warmth by placing yourself a comfortable distance from the stove.
• Generally, central heaters consume more fuel than a single stove because they heat more of the house. (Their energy efficiencies are roughly the same as those of stoves.) A stove often heats only part of a house, and the homeowner lets the rest of the house be relatively cool, thus saving on fuel. The savings is mostly because part of the house is unheated, an option not available with many central heaters.
• Central heaters are virtually always more expensive to buy than a single stove, and usually more expensive to install. Even compared to the cost of several installed stoves, installed central heaters sometimes are more expensive.
• Most central heaters are useless during an electrical power failure.
• Central heaters do not contribute to the atmosphere of a home as a stove or fireplace can.
• The safety aspects of central heater installations are more complex. Thus, do-it-yourself installations have more risk.
• Creosote is usually more of a problem with central heaters because of their large fuel capacities and the automatic control of the combustion.

TYPES OF CENTRAL HEATERS

In almost all cases, a solid fuel central heater supplements the heat from a conventional source, such as oil or natural gas. The whole system generally is arranged so that the conventional fuel source kicks in whenever the solid fuel system is not meeting the demand.

Unless you are building a new house or are replacing your entire heating system, your choice between a furnace and a boiler will be made to match your existing conventional heating system. Most solid fuel boilers are hot water boilers —designed for hydronic heat distribution systems, such as hot-water baseboards. However, steam boilers are also available.

If you have a forced-air heating system, you have a wide choice of solid fuel central heaters. Solid fuel furnaces tend to be less expensive to buy than boilers. Installation costs often are less also, unless the existing duct systems need to be modified.

A gravity flow (or natural convection) furnace has the advantage that it requires no electrical power. Such units are also very quiet. Two-story or taller houses are ideal for such systems since hot air rises. Oversized duct work with no sharp corners is also necessary. The disadvantage to gravity flow furnaces is that some rooms will be cooler than others; but many people do not find this a disadvantage.

Add-on or supplemental furnaces and boilers are little different from regular central heaters except for their smaller capacities and, often, simpler controls. Add-on units are intended only to supplement the existing conventional system.

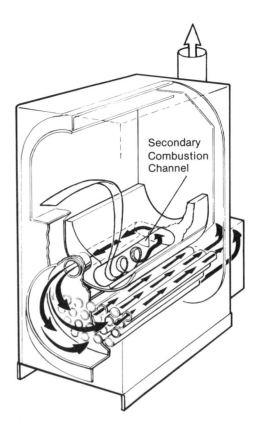

Figure 13–2. A wood-fired boiler with a highly turbulent and well-insulated secondary combustion channel at the base of the fuel chamber. Smoke enters the channel through a slot in the cast refractory. (The unit is an Essex boiler.)

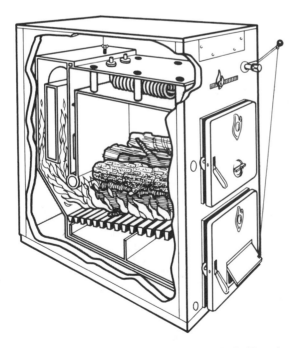

Figure 13–1. A wood-fired boiler (an H. S. Tarm).

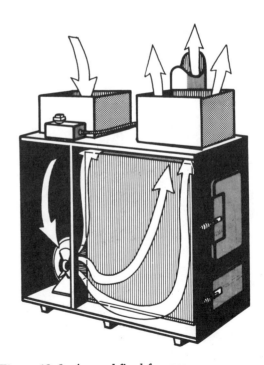

Figure 13–3. A wood-fired furnace.

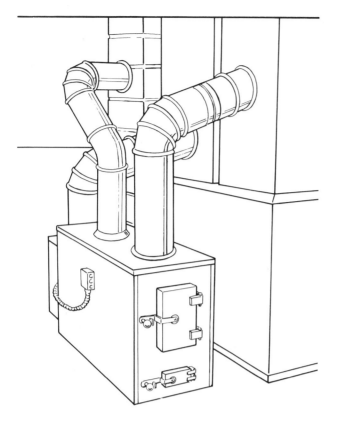

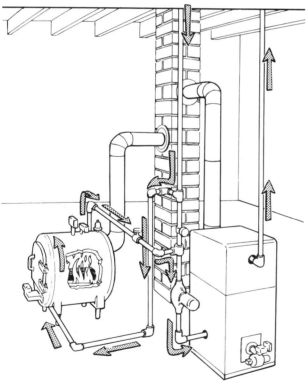

Figure 13-4. A supplemental furnace (top) and a supplemental boiler (bottom).

If your existing conventional central heater needs replacing, then a *dual fuel system* may be appropriate. Dual fuel usually means a solid fuel plus a fluid fuel (oil or gas) and is used in this sense in this book, although sometimes the term is used to mean wood and coal. (Actually wood plus electric systems are also available, but are very rare.)

Dual fuel units come with either single or separate combustion chambers (Figure 13-5). Separate combustion chambers can decrease the possibility of damage to the fossil fuel burner from logs, and lessen the chance of creosote fouling the fossil fuel burner. However, with proper engineering, these problems can be managed in a single combustion chamber unit. Separate combustion chamber units tend to be more expensive.

Before buying a dual fuel heater, compare prices to those of separate solid fuel and fossil fuel systems. It is not always cheaper to buy the dual fuel system. Keep in mind, however, that installation costs are usually less for the combined dual fuel units.

Most central heaters are manually fed, but stokers may be having a revival (Figure 13-6). Automatic coal stokers used to be fairly common. Typically, an auger powered by an electric motor feeds the fuel to the combustion chamber in response to a thermostat. The system can be fully automatic, except the ashes must be removed manually (and the grates shaken, in the case of

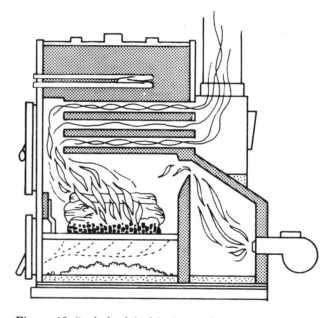

Figure 13-5. A dual fuel boiler with separate chambers for the wood and oil fuels.

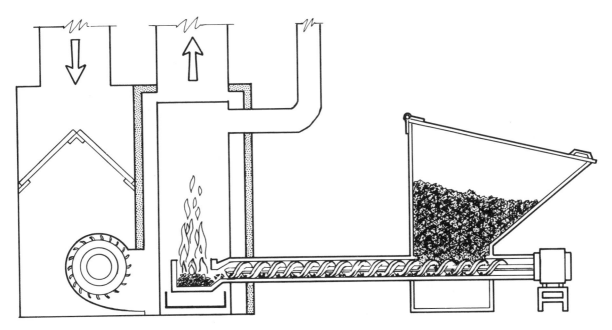

Figure 13-6. A coal stoker furnace.

some coal systems) and the fuel bin must be filled by the operator occasionally.

Although coal is the traditional fuel used in stokers, units are made that use wood chips and pellets.

The reason automatic stoking generally is associated with central heaters and not room heaters is that the complexity and bulkiness of such systems make them inappropriate for installations in prime living space.

EFFICIENCY

Central heaters have the potential for high energy efficiencies. They are not as constrained by size and aesthetics as are stoves. There is room for extra heat exchangers, separation of combustion and heat transfer, liners, baffles, and ungainly protrusions, if necessary.

Since most central heaters need electricity for heat distribution, they can take advantage of a wide range of electrically powered design options with the potential for improving energy efficiency, such as sensors and controls, forced convection heat transfer, forced draft for increased turbulence in the combustion chamber, and induced draft to help the flue gases move after being cooled in an efficient heat exchanger.

Most manufacturers are aware of these possi-

bilities. Some have taken advantage of them. But design improvements usually make appliances more expensive, so more efficient designs may not always be economically practical.

Very few central heaters have been tested for efficiency by independent laboratories using methods that are reliable and comparable to those used for stoves. There is not enough data to make comparisons. Even if the data were available, comparisons must be done very carefully, for the measured efficiency of the appliance in a laboratory and the efficiency of an actual installation may be quite different, for 4 reasons.

• Central heaters are not usually installed in prime living space. Thus the *jacket losses*—the heat given off by the exterior of the unit to the surroundings—usually are not fully utilized. Stack-loss methods of measuring heat output and efficiency presume the jacket losses are fully utilized; so efficiencies and outputs measured this way tend to be overestimated.

• The network of ducts or pipes that distribute the heat throughout the house loses heat. These duct or piping losses, again, are not all used effectively for house heating. Most laboratory measuring methods for efficiency and output presume there are no net losses from the distribution system; so again, laboratory efficiencies may exceed field efficiencies.

• All of the heat given off by the stovepipe connector in a room heater contributes to heating the room, and the contribution is significant. Again, since central heaters are not located in prime living space, not all of the heat output from connectors will be utilized. Laboratory efficiency measuring methods differ in whether or not the connector heat output is included.

• Finally, central heaters usually have *standby losses*; room heaters do not. Each time the demand for heat in the house has been satisfied, there is a period of time when the blower or pump is off. The fire is still on in the appliance, and there is residual heat in the pipes or ducts. Again, not all of this heat ultimately finds its way up into the living spaces of the house.

Considering these factors, the designs of most central heaters, and what little data is available, I expect that the net energy efficiencies of most central heating systems covers about the same range as the efficiencies for stoves.

INSTALLATION AND SAFETY

The installation of central heaters is considerably more complex than that of stoves or inserts; and there are additional safety aspects of the installation which are critical. Installations should be attempted only by the experienced, and then only in conjunction with a very careful reading of the instructions. I recommend reading chapter 5 in

Wood Heat Safety (Garden Way Publishing), which goes into more detail than this book and explains some precautions which are either not explained or not even mentioned in the instructions for some central heaters. Below are some of the most important safety concerns which are unique to or more important with central heaters. All the normal issues associated with stoves and discussed elsewhere in this book are also important with central heaters.

Plenum and Duct Clearances

Do not presume that you can use your existing duct system for the heat from a solid fuel furnace (either full size or supplemental). Many solid fuel furnaces are capable of generating much hotter air than are ordinary fossil fuel systems. And even if this is not true for a particular system under normal operating conditions, it is likely to be true during a power failure.

Even though power failures may not be common, it is vital that heating systems be safe if and when electric power is lost. Without electric power, the blower stops and the heat is not carried away very effectively. Temperatures in the plenum and duct work can rise very quickly and very high.

Mostly to handle this worst-case situation, required clearances between the plenum/duct system and combustibles are large (Figure 13–7). Very few existing heat distribution systems de-

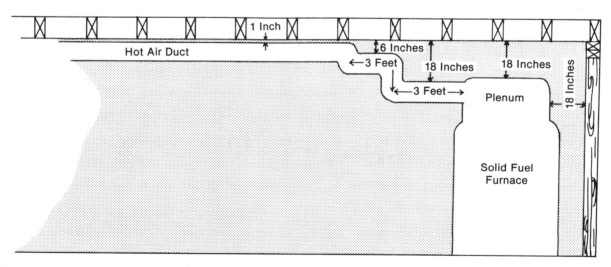

Figure 13–7. Recommended clearances to combustibles from the plenum and horizontal ducts of a hand-fired solid fuel furnace, according to the NFPA.

signed for gas, oil, or electric furnaces have (or need) these clearances because in conventional appliances, the heat source is usually turned off as soon as power is lost. Solid fuel fires usually cannot be cut off so easily, and there is more residual stored heat in any case.

Thus, existing duct systems usually must be modified to be safe with solid fuel furnaces. This may involve lowering the ducts, adding ventilated sheet metal shielding, adding insulation, or replacing the whole duct system.

Avoiding Negative Pressure In the Heat Exchanger

Few heating systems are perfectly airtight. This is of little consequence normally because the draft of the chimney provides suction to the appliance, pulling in air through leaks rather than allowing fumes to escape.

Furnaces use blowers to circulate air around the combustion chamber and/or its heat exchanger. The normal practice is to locate the blower before the heat exchanger, so that it *pushes* air through. This results in slight pressurization of the air passages around the combustion chamber and/or heat exchanger, which reinforces the tendency of the natural draft to force air in any leaks.

If a solid fuel furnace is installed in such a way as to create suction in the air around the combustion chamber and/or heat exchanger, fumes, including carbon monoxide, may be sucked out of the furnace and into the air stream which is being sent into the house.

Using the solid fuel furnace as a preheater for the return air to the conventional furnace is not appropriate, for then the blower in the conventional furnace will be sucking air from the solid fuel unit. Usually the solid fuel system should be on the hot-air side of the conventional furnace, or it should be parallel with the conventional furnace and be equipped with its own blower (Figure 13–8).

Again, consult the instructions and *Wood Heat Safety* carefully when installing solid fuel furnaces.

Avoiding Steam Explosions

Installing boilers is even more complex than installing furnaces. With boilers, the number 1

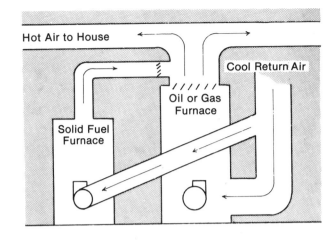

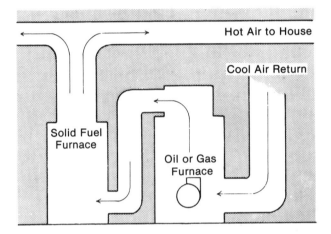

Figure 13–8. Two satisfactory ways to install a full-sized solid fuel furnace in association with an existing fossil fuel furnace.

boiler-specific safety concern is to avoid the explosion of the system due to the buildup of steam pressure.

This is so important that there are usually several lines of defense, all of which would have to fail before an explosion could occur. Below is a brief description of each.

• The fire in a boiler normally is thermostatically controlled with a temperature sensor in the boiler water (an *aquastat*). Before the boiler water becomes dangerously hot, the aquastat slows or stops the air supply to the fire.

• Many boilers are installed with a provision to dump heat if the boiler overheats. This overheat control system overrides the house thermostat and sends heat into the house regardless of whether the house needs heat.

• All boilers are equipped with at least 1 pressure and temperature relief valve. If the previous 2 systems have not prevented the boiler temperature from rising, then either excessive pressure or temperature will cause a valve to open which will release steam and/or water from the boiler. The discharge is usually near the floor level beside the boiler, although increasingly, the discharge is directed outdoors.

These are only some of the principles involved in boiler safety. I do not recommend that boilers or furnaces be owner-installed.

CHAPTER 14

OPERATING SOLID FUEL HEATERS

Both wood fires and coal fires are started by establishing a healthy wood fire.

Starting fires from scratch is not difficult if you have the right ingredients: dry wood in small pieces. Dry wood is preferable because moisture retards combustion, and small pieces help because of their larger surface area relative to their weights. Sufficiently small and dry pieces can burn all by themselves, singly, without any other burning material nearby.

Fires start and burn most vigorously in the regions *between* pieces of wood. When the heat from 1 piece of burning wood contributes to heating a second, the second burns more vigorously. The burning of the second, of course, enhances the burning of the first. The mutual heating between pieces of burning wood occurs by flames from one piece licking another, and, just as important, if not more so, by radiation both from the glowing charcoal and the flames from one piece being absorbed by another.

When establishing a fire, the pieces should be arranged to maximize both this heat feedback between pieces and to assure an adequate air supply between the pieces. Some effective geometries are illustrated in Figure 14-1. Details, such as piece sizes and the number of pieces needed, depend on how dry the wood is, on the type of wood, and on the experience of the fire builder.

Using wood with a high pitch content makes it easier to start a fire, which is why softwoods are sometimes preferred over hardwoods for kindling. Wood naturally saturated with pitch is sold as a fire-starting aid. It is called *fatwood* in some parts of the country. One or two pencil-size pieces are used in place of, or in addition to, kindling. The high pitch content results in vigorous flaming even in the absence of any surrounding burning material.

A number of manufactured fire-starting aids are also available, often in the form of packets or pellets; they are used and perform much like fatwood.

Sawdust and ashes, or a porous high-temperature clay ball, saturated with kerosene, make a traditional but somewhat dangerous, fire-starter. About a cupful of the kerosene-saturated ash or sawdust, or the saturated clay ball (which is attached to a metal rod handle), is placed under the wood and ignited. After the fire is established and the kerosene has burned out of the ball, the ball is removed. It is disastrous not to let the ball cool before placing it back in the kerosene pot! Although effective, using liquid fuels, even in these absorbed forms, is not considered safe. I do not recommend using these methods. Nor do I recommend using charcoal lighter fluid, which is intended for outdoor use only.

Since the use of *dry* kindling is generally the single most important factor in easy fire starting, it is sometimes useful to store a small amount of easy to split (or already finely split) wood in a warm, dry indoor location months in advance of the heating season. Such storage usually will accelerate drying.

Kindling geometries. Don't use too
many sticks too close together or
not enough air will circulate. Smaller
pieces ignite more easily, but take
more time to prepare.

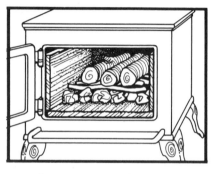

A common stove fuel load
ready for ignition.

A common fireplace fuel load
ready for ignition. The logs
should be selected and laid so
the flames can penetrate
between them.

A fuel loading for fireplaces
with neither andirons nor
grate.

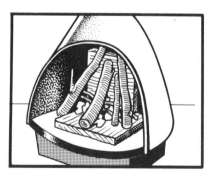

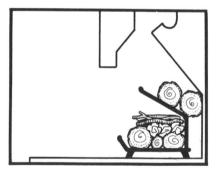

Some fuel arrangements that often require less kindling because
the fuel more completely surrounds the ignition location. In the
last two, lumber scraps placed below and/or beside the kindling
can increase the ease of fire starting. The last drawing
illustrates a special kind of grate that naturally creates a cavity
partially surrounded by wood.

Figure 14-1. Some useful fuel geometries for starting fires.

Starting and Maintaining Coal Fires

Most types of coal are harder to ignite than wood for a number of reasons. Many coals have less flaming tendency than does wood—their volatile content is smaller; this makes them more difficult to ignite. (Raw coal and wood do not *contain* volatile matter. Rather, when heated, the coal is chemically transformed into gases and aerosols, or volatile matter, and residual coke. The "volatile content" does not refer to contained volatiles, but rather to the potential for their production when the coal is heated.)

Coal does not lend itself to the favorable geometry of small pieces with adequate air spaces between, as with wood kindling. Small pieces can be used, but then the packing is too close; not enough air can get through. Finally, coal is denser than wood. This makes it easier for heat to conduct away from the surface into the interior of a piece of coal that you may be trying to ignite. This makes it harder to get the surface to ignite.

For all these reasons, coal fires generally are started with wood fires. Once a strong wood fire is established, coal can be added, but only a little at a time: 1–2 inches. The primary air inlet should be wide open. Even the ash door may need to be open, particularly when burning anthracite, and if the chimney generates little draft. After this first addition of coal has been burning for 5–10 minutes, a little more coal can be added. After a few such additions, with 5 or 10 minutes in between, a vigorously burning coal bed should be established. Only then can the stove be filled up to the recommended level. If the stove is filled with coal prematurely, the whole fire may go out. Then you have to wait for the stove to cool down, unload the fuel, and start the process all over again.

The primary air control and the ash door may be left open for a few minutes after the stove is filled. When the fire is firmly established, then the ash door must be closed, or the stove may overheat dangerously. The primary air control (or thermostat) is then also set to the desired position.

If you have the choice, particularly with anthracite, use smaller pieces of coal for starting coal fires; they ignite more easily as long as not too much is added at one time, and there are not too many *fines* (small pieces of coal that break or are ground off during shipping and handling).

Many people successfully use self-starting charcoal briquettes to start coal fires. Note that these do *not* need lighter fluid.

Reloading a coal stove is best done *before* the fuel bed begins to cool off. Then full loads can be added. If the fire has started to die, supply maximum air (ash door open if necessary) and add only small quantities of coal until the bed is vigorous again. Often when a full fuel load is added, it is advantageous to open the primary air control for a few minutes until the coal is clearly burning, then return the air control to the desired setting.

Fuel piece size affects coal fires. For the hottest fires, use larger pieces; they let more air through, which promotes higher combustion rates. For the longest-lasting fires, use smaller pieces; they restrict the air.

These generalizations apply primarily to anthracite and low-volatile bituminous coals. The lower rank coals are less touchy. Fuel beds need not be deep, and the fire has less tendency to go out. The lower the rank, the more coal behaves like wood.

Shaking Coal Grates

The accumulating ash at the bottom of the coal bed does not readily fall through the grates. Approximately 2–4 times a day, the grates need to be shaken (or the ashes sliced). If this is not done, the ashes will impede the air flow, and the fire eventually will go out.

Shaking the grates too much or too often is not desirable for 3 reasons.

• The ash bed keeps the burning coal from direct contact with grates, which preserves the grates.

• Overshaking wastes fuel. Unburned (or partially burned) coal will drop through the grates if shaking is excessive.

• Shaking too often encourages the formation of clinkers, as does poking or stirring the fire. These actions tend to let more air into the fire, making it hotter and thus bringing it closer to the ash-fusion temperature, and to bring the ash and the hottest parts of the fire in closer proximity, which facilitates the fusion of the ash into clinkers.

How often shaking is needed depends on the ash content of the coal, the burn rate, and the design of the appliance. Twice a day is typical. How much shaking is needed each time?

A rough guide is to shake until some substantial size pieces of live coals come through, and until the entire fuel bed is essentially free of ash. In practice, under-shaking rather than over-shaking is the more common problem. Under-shaking leaves too much ash, which then impedes air flow and slows the fire down. You will need to experiment with your stove to determine the appropriate amount of shaking.

Some coal appliances have an ash-slicing mechanism instead of grates. The effect and use are similar. In units with neither mechanism, a poker or similar tool can be moved around at the bottom of the fuel bed to shake down the ashes.

Slicing does not disturb the fuel bed as much as shaking. Slicing works well with coals with lower ash contents and where the ash is soft. Shaking is better for coals with coarser, harder ash.

Coal Clinkers

Clinkers are hard chunks of fused ash that form in the fuel bed. They are a problem when they are too large to fall through the grate. As they accumulate on the grate, they impede the flow of combustion air. The fire becomes sluggish and eventually may go out.

The only way to get these large clinkers out is from above. The fire must be out, and most of the unburned fuel removed to get access to the clinkers—a messy and frustrating job. Needless to say, large clinker formation is to be avoided.

Using coal with a low ash fusion temperature and a high ash content is the primary cause of clinkering, but other contributing causes include poking the fuel bed, burning coal with impurities, such as rocks and sand, and overfiring, such as by leaving the ash door open.

Coal Ash Removal

You can expect to remove coal ashes once a day. The ashes can contain red hot coals, so handle carefully. Put the ashes in a metal container with a tight metal lid outside the house, pending final disposal. Coal ashes are not considered a good soil additive.

Banking Wood and Coal Fires

To bank a fire is to extend the burn time. This may involve adding fuel, or burying glowing coals in ash or coal fires, and/or reducing the supply of combustion air. The objective is to assure easy revival of the fire when large amounts of heat are next needed. Fires often are banked in the evening for quick revival in the morning. With the right appliance and the right fuel, fires can last in a semidormant state for 2 days or more.

Details of banking depend on the appliance and the type of fuel. In an open fireplace, there is no chance of restricting the air supply. The most effective banking procedure is to bury the hot coal bed in ash a few inches deep by shoveling ash from the front and sides of the fireplace on top of the coals concentrated in the back. In the morning, the coals can be raked out of the ash, and a new fire can be established merely by throwing on more wood—paper and kindling are not necessary. Coals stay hot for more than 24 hours using this technique.

In an airtight wood stove, adding a large load of fuel and restricting the combustion air usually will result in an overnight burn. However, such operation usually results in considerable creosote and air pollution and is not recommended.

Methods for banking a coal fire depend on the design of the appliance and type of coal being used. Banking a coal fire merely may require placing a quantity of fuel in the combustion chamber. Sometimes piling the fuel higher around the sides than in the center is helpful. In hopper-fed coal burners, no special banking procedure is necessary other than filling the hopper. These units are specifically designed for long duration burns.

In any closed burner, wood or coal, reducing the combustion air flow extends the burn time. This can be accomplished by restricting the air inlet; by adjusting a barometric draft control; or by closing a stovepipe, flue collar, or other downstream flue-gas damper.

FUEL VARIABLES AND OPERATION

The moisture content of wood, size of the wood pieces, type of wood, and the size of the total fuel load all affect efficiency, creosote, emissions, and ease of operation.

Wood Moisture Content

The conventional wisdom is that the drier wood is, the less creosote is generated, the higher the stove efficiency, and the easier the fire is to start and maintain. Much of this conventional wisdom is true, but there are some interesting exceptions.

Some effects of moisture content are indisputable. Green wood is heavier than dry wood, sometimes twice as heavy, and so is harder to handle. Very dry wood is superior as kindling. Green wood is harder to burn; wood can be so moist that it cannot be burned by itself in most heaters.

Because green wood is harder to burn, it can be used to slow down the burn rate of a fire. Particularly in stoves without good air control, mixing some green wood with seasoned wood can extend the burn duration. The best proportion of green wood must be learned from experience. It depends critically on how moist the green wood is and the type of stove it is burned in. If too much green wood is used, the fire can die.

Using green wood rather than seasoned wood almost always results in a lower overall energy efficiency for 3 primary reasons.

• Some of the chemical energy of the wood is used to evaporate the extra moisture. Since in most systems the moisture is not condensed, this latent heat is lost. This decreases the heat transfer efficiency.

• Green wood requires more air to burn at the same rate. Thus, green wood burns with more excess air, which results in decreased heat transfer efficiency.

• Too much moisture in wood fuel decreases its combustion efficiency. Temperatures are lower and the combustible gases are more dilute because of the extra water vapor present.

In a stove, overall energy efficiency tends to be highest for seasoned wood (lower curve in Figure 14-2). Using green wood substantially lowers overall energy efficiency, by as much as 10 percentage points. The principal reason is poor heat transfer, not poor combustion. Combustion efficiency (the top curve in the figure) falls somewhat when wood is very green, but the larger cause for the drop in overall energy efficiency is poor heat transfer (the middle curve in the figure). The energy required to evaporate or boil the extra water in 35 percent versus 20 percent moisture content wood is 3.5 percent of the heating value of wood, which would cause a drop of only 2 percentage points in overall energy efficiency. More important is the effect of the extra excess air on heat

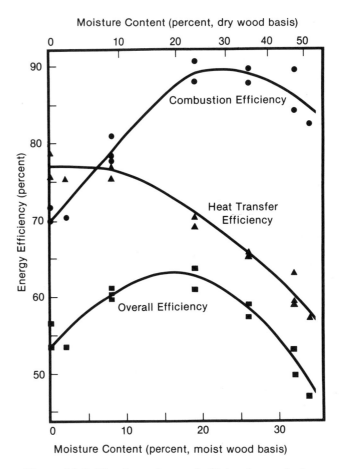

Figure 14-2. The dependence of efficiencies on fuel moisture content in an airtight stove. The air inlet setting was varied to maintain an average power output of about 17,000 Btu. per hour for all moisture contents. The fuel load volume was approximately constant. This figure illustrates test results for a particular stove.

transfer. Because green wood does not burn very easily, considerable air must be supplied, much of which blows by and around the fire, sweeping much of the heat up the chimney.

Perhaps more surprising is the effect of very dry wood on efficiency. In an airtight stove, using wood with a moisture content of less than about 15 percent decreases overall energy efficiency, compared to using wood at a more normal moisture content of 15–25 percent.

Why does using very dry fuel decrease combustion efficiency in an airtight stove or, for that matter, in any air-limited combustion system? Basically, the natural burning rate of very dry wood is very high. When this natural tendency is held in check by the limited air supply in a wood stove, large amounts of unburned smoke rise up the chimney. Since the smoke is gasified but unburned fuel, combustion efficiencies are low.

The smaller amount of moisture in very dry wood results in faster pyrolysis. Less heat is needed to evaporate or boil the moisture. When a very dry log is placed on a hot bed of coals, the heat penetrates to the core of the log much sooner than occurs with green wood. In fact, green wood can still be cool to the touch in its interior long after it has been placed in a fire, whereas very dry wood is soon charred to the core.

This quicker heat penetration allows the pyrolysis reactions to occur more nearly simultaneously throughout the whole fuel mass near the beginning of the burn cycle. A large amount of smoke is generated, much more than can be burned with the limited air available in an airtight stove. This chemical energy loss near the beginning of the fuel cycle causes the average combustion efficiency to be low. In contrast, with seasoned or green wood, the smoke evolution is more gradual so more of the smoke can be burned by the limited air supply.

Some other factors also come into play. Because very dry wood burns so easily, a switch to very dry wood without changing the air-inlet setting will result in an overheated house. To maintain thermal comfort in the home, manual air-inlet dampers must be turned down; thermostatically controlled appliances will automatically restrict their combustion air supplies. This exacerbates the problem; even less of the smoke surge will be burned if the air supply is more restricted.

The effect of fuel moisture content is the oppo-

site in open burners compared to closed burners. In an open fireplace or an open fireplace-stove, using very dry wood enhances combustion efficiency (Figure 14–3). There is always enough air available to burn any amount of smoke. In fact, the high rate of smoke evolution results in a larger, *hotter* fire, which results in better combustion. The net effect is that using very dry wood results in higher overall energy efficiencies.

Piece Size

When the wood is split very small so that a large number of pieces is needed to fill the stove, combustion efficiency is at its lowest. When 2–4 pieces constitute a full load, combustion efficiency is highest. Since heat transfer efficiency does not vary much over this range, the highest overall energy efficiency also occurs when larger pieces are used.

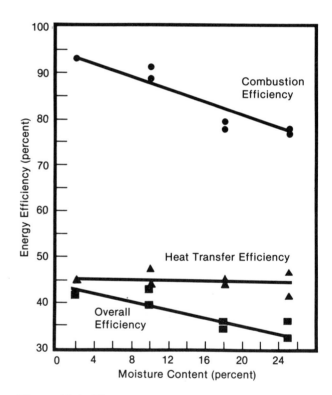

Figure 14–3. The dependence of efficiencies on moisture content in an open fireplace stove. Here power output could not be held constant but increased with decreasing moisture content. Fuel loads were of uniform size and were added when the previous load had been reduced to charcoal. (Moisture contents are on the moist wood basis).

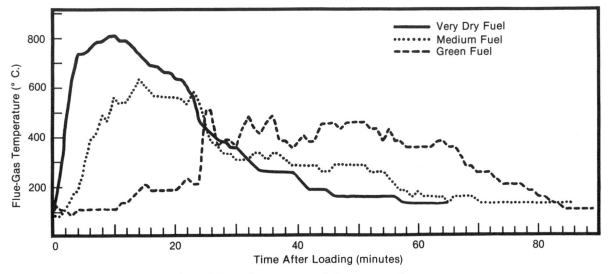

Figure 14-4. Flue gas temperatures about 1 foot above a stove being operated with its door open. Each curve represents the full burning cycle of a single load of fuel. The very dry fuel starts up more quickly. The green fuel took about 25 minutes to start burning vigorously. Pyrolysis and combustion are steadier with green wood.

A single, very large piece of wood in a stove cannot do much more than smolder, if that. The tests illustrated in Figure 14-5 used wood at about 15 percent moisture content. This is dryer than most people have available. With 20-25 percent moisture content fuel, the optimum piece size probably would be slightly smaller. But, the optimum piece size also may depend on the power output (which was held constant at a medium level for the tests illustrated in Figure 14-5). However, overall, you should not split pieces smaller than is necessary to obtain a reliable fire yielding adequate heat output.

Why do smaller pieces yield poorer performance? If the same total weight of fuel is to be used, then using smaller pieces results in more surface area for the fuel. There is quicker heating and quicker pyrolysis of the whole fuel mass. The consequences are then the same as in the case of very dry fuel: The resulting surge of smoke cannot be burned with available air supply, and much of it passes up the chimney unburned.

The effect of piece size is again the opposite in open burners from closed burners. Smaller pieces in an open burner burn with higher combustion rates, cleaner combustion, and higher overall energy efficiency.

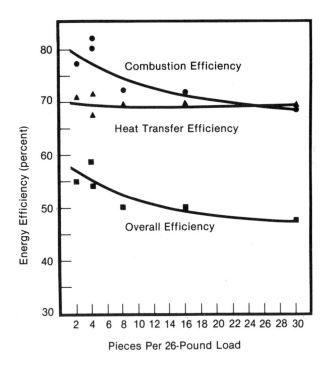

Figure 14-5. The dependence of efficiencies on fuel piece size in an airtight stove. The air inlet setting was varied to maintain a uniform average power output. Moisture content was uniform at around 15 percent.

Total Load Size

The effect of total load size on energy efficiencies has not been as well researched as some of the other fuel variables. In 1 experiment, smaller loads in an airtight stove resulted in about the same overall energy efficiency due to the combination of a higher combustion efficiency and a lower heat transfer efficiency. The higher combustion efficiency probably results from the smaller surge of smoke due to the smaller load, and the lower heat transfer efficiency probably results from higher excess air with smaller loads; more air tends to bypass a smaller fire than a large fire. Thus, in this experiment, smaller loads did not change the amount of fuel needed to produce a given amount of heat, and the increased combustion efficiency would doubtless reduce both emissions and creosote. But tests on a variety of appliances are necessary to determine if the result is general.

In a fireplace, load size has a different effect. Larger loads, of course, result in bigger fires and a higher power output, but the overall efficiency is also higher—more useful heat is derived from each log burned. Again, the key difference between closed and open burners is the unlimited air supply to an open burner. As a consequence, the higher peak smoke output resulting from larger fuel loads is not a liability; in fact, not only is there enough air to burn it, but the resulting higher intensity of the fire overall results in improved combustion with larger loads.

Wood Species

Efficiencies also depend on wood species, but again not enough experiments have been done to be able to predict results in all cases. However, it appears that when airtight stoves are fueled up to the same level (the same *volume* of fuel), then the less dense fuels (softwoods) tend to give higher energy efficiencies. This, again, can be understood on the basis of the size of the initial smoke surge. With equal fuel volumes, denser fuels will weigh more. More weight means a large surge of smoke simply because there is more fuel. Using low-density softwoods results in smaller loads on a weight basis.

In practice, the more important effect of species is on burn duration or refueling frequency. Since the denser hardwoods contain more fuel energy in a given volume load, a hardwood load will last longer. You can maintain the same average heat output rate with less frequent refueling. This is the primary reason for the general preference for hardwoods if there is a choice.

Fuel Variables and Creosote

What are the effects of these basic fuel variables on creosote? The long-standing conventional wisdom is that both green wood and softwoods result in more creosote compared to seasoned wood and hardwoods respectively. The truth is that there are some important exceptions to these notions.

Green wood *can* result in *less* creosote than seasoned wood, and very dry wood can result in more creosote than normal seasoned wood. Appliance type and operation can both have effects.

Carefully controlled experiments at Shelton Energy Research (SER) laboratories and elsewhere have clearly shown that in *stoves* or any other air-limited burners, greener wood results in less creosote under many operating conditions.

The results of an SER experiment are illustrated in Figure 14-6. Six identical, small airtight stoves were used, connected to six identical chimneys. All systems were operated simultaneously with the moisture content of the fuel being the principal variable. The chimneys were weighed before and after each experiment to determine the weight of the creosote deposited.

With the doors of the stoves left open to approximate open fireplaces, creosote increased with moisture content. Fuel at 34 percent moisture content resulted in about 4 times as much creosote per kilogram of fuel burned than did either 17 percent or 5 percent moisture fuel.[1] Open door operation confirmed the conventional

---

[1] The qualitative nature of this result is unchanged when compared on other bases. On a per-log basis (or oven-dry fuel weight basis), the 34 percent fuel still resulted in about 3 times as much creosote as the other drier fuels. Since overall energy efficiencies are lower in open burners with green wood, when creosote is compared on a per-unit-heat-output basis, the green wood results in more than 4 times as much creosote compared to the medium and dry wood. Finally, dry wood burns more quickly than green in an open burner. Thus, in a given time, more dry wood will be burned. But even on an equal time basis, the green wood resulted in about 3.7 times as much creosote as the very dry wood, despite the fact that about 46 percent more of the dry wood (in terms of number of equal volume logs) was burned.

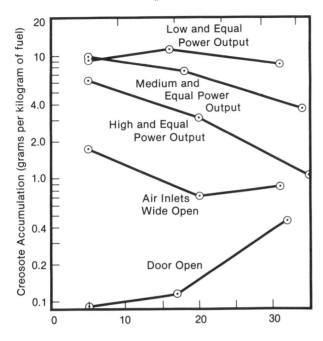

Moisture Content (percent, dry wood basis)

Moisture Content (percent, moist wood basis)

Figure 14-6. Creosote accumulation as a function of moisture content, using piñon as the fuel (a relatively pitchy pine).

wisdom, probably because it simulates the conventional appliance, an open fireplace.

In an open fireplace with an essentially unlimited air supply, the high pyrolysis rates with drier wood are not a liability—most of the smoke burns, and the resulting higher average temperature of the combustion chamber probably increases combustion efficiency generally, and hence, decreases creosote. Green wood fires are cooler, which is detrimental to combustion efficiency. Combustion efficiency may also be decreased due to the dilution of combustible gases by the extra water vapor from the fuel.

The upper 3 curves in Figure 14-6 correspond to typical airtight stove operation. Since the objective in operating the stoves was to be as realistic as reasonably possible, the air inlets on the stoves were not the same. Dry wood burns much more easily than does green (Figure 14-4). With roughly equal volume loads, the typical operator (or thermostatic control) will restrict the combustion air supply with the very dry wood to avoid overheating the house. Similarly, green wood requires more air to burn at about the same

rate as seasoned wood. Thus, during the controlled low, medium, and high-heat-output runs, the air inlets on each of the 6 stoves were adjusted to different settings to achieve the same average heat output rate.

For a *low-heat-output smoldering burn*, as is typical of overnight burns (the topmost curve), *roughly* the same amount of creosote was observed for all 3 moisture contents tested.

The results that are most startling in light of the conventional wisdom are represented by the second and third curves in Figure 14-6. With the stoves operated at medium or high power outputs, creosote accumulation was the highest for the driest wood and least for the greenest. At the high power output condition, burning 5 percent moisture content piñon resulted in about 6 times as much creosote as 35 percent moisture content piñon. The medium moisture content fuel (20 percent) resulted in 2.9 times as much creosote as the green (35 percent moisture) wood.

These results for *medium and high fires* in airtight burners are consistent with measured effects of moisture content on combustion efficiency discussed previously in this chapter. How can drier wood result in more creosote? Drier wood burns so easily that the air supply must be more limited to maintain the same heat output. Less air generally results in lower combustion efficiencies. In addition, the whole fuel load tends to undergo pyrolysis simultaneously and more quickly than with moister wood. This results in a larger surge of smoke, and since combustion tends to be air-limited under these circumstances, most of the smoke cannot burn for lack of oxygen. The result is more creosote with the drier wood.

There is another important factor as well. Dilution air in chimneys reduces creosote accumulation. The more air is mixed with the combustion products in the flue, the cooler the temperature, but the lower the concentration and the higher the velocity. Empirically, the net effect of dilution air is a reduction in creosote accumulation. The excess or dilution air was the least with the very dry wood (because the air inlet had to be more restricted to achieve the same power output); hence, the very dry wood gave more creosote.

Other mechanisms may be contributing also. These are probably less important and are even uncertain as to the direction of the effects. The amount of moisture in wood may affect the pyrolysis and/or combustion reactions, resulting in a different total amount of condensible products and a different composition for the product mix. Extra water vapor in the flue could influence the amount of organic material that condenses.

No large moisture content dependence was observed at the *low power output* firing condition. Under these conditions combustion was essentially flameless—only the charcoal burned and essentially all the smoke evolved in pyrolysis was vented into the chimney. With no significant combustion of the smoke, the roughly equal creosote deposits are not surprising.

The effects of species on creosote were investigated by repeating the equal-power-output tests burning oak instead of pine in airtight stoves. The results (Table 14-1) indicate more creosote from the pine for the medium and high power output conditions. The pine used was piñon pine, a relatively pitchy wood.

Practical Implications

The practical implications of all these results is not that woodpiles should be wetted down to keep them green. Although creosote may be reduced by not letting wood get too dry, energy efficiency

TABLE 14-1
SPECIES DIFFERENCES IN CREOSOTE ACCUMULATION

Fire	Fuel Moisture Content, Moist Basis (percent)	Creosote Accumulation (piñon as fuel) (gm/kg.)	Creosote Accumulation (oak as fuel) (gm/kg.)	Percentage Difference: Piñon Relative to Oak
Low	5	9.2	16.5	−44*
	15	11.0	16.6	−40*
	25	9.5	19.4	−52*
Medium	5	9.8	4.1	+139
	15	8.0	2.9	+176
	25	5.4	2.3	+134
High	5	6.4	2.7	+137
	15	3.9	0.88	+333
	25	2.2	0.40	+450

*The observed decreases may not be generally applicable.

is generally highest for normal seasoned wood—wood with a moisture content of 15–25 percent.

All things considered—creosote, energy efficiency, emissions, and ease of operation—I recommend seasoned wood for stoves and other closed burners. For use in open fireplaces and other essentially air-unlimited burners, I recommend wood that is as dry as possible (Table 14-2).

But for controlling creosote in a stove, there is something much more important than either moisture content or species, and that is combustion conditions, as affected by air settings and load size.

With oak at 25 percent moisture content as the fuel, we observed 7 times as much creosote with a medium compared to a high fire, and 48 times as much creosote with a smoldering versus a high

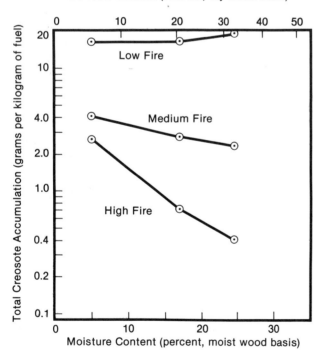

Figure 14-7. Creosote accumulation as a function of moisture content, using oak as the fuel.

TABLE 14-2
OPTIMUM MOISTURE CONTENTS

Type of Appliance	Overall Energy Efficiency	Creosote	Emissions
Open	very dry	very dry	very dry
Closed	seasoned	green	seasoned

Estimates for optimum moisture contents for maximizing overall energy efficiency and for minimizing creosote and emissions. It is presumed that in closed appliances, air inlets are set to achieve the same average power output for each fuel condition.

fire (Figure 14-7). Over the range of moisture contents usually available (about 20–35 percent on the moist basis), the difference in creosote accumulation ranged from negligible for a smoldering fire to a factor of 3 for high-power-output burns. The species dependence over the same moisture content and firing conditions ranged from very small to a factor of about 5. These are not negligible effects, but they are much smaller effects than the maximum factor of 48 due to firing conditions. As a practical matter, a hot, flaming fire is much more useful for reducing creosote than a particular species of wood or its moisture content.

To control creosote in closed burners, it is best to use relatively small fuel loads (typically 2–4 sticks of wood) and to adjust the air-inlet control for moderately large amounts of air, particularly during the initial (20 minutes to 1 hour) portion of the burn when most of the smoke-generating pyrolysis reactions are occurring. (Thermostats are often counterproductive for this purpose.) Using smaller fuel loads means that more frequent refueling will be necessary, but tests at SER suggest that no more fuel in total will be burned. The amount of visible smoke coming out of the chimney is a good creosote-potential indicator—operating stoves to minimize the smoke density will minimize the creosote buildup also.

Fuels Unsuitable for Solid Fuel Heaters

Along ocean coasts, driftwood can be an abundant and free source of fuel. The temptation is large to use it. Don't.

Ocean driftwood contains salt and many other chemicals that remain in the wood as the water evaporates. Salt is corrosive to all materials commonly used in stove and chimney construction: steel, cast iron, stainless steel, firebrick, and fireclay flue liners.

Figure 14-8. A cast-iron stove before and after being used for a season or two with driftwood for fuel.

TABLE 14-3
A SUMMARY OF SOLID FUELS

Chunk wood (cordwood). Usually the least expensive and most practical wood-based fuel.

Wood chips and sawdust. Generally used only in stokers; rare as residential fuel.

Densified wood logs, sawdust/wax logs, and cardboard logs. Common fireplace fuels; easy lighting, clean to store and handle, available in grocery stores, expensive compared to regular wood logs.

Densified wood pellets and briquettes. Used in some industrial systems; not a common residential fuel.

Rolled newspaper logs. Usually used in conjunction with wood. It is difficult to maintain a pure newspaper-log fire.

Densified agricultural waste. Rare; agricultural waste is generally more useful for soil conditioning and/or conversion to gaseous or liquid fuels.

Other densified biomass. Materials such as paper and food garbage can be made into fuel pellets.

Charcoal. Very clean burning.

Coal, natural. The least expensive of the coal-based fuels.

Coal, blocks. Processed coal in paper wrapping; features easier lighting and cleaner handling compared to natural coal.

Coke. Very clean burning.

Peat. Usable where available; substantial drying needed due to high natural moisture content.

Railroad ties and similarly treated wood, such as utility poles, are not recommended as fuels. Many decay-resisting treatments for wood contain metals and other ingredients that when burned could constitute a serious emissions problem. Creosote and odor are additional problems.

BACKFLASHING

Airtight solid fuel heaters are capable of spitting fire and/or smoke back at the operator when the door is opened (Figure 14-9). Such occurrences are clearly to be avoided!

If air is suddenly admitted in an airtight burner by opening the door, there can be a combustion surge of the gases that is so quick the resulting pressure can force smoke and flames out any available openings – the loading door, the air inlet, and cracks or leaky joints in both the stove and stovepipe. The phenomenon could be called a small explosion.

Backflashing happens when a substantial amount of smoke and air becomes premixed before ignition occurs; this premixing allows the flame to spread fast and create the mini-explosion. Thus, there is often a delay of up to a few seconds after opening the door before the backflash occurs. It is most likely to occur after a load

Figure 14-9. Backflashing from a stove, in this case induced in SER's laboratory.

of wood is added to a bed of hot coals and the air inlet is turned down.

Backflashing is dangerous; the emerging gases may be burning and can burn people nearby. To avoid backflashing and momentary leakage of smoke, fully open the air inlet for half a minute before opening the door. Then open the door slowly. Keep your face well back until the door has been fully open for a few seconds.

Backflashing can occur spontaneously, without the operator doing anything. On rare occasions, the pressure within the stove will blow doors open or blow cooking plates off. The ash and charcoal on the apron of the stove may be blown off by a flash from the air inlet above. In some cases, live coals or sparks may come out of the air inlet. This is 1 reason to have floor protectors of the full extent recommended by NFPA.

Spontaneous pulsating backflashing – small flashes occurring consecutively, separated by as little as a half second or as much as 5 seconds – is another possible occurrence. Each flash consumes the available oxygen and pressurizes the stove, momentarily preventing more oxygen from getting in while the smoke and/or flames are coming out. The next flash will occur when enough oxygen has reentered the stove.

The phenomenon of backflashing deserves respect and perhaps some research effort. Particular stove designs may be more or less susceptible than others.

ASH DISPOSAL

Every year many fires are caused by ashes emptied into cardboard boxes or paper bags. My uneducated intuition used to be that if there had been no fire in a stove or fireplace for at least a day or two, the ashes would not be hot enough to ignite paper or cardboard. Wrong! Coals buried in ash can stay red hot for days! Ash is a good thermal insulator and keeps enough oxygen away so the charcoal does not burn up. This is why it is important to empty the ashes from a stove or fireplace into a noncombustible container, such as a metal container.

Additional precautions are to keep a tight-fitting metal cover over the ash container and to keep the ash container itself off combustible floors and not touching any other combustible materials. The cover will keep any noxious fumes inside and will keep oxygen out, thus inhibiting combustion of the coals. Place the ash container outdoors.

If the stove does not have an ash drawer and a firebrick floor, leave about an inch of ash when cleaning to insulate the bottom. This prevents the stove bottom from overheating and burning out.

POWER FAILURES

What should be done with a solid fuel heating system if there is an electrical power failure? If the solid fuel heating system does not use electricity, go ahead and use it. But what if the solid fuel heater includes an electric fan, blower, pump, or controls? Will the system be safe without the electric power that is used to move heat via a fan or pump?

Without air (or water) circulating over heat transfer surfaces, those surfaces (such as the tubes in a tubular heat extractor) will get hotter. With a large fire in the appliance, as is likely if the central heating system is out, the high temperatures may burn out parts of the system. This then creates a hazard.

Most of the more sophisticated electrically assisted wood heating systems, such as furnaces and boilers, have an automatic shutdown feature; when power fails, the combustion air inlet closes, thus cooling off the fire.

Some systems are either designed so ruggedly or have such ineffectual fan assists that power failures need be of no concern. It is not easy to predict this in advance. The best general procedure is to check the manufacturer's instructions carefully for cautions about firing the unit when the fan or blower is not operational.

Many of the fans used in stoves and accessories are not designed to withstand high temperatures. If the temperature of the fan exceeds roughly 150–200° F., the unit may be damaged. Normally, the air pulled through the fan prevents the overheating. But during a power failure, or if the fan is not plugged in, the heat from the fire can overheat the fan. This is another reason to be cautious about using heaters with fans during power failures.

PERIODIC MAINTENANCE

Some maintenance is so obvious that no reminders are necessary; replacing broken glass and burned-out motors or switches, are examples. Some maintenance chores are easily forgotten, but they shouldn't be.

Door gaskets need occasional replacement to maintain peak effectiveness. Some people find that asbestos gaskets work best. Unfortunately, asbestos fibers are known to cause lung cancer, but it is not known whether the amount of exposure to asbestos from door gaskets presents a significant health risk. Alternative materials—glass and ceramic-wool gasketing—are available.

Clearances to combustibles need to be maintained. An initial installation with all the proper clearances does not guarantee safe clearances to everything forever. Rugs, furniture, newspaper, wood fuel, and matches should all be kept the required distance from the appliance. Growing vines, shrubs, and trees may need occasional pruning to keep them a safe distance away from a chimney.

Stovepipe connectors do not last forever. Corrosion is particularly a problem if general household trash is burned in the appliance (it shouldn't be), or if the inside of the pipe tends to get damp from condensation of flue gases or from rain or snow getting into the chimney, or if you burn coal. Burning driftwood, using salt-based chemical chimney cleaners, or just being on an ocean coast can also accelerate corrosion.

Stovepipe replacement may be necessary more than once a season, but more typical is once every few years. At least once a year, the strength and integrity of the pipe should be checked. Tap lightly with a small hammer or poke with a screwdriver to reveal where the metal is getting thin due to corrosion on the inside. Elbows usually give out first.

In climates where summers are very humid, careful late-spring cleaning of the whole system—chimney, stovepipe connector, and the in-side of the appliance—may help reduce corrosion. The need for such spring cleaning has not been proven, but it is plausible that it is better than fall cleaning for controlling corrosion. Chimney sweeps should be able to serve you better in the spring because their work load is less than in the fall.

Some sweeps offer an acid-neutralizing wash or spray that is applied after sweeping. Again, it is plausible, but unproven, that such treatment is helpful. Some people have suggested removing the ashes in the appliance and coating all interior metal surfaces with oil. This procedure cannot hurt, but evidence that it is necessary is lacking.

Last but not least, the whole chimney system must be inspected for creosote frequently—typically once a month. The *only* way to know for sure if a chimney needs cleaning is to look inside it. Read the next chapter!

CHAPTER 15

CREOSOTE, CHIMNEY FIRES, AND CHIMNEY CLEANING

Combustion of solid fuels in residential heaters is never perfectly complete. Flue gases always contain a combustible component of vapors and particles. When these ingredients of wood smoke condense or stick to a surface such as the flue, the resulting deposit is called creosote.

Creosote is combustible. When the creosote deposits inside a flue burn, the resulting fire is called a chimney fire, or a creosote fire. Under some circumstances, chimney fires cause house fires. Short of causing a house fire, a chimney fire can damage both prefabricated and masonry chimneys.

Wet creosote contains acids that can eat through stovepipe connectors in a season or less. Some masonry materials, particularly concrete, are susceptible to corrosion by creosote.

Some creosote has an unpleasant acrid odor. Creosote can stain roofs, walls, fabrics, and floors. Thus, the leaking of fluid creosote out of the flue into the living space of a house is highly undesirable. The dry residue after the water evaporates is also flammable.

Creosote deposits can build up to the point where the flue is severely restricted. This condition limits the heat output from the connected appliance and can result in noxious fumes spilling out of open appliance.

Creosote can act as a thermal insulator. Deposits inside a stove, connector, or interior exposed chimney can reduce heat transfer, which lowers the overall efficiency of the system. (Some creosote deposits have the opposite effect; they *increase* heat transfer.) (See "The Effects of Creosote on Stovepipe Heat Transfer," an SER publication. See Appendix 7.)

Creosote is a fact of life when heating with solid fuels. It is desirable to minimize its buildup, clean it out when necessary, and prevent, or at least control, its combustion.

DEFINITION OF CREOSOTE

As used in this book, creosote means the combustible component of deposits originating as condensed smoke from solid fuels. Vapors, soot particles, and tar particles are the ingredients of smoke which become creosote when they condense on or stick to a surface.

Water vapor may condense with the creosote, although the water itself is not considered part of the creosote. The wetness it contributes is part of the creosote problem since it contributes to the corrosiveness and volume of the total condensate. Little of this wet creosote remains in the flue.

Ashes, which are carried up the flue with the smoke ("fly ash"), may become lodged in the creosote deposits, but are not considered part of the creosote because they are not combustible.

Creosote can condense anywhere in a solid fuel system from the inside of the appliance itself up to the chimney cap.

Distinctions can be made among various types of creosote. Here are some of the forms.

TABLE 15–1
ORIGINS OF CREOSOTE

Origin in Fire	Smoke Ingredients	Initial Type of Creosote
Smoldering fuel → Tar mist →		Tar and other fluid creosote*
Flames → Vapors →		
→ Soot particles →		Soot

*Although initially fluid, this condensate does not necessarily flow down the flue, either because of its high viscosity or because it soon becomes pyrolyzed into a solid.

• Shiny tar glaze. This kind of creosote is largely wood tar and will melt when it gets hot, just as does roofing tar; it is one of the most difficult to clean. (This is what some call "third degree" creosote. I prefer descriptive terminology; it is more useful than jargon.)

• Slag. A very hard, thick, somewhat porous and brittle deposit that is predominantly partially pyrolyzed tar. Not cleanable with brushes.

• Flaky creosote. Layered flakes, sometimes glossy, sometimes matte black or brown. Usually cleanable with brushes.

• Soot. Velvety soft agglomeration of soot par-

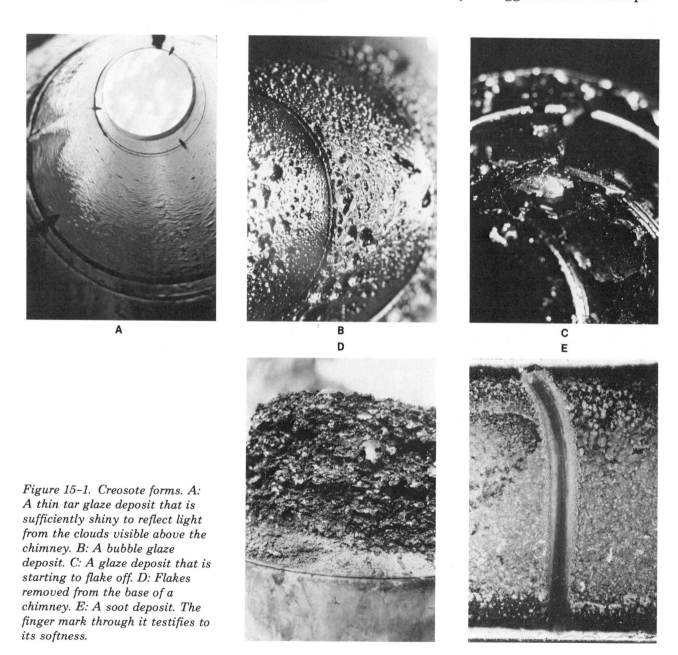

Figure 15–1. Creosote forms. A: A thin tar glaze deposit that is sufficiently shiny to reflect light from the clouds visible above the chimney. B: A bubble glaze deposit. C: A glaze deposit that is starting to flake off. D: Flakes removed from the base of a chimney. E: A soot deposit. The finger mark through it testifies to its softness.

ticles; easily cleaned. Coal appliances and open wood fires tend to produce more soot than tar.

Many actual creosote deposits are combinations of the above forms.

Some people use the term creosote only for tar-based deposits, such as glaze and slag deposits. I prefer to include soot as a form of creosote, partly for simplicity, since it also can fuel a chimney fire and needs to be cleaned from flues, and also to avoid semantic difficulties for deposits that are a combination of tar and soot particles. It is only a matter of semantics. But as the term is used in this book, creosote refers to any combustible deposits that condensed out of solid fuel smoke.

CHIMNEY FIRES AND THEIR CONTROL

Chimney fire, creosote fire, and flue fire are all synonymous; they all mean the burning of creosote deposits inside the system—typically inside the chimney and the stovepipe connector. Sometimes deposits in the appliance may also be involved. If creosote has leaked to the outside of the flue, it may also burn.

Ignition of the creosote deposits inside a flue requires adequate oxygen, which is usually available, and sufficiently high temperatures—the same conditions for the ignition and combustion of any fuel. Chimney fires are most likely to occur during a very hot fire, as when cardboard or Christmas tree branches are burned, or when the appliance burns normal fuel, but at a higher than normal rate. As is true for all solid fuels, creosote has no one meaningful and simply defined ignition temperature.

Chimney fires also may occur at unexpected times. Very intense fires in the appliance will sometimes fail to ignite the creosote in the chimney because of a lack of oxygen in the flue. Intense appliance fires can use up most of the entering oxygen. Low fires usually result in adequate oxygen concentrating in the flue. A spark can ignite the creosote. Laboratory experience indicates chimney fires can start almost any time.

Chimney fires can generate very high temperatures inside the flue. In tests at Shelton Energy Research (SER), peak flue-gas temperatures of over 2,000° F. have been measured.

Chimney fires generally last 10–30 minutes, al-though both shorter and longer chimney fires may occur.

In many chimney fires, the location of the peak temperature moves up the chimney; but occasionally the location of most intense burning will start high in the chimney and slowly work its way down. In some cases there is more than 1 stage or wave of burning. Some chimney fires are concentrated at the base of the chimney near the horizontal breaching, apparently due in part to the combustion of large amounts of creosote debris which has dripped or fallen down from above and accumulated in the chimney cleanout area.

In some chimney fires, the pressure of the burning gases in the upper portions of the flue can be higher than the surrounding air pressure. The result is that flames will penetrate through any available openings or cracks.

RECOGNIZING CHIMNEY FIRES

Once started, there is little mistaking a chimney fire. The sound is usually awesome, often described as a jet airplane taking off; chimney fires usually roar! This sound probably is caused by the very high velocity of the flue gases and of the air being sucked into the air inlet of the appliance and through leaks in the lower parts of the system.

If the fire is in a stovepipe connector, the stovepipe usually glows red hot. The stovepipe may also shake or vibrate.

The most spectacular part of the system during a chimney fire is usually the top of the chimney. Very dense and dark smoke is sometimes seen, but even more impressive is the plume of flame and sparks. In chimneys without caps, the plume can be well over 10 feet high. The sparks can also consist of substantial chunks of burning creosote.

THE HAZARDS OF CHIMNEY FIRES

The most obvious hazard of a chimney fire is starting a house fire. This can happen via several routes.

• Heat conduction through the chimney to nearby wood, combustible insulation, and so on.

- Flames from the flue penetrating through cracks in the chimney.
- Sparks and burning chunks of creosote falling on the house and its surroundings.

Sound Class A (or Solid Fuel or All Fuel) prefabricated chimneys *properly installed* and sound, *code-consistent* masonry chimneys are designed to withstand an occasional chimney fire. They are unlikely to be implicated in a house fire caused by the conduction of heat through the chimney walls or via flame penetration through cracks or joints. However, sparks coming out the top of any chimney can cause fires. Do not be complacent if you have a sound chimney.

Despite the risks, some people recommend letting a chimney fire burn itself out—of course keeping watch that the house not be ignited—in order to clean the creosote out, or at least render the remaining creosote easier to clean. Typically, some creosote will remain, but the high temperatures of the chimney fire will make it much easier to clean out.

This I cannot recommend. Even if no house fire results, chimney fires are not desirable. The intense and sudden heat can warp the flues in prefabricated chimneys and can crack masonry chimneys. The damage is serious and requires replacement of damaged prefabricated chimney sections, and, usually, replacement of most of the masonry chimney.

And there is more. Even if there is no visible damage, there is likely to be some invisible wear and tear. The thermal stress of a chimney fire usually increases the number of and extent of hairline cracks in the masonry. Prefabricated metal chimneys may suffer slight deterioration of the flue. Thus, chimney fires should not be allowed to run their course, but should be put out or at least held back as much as possible.

EMERGENCY ACTIONS IN CASE OF A CHIMNEY FIRE

What should you do if you have a chimney fire? Here are traditional recommendations for any house fire in order of importance.

1. Alert all people in the house. Either have them leave, or be ready to leave.
2. Call the fire department.

3. Suppress the fire as best you can until the fire department arrives, being careful of your own safety. Be sure you always have a way out of the house should the fire get out of hand.

In my opinion, deviation from this order of action is reasonable in some cases. Because a chimney fire is initially contained within a structure designed to withstand such an event, sometimes it makes sense to take simple, *quick* measures to control the chimney fire, *before* calling the fire department. Such activity should take no more than 30 seconds.

There are 3 quick measures you can use to control the chimney fire.

- Close the air inlet(s) of the appliance.
- Discharge a dry chemical household fire extinguisher into the appliance.
- Use a chimney fire extinguishing product.

Be watchful for a house fire; check the roof, shrubbery, and all accessible wood and other combustibles near the chimney, including ceiling and roof penetrations, chimney chases, and the house adjacent to exterior masonry chimneys. Water is very effective at extinguishing smoldering wood and keeping wood cool near a hot chimney. A garden hose with a spray nozzle can deliver controlled amounts of water to otherwise awkward locations.

SUPPRESSING CHIMNEY FIRES

It is *not* necessary to put out the fire in the appliance to control a chimney fire. Appliances are intended to have fires in them. Once started, a chimney fire is not greatly affected by the condition of the fire in the appliance. Chimney fire control measures should be aimed at the fire in the chimney.

With sufficient foresight, one of the simplest, quickest, and most effective ways to suppress a chimney fire is to deprive it of air merely by closing the air inlet(s) on the appliance. The foresight is to have no other way for air to enter the chimney. Thus the appliance, stovepipe connector, and chimney (including its cleanout door), must all be relatively airtight. There must be no barometric draft control, or the control must also be closed and kept closed (the draft in the flue will tend to pull it open). There must be no other appliance connected to the same flue. Almost all gas and oil appliance installations intentionally admit substantial amounts of air into the flue. A chimney fire will be extinguished within seconds after it is denied air.

Fire Extinguishers

You can use a dry-chemical household fire extinguisher to dampen the spirits of a chimney fire. Discharge the extinguisher into the appliance above the fire; the extinguishing agent will be carried up into the flue. If applicable, close the door of the appliance and shut the air inlet immediately afterwards. No experiments have been done to determine how much extinguishing agent is usually necessary. Even if the entire contents are discharged (which will take only 10–15 seconds), the flue fire may not be completely extinguished, but it will probably be slowed down.

There are fire extinguishers designed for chimney fires. One product looks like an emergency roadside flare, but in fact is quite different. It generates large volumes of smoke which the manufacturer claims suffocates chimney fires. The flare is ignited by striking the end with a special cap, just as are roadside flares. It is then placed in the appliance or just above it (on the smoke shelf of a fireplace if convenient). Placement is not critical. The draft will sweep the smoke up into the flue.

The product is not universally recognized as being effective. I am aware of no independent laboratory tests, and field reports are mixed. My interpretation of the field reports is that in some cases more than 1 flare going at the same time may be necessary for large flues, and that a second application of the flare(s) may be necessary after the first 1, or set, has burned out. However, be careful about using more than 1 at a time; the smoke is probably not healthy to breathe. Virtually all the

Figure 15-2. A chimney fire extinguisher (Chimfex).

smoke from the flares should go up the flue and not spill out of the appliance into the house.

Water can help control a chimney fire, but should be used very sparingly, because sudden cooling of the stove, fireplace, or chimney can cause serious damage. Also, a large quantity of steam entering the house is dangerous. Tossing small amounts of water, about a half cup at a time, on the fire can generate enough steam to help suffocate the chimney fire. A side effect of this action may be putting out the fire in the wood heater, but this is not the primary objective in controlling chimney fires.

A more down-home remedy often suggested is to empty a large box of baking soda on the fire in the appliance. This certainly will slow down the fire in the appliance. Sodium bicarbonate is a traditional fire-extinguishing chemical. But its effectiveness at suppressing fires in chimneys when it is dumped in the appliance is not well established. It certainly can't hurt.

Salt is sometimes suggested as a chimney fire suppressant. Its effectiveness has not been demonstrated in laboratory tests.

Baking soda and salt may be effective, but if so, it would be useful to know how much is needed. If 2 pounds is needed, then a 6-ounce box will not help much. Some laboratory tests on the relative effectiveness of all chimney fire suppressing substances in various quantities would be useful.

CHIMNEY FIRE ALARMS

There are a number of products on the market intended to warn the occupants of a chimney fire by sensing the temperature in the flue, then sounding an alarm when the temperature exceeds a particular value. If the occupants are home, the alarm can alert them in time to attempt to control the chimney fire or to get out of the house. A chimney fire alarm will go off long before a smoke or heat detector in a house would detect a resulting house fire.

Designing and installing chimney fire alarms can be tricky. Where should the sensor be placed, and what should the temperature be, so that most chimney fires will be detected, false alarms will be rare, and the installation will not be too difficult? For simplicity in installation, some put the sensor in the stovepipe connector, not the chimney. However, some creosote fires are restricted to the

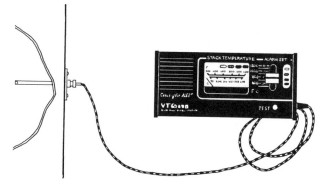

Figure 15-3. A chimney fire alarm. (This one is manufactured by Vermont Technology Group, Inc.)

chimney; a sensor in the connector misses these. A better location is the top of the chimney, since almost all serious chimney fires result in very high temperatures there. However, this location is awkward. All things considered, high in the stovepipe connector is the most reasonable location.

Wherever the sensor is located, it is important to have the alarm-triggering temperature low enough to detect any serious creosote fire, and high enough not to be triggered by hot flue gases from the appliance. The closer the sensor is to the appliance, the more difficult this is.

A highly desirable feature of some chimney fire alarms is an adjustable set-point temperature. This allows the alarm to be tuned for each system to eliminate most false alarms and still be sensitive to unusually high temperatures.

An adjustable alarm can be used intentionally to warn of an unusually hot fire in the appliance, in addition to chimney fires. An overheated appliance is itself dangerous, and could be the prelude to a creosote fire.

Another feature of some alarms on the market is that they double as a flue-gas temperature *indicator*—you can read the temperature at any time on the meter. It is superior to a surface-mounted temperature indicator in that the temperature probe penetrates into the flue and reads actual flue-gas temperature. Surface indicators give distorted temperatures when there is creosote in the pipe.

Keep in mind that a chimney fire alarm is a second line of defense. The primary defense against chimney fire hazards is prevention. If you have a chimney fire alarm, you should still inspect and clean as often as if the alarm were not there.

FACTORS AFFECTING THE CONDENSATION OF SMOKE

There are 4 fundamental factors affecting the rate of creosote buildup.

- Smoke density.
- Temperature of the flue wall.
- Residence time of the smoke in the flue.
- Turbulence of the smoke.

Clearly denser smoke has the potential for resulting in more creosote. This is probably the single most important factor. Thus, the single most effective preventative measure is to achieve relatively complete combustion in order to minimize the smoke density.

The temperature of the flue gas itself is relevant, but less important than temperature of the surrounding surfaces, such as the flue wall. The cooler these surfaces are, the higher will be the rate of flue-gas condensation.

There is no one critical temperature above which condensation will not occur. Smoke contains hundreds of condensible compounds, not just one, and each has its own concentration-dependent condensation temperature.

In addition, the particle component of smoke is not subject to the principles of condensation in the first place. They are already liquid or solid. Hence, they stick to the flue walls rather than condense in the technical sense of the word. There may be no practically achievable temperature that will prevent the accumulation of these particles on the flue walls.

Nonetheless, if flue-gas temperatures throughout the flue are kept above about 250-300° F., creosote accumulation will be substantially reduced. See chapter 11 for more discussion of monitoring flue-gas temperatures.

Both vapor condensation and particle sticking processes take time. All things being equal, the slower the smoke moves up the flue, the more creosote will condense out of it. This is one reason oversize flues are not good; the smoke moves more slowly up the flue. Oversize flues also allow more cooling of the smoke.

The effect turbulence has on creosote buildup is not certain. Turbulence enhances heat transfer by bringing the hot core of flue gas at the center of a flue in contact with the flue wall. Turbulence may help bring fresh dense smoke from the center to the flue wall. Sudden changes in smoke velocity, such as at elbows, may result in some of the larger particles "falling out" to the outer surface of the curve. None of these possible turbulence effects has been well documented.

HOW OFTEN IS CHIMNEY CLEANING NECESSARY?

Opinions vary, but there is only 1 truthful answer: A chimney should be cleaned whenever it needs it. Let me explain.

Conventional wisdom is that 1 cleaning a year is reasonable. But is it? Clearly it makes more sense to take into account the amount of use of the appliance. Some people recommend a cleaning for every 3 cords of fuel consumed. But this does not make much sense either, for, in fact, creosote accumulation also depends on *how* the wood is burned. The accumulation from smoldering combustion in a closed stove can be 100 times that from flaming combustion in an open burner! The type, location, and size of chimney also have a large effect.

There is no doubt that some systems need weekly cleaning. The creosote shown in Figure 15–4 was, in most cases, accumulated in less than the equivalent of 5 weeks of full-time operation of wood stoves at SER. Many such deposits were intentionally ignited, and the resulting chimney fires reached peak flue-gas temperatures of over 2,000° F. and lasted for about 15 minutes.

What is the conclusion? No general recommendation based on a time period or an amount of fuel consumed is worth very much. There are too many variables. The only safe advice is to clean the flue when it needs it, and the only way to know if the flue needs cleaning is to look in it.

How dirty is too dirty? The rule of thumb used by most people is ¼ inch. A ¼-inch buildup of tar can result in a very substantial chimney fire. A ¼-inch buildup of soot or cobweblike creosote contains much less weight and hence fuel energy, but can still result in a chimney fire.

The ¼-inch rule is somewhat arbitrary, but is nonetheless useful because it is in the right ballpark. In fact, there have not been enough laboratory tests of chimney fires to justify that ¼ inch is preferable to ⅛ or ³⁄₁₆ inch.

My recommendation is that if there is a ¼-inch or more buildup of creosote anywhere in the system, the system should be cleaned.

The question then becomes, how often is it necessary to inspect the system to have reasonable assurance that the buildup will not exceed ¼ inch between inspections? I recommend inspecting once a week for the first month, then once a month for a few months, until you find out how quickly creosote is likely to build up with your ap-

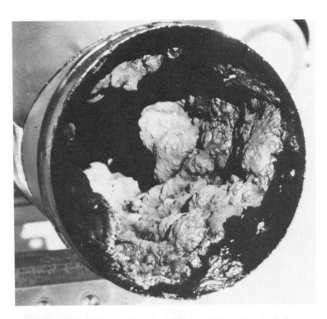

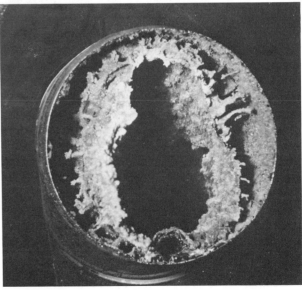

Figure 15–4. Two-week creosote deposits. These deposits all accumulated in about two weeks of equivalent full-time firing of the stoves in SER's laboratory.

pliance, your chimney, your operating habits, and your fuel.

Because inspection is the only way to know when cleaning is necessary, there is clearly a great advantage to installations wherein inspection is easy. See chapter 6 for suggestions.

HOW TO CLEAN

If you clean your own chimney and stovepipe, use the equipment professionals use. Wire brushes are available in enough sizes and shapes to allow a snug fit inside any common flue. Screw-together lightweight fiberglass rods are used to work the brush up and down the flue.

While cleaning a chimney from the roof, leave the appliance connected and all openings closed off. Put a cover over the fireplace opening. This keeps the creosote and dust contained and out of the house. After brushing, the loose, fallen creosote can be removed from the cleanout area, stove, or fireplace.

When cleaning a chimney from below, most professionals use a specially designed powerful vacuum cleaner to keep the house clean. Home

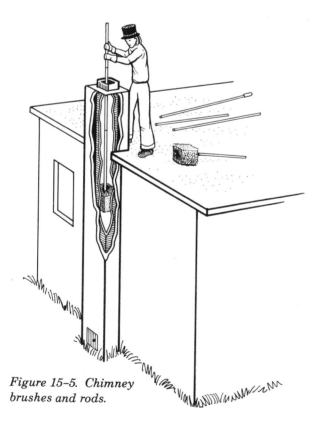

Figure 15-5. Chimney brushes and rods.

vacuum cleaners will not do the job. If you don't have access to a professional vacuum cleaner, stuff a rag into the bottom of the chimney or hold it across the chimney bottom in such a way that a rod can penetrate through or slip by the rag, but the falling creosote cannot.

If inspection of a chimney reveals a thin (1/16 inch or less) tarlike deposit, cleaning is not worth the effort. If the tar is soft, the brush may be fouled. If the tar is hard, the brush will not clean it. There is no safe solvent to help clean this kind of deposit. The best bet is that a hot fire will dry, crack, and loosen these deposits so that cleaning is easier.

Some deposits are very difficult to clean. If the deposit is a thick, hard glaze or slag, cleaning is extremely difficult. I recommend calling a professional chimney sweep and changing how you operate the system so as to avoid this type of deposit in the future.

CREOSOTE CONTROL

By installing your stove properly and operating it effectively, you can control the amount of creosote that accumulates in your system.

Chimney Type, Location, and Size

The way an appliance is installed has a substantial effect on the rate of creosote buildup. To minimize creosote, the flue gases should be kept as warm as possible throughout the venting system. Listed below are the features of installations which help reduce creosote. Each is discussed in more detail elsewhere in this book. The list is roughly ordered with the more important features listed first.

- An appropriately sized flue, not oversized.
- An interior chimney.
- A chimney with good thermal insulating properties, such as mass-insulated or air-insulated prefabricated chimneys, or insulated masonry chimneys.
- A short stovepipe connector.

Appliance Operation

Once an appliance is selected and installed, the single most effective way to minimize creosote is to burn the fuel, not smoke it! Wood and coal were never intended to be smoked; yet that is what many people do. Misled that long burn durations are desirable, homeowners fill their appliances with fuel and then starve them of air, either by setting the manual air-inlet control too low, or by letting the thermostat of the appliance take over.

Although this stuff-it-and-starve-it routine yields long times between refueling, the result is that the fuel smokes rather than burns. Here I use the term "burn" to mean sufficiently complete combustion to burn most of the smoke.

In fact, most airtight wood appliances can be operated to burn rather than smoke the fuel. My comments on operation apply to the vast majority of reasonably tight wood burning appliances. Appliances with radically different designs may behave differently.

Typically, smokefree, clean burning requires small fuel loads—2–4 logs at a time, or ¼–½ of a full load—and leaving the air inlet relatively wide open, especially during the first 10–30 minutes after each loading, when most of the smoke-generating pyrolysis reactions are occurring. Towards the end of each burning cycle, the air inlet can be turned down substantially without fear of smoke generation; charcoal cannot generate creosote-producing smoke.

In my own home, I try to use the appearance of the flue gases coming out the top of the chimney as a guide. If I see much smoke, I open the air inlet further, or remember to use less fuel next time. If the flue gases are clear, I cut back on the combustion air. In this mode of operation, the rate of heat output essentially is determined by the size of the fuel load. In mild weather, small fuel loads must be used; otherwise too much heat will be given off when enough air is admitted to achieve clean combustion. In cold weather, substantial loads, perhaps even full loads can be used. Because many people tend to use large fuel loads all the time, some manufacturers of large central wood-heating systems recommend that their appliances not be used at all in spring and fall, but that only the backup gas, oil, or electric systems be used. This is one way to decrease creosote, but another, of course, is to use smaller and more frequent loadings in the spring and fall.

Constant checking and adjusting is, of course, impractical and unnecessary. The real and necessary sacrifice is more frequent refueling of the wood heater—perhaps every 1–3 hours instead of every 3–9 hours. If the long duration, large load,

unattended burn is important, more frequent chimney cleaning will be required. On the other hand, if you are at home most of the time and do not mind tending the stove every 1–3 hours, using smaller, more frequent loadings usually will decrease creosote accumulation significantly.

In the spring and fall when not much heat is needed, I suggest making small occasional hot fires, leaving the air inlet wide open throughout. Open a window if the house becomes too warm, and let the fire go out when the heat is not needed.

The effect on overall energy efficiency of this operating mode has not been measured over a range of appliance types or heat output rates. Probably the same approximate amount of fuel is consumed to produce the same amount of useful heat. The clear improvement in combustion efficiency may be compensated by a decrease in heat transfer efficiency. Testing is needed to clarify the net effect.

Fuel: Moisture Content, Species, Piece Size

In the past, too much emphasis has been put on the effect of fuel on creosote. Heavy creosote was most often blamed on the fuel being too wet or the wrong species. Using "seasoned hardwood" was supposed to solve most creosote problems.

Fuel variables are relevant, but in fact they are less important than a number of other factors

TABLE 15–2
EFFECTS OF FUEL ON CREOSOTE ACCUMULATION

	Creosote Accumulation	
	Closed Appliances	Open Appliances
Fuel Variables	(stoves)	(fireplaces)
Drier fuel	more	less
Pitchier fuel	more	more*
Smaller pieces	more	less†

*This is a prediction, not a measurement.

†This trend is based on indirect evidence—measurement of efficiencies, not creosote.

Summary of laboratory-measured effects of fuel variables on creosote accumulation. See chapter 14 for details. In the real world, there will be exceptions to these results because of differences in other conditions. However, I believe the above trends are probably applicable in most cases.

such as how the appliance is operated, and the chimney type, size, and location.

The effects of fuel variables are discussed extensively in chapter 14 and summarized in Table 15-2. These comments on fuel variables apply to the vast majority of appliances. Some appliances with radically different designs may respond differently.

Piece size should be whatever is most convenient, but keep in mind that the smaller the pieces are, the more creosote will accumulate in closed appliances.

Intentionally Hot Daily Fire

A daily (or weekly) short, hot fire is often recommended as a way to help control creosote. As usually explained, the intent is to have a chimney fire, but only a small one. If only a day's accumulation of creosote is involved, the chimney fire will not be very dangerous and will clean out the chimney. Thus, allegedly, both brushing and a dangerously intense chimney fire are avoided.

I cannot recommend this practice. It is critical that there never be a large creosote buildup when an unusually hot fire is in the stove; a real and substantial chimney fire is not desirable. The key to reasonably safe use of this method is to be sure it is working—to be sure large creosote deposits are not building up.

How can you be sure? There is only one way—by inspecting the flue. If you try this method, I recommend starting with a clean chimney, and then carefully inspecting every day for the first week, every week for the next month, every month for the next year. If inspection ever indicates creosote is still building up, stop using the method. If inspections indicate success, then the frequency of inspection can be decreased gradually.

In many, if not most cases, a hot daily fire does not *burn out* the creosote. The heat in the flue gases dries, shrinks, and cracks the thin tar glaze creosote layer which has accumulated since the previous hot fire, and this loosened creosote then either blows out the top of the chimney or falls down to the bottom of the chimney.

In prefabricated chimneys, only a few minutes with a hot fire in the appliance will do the job. However, the much larger mass of masonry chimneys necessitates a longer fire—perhaps up to half an hour.

Note that although this technique can eliminate the need for brushing a flue, usually it does not eliminate the need to clean out the creosote. The fallen creosote flakes can build up quickly in a pile at the bottom of the chimney. This material needs to be cleaned out; it is combustible, and it can block the breachings where appliances are connected into the chimney. If the venting system is vertical throughout, the flakes can fall back into the appliance. In some appliances, the flakes will land in the fire and be burned. In others, they will land on a baffle plate and may need periodic removal.

One final precaution. Keep in mind that even small chimney fires can result in burning material coming out the top of the chimney. Thus, the method may not be advisable if your roof or surroundings are especially combustible. A spark screen would, of course, help.

Additives

There is a long history of adding chemicals and other materials to wood and coal fires to help control creosote. Often mentioned in older writings are salt, used batteries (do not try this one—most contemporary batteries will explode), old tin cans, and potato peels. Added once a week to a fire, these materials were claimed to help reduce creosote.

Most contemporary additives are chemicals or mixtures of chemicals. Common ingredients include trisodium phosphate, bentonite (clay), sodium chloride (common salt), cupric chloride, sodium carbonate, sodium acetate, and zinc oxide.

It is interesting that batteries, tin cans, and, of course, salt are made of some of the same elements that are found in some modern chemical treatments. To the extent that the newer chemical formulations work, this could be an explanation for some of the down-home remedies. However, I see no mechanism yet by which potato peels could be effective.

What are the claimed effects for the chemical formulations? They include:

• Cleaning dirty chimneys with complete disappearance of the creosote.
• Cleaning dirty chimneys by loosening the creosote, which then falls off and down the chimney.
• A change in the creosote itself or in its adhe-

sion to the flue, which makes the creosote come loose more easily when cleaned mechanically.

• Decreasing or eliminating the rate of creosote buildup.

• Decreasing the heating value or raising the ignition temperature or other properties of the creosote which make chimney fires either less likely or less intense.

• Decreasing flue corrosion due to the creosote and/or the flue gases.

How could chemicals have some of these effects? One mechanism is catalytic combustion. Many metals have a catalytic effect, and most of the chemical formulations include some of these metallic elements.

When a chemical cleaner is applied to a fire, some of it, usually after being burned or vaporized, is transported through the system along with the smoke and other by-products of combustion. It can then condense on or adhere to the creosote on the walls of the appliance or its chimney.

It has been shown in laboratory tests that such dustings of certain metals on *coal soot* reduce the ignition temperature of the deposits.[1] The same study demonstrated that the catalytic action can result in slow nonflaming and nonglowing oxidation of the soot — what would appear as a gradual disappearance of the soot. However, field tests of the same compounds showed little if any effect in chimneys, although there did sometimes seem to be effects in the appliances themselves.

The explanation, according to the researchers, is that rarely are temperatures high enough in the flue even for catalytically assisted combustion.

Other mechanisms may exist, but are not generally discussed openly by the manufacturers and are thus not possible to discuss here. In my opinion, it is plausible scientifically that chemicals could be effective.

Most of the products on the market today are intended for *wood* creosote. Do they work? Strong and numerous personal testimonials abound on both sides of the issue. Many users feel the chemicals are very effective; but many other people have tried chemical cleaners and find them to have no effect. Thus, it is at least clear that they do not always work.

Mixed personal reports such as these suggest that scientific evidence would be useful to help resolve whether or not, and under what circumstances, chemical cleaners in fact help reduce creosote in chimneys.

SER conducted a series of tests in 1980. The following is quoted from the summary and the conclusions of the report.

". . . To test the effectiveness of chemical chimney cleaners, four representative name-brand cleaners were simultaneously tested in six identical stoves (one for each cleaner plus two controls), and the weight changes in the six chimneys were carefully monitored. Power output, fuel species and moisture content, cleaner dosage, and length of test series were varied over seven months of testing.

"The particular brands of chemical cleaners we tested did not exhibit any substantial effectiveness in our tests. We used both oak and pine, seasoned wood and green wood, cool and hot fires, and normal and high doses. We looked for prevention of creosote buildup, the disappearance of creosote, the falling down of creosote flakes, changes in the creosote's brushability, and changes in the creosote's flammability. Although some of these phenomena were observed, they were equally strong in the untreated systems as in the treated ones. Thus the effects were not attributable to the chemicals but rather to factors common to all the systems, such as temperature. Hot fires tend to dry and loosen many creosote deposits."[2]

The research has been criticized by the chemical chimney cleaner industry. One manufacturer's rebuttal, and SER's reply, as well as the original report, are both available from SER.

More recently, a test was conducted in Canada by an agency of the Canadian government. Some revised formulations were used in the test. Again, the results were null: no significant effects were observed.

It is notable that most manufacturers base their claims on personal testimonials, and at least some of those who claim to have scientific evidence of effectiveness will not provide any report on the tests.

None of this research proved that chemical chimney cleaners cannot work. That is impossible for 2 reasons. The products may work under some field conditions which were not replicated in the

---

1. P. Nicholls and C. W. Staples, "Removal of Soot from Furnaces and Flues by the Use of Salts or Compounds," U.S. Dept. of Interior, Bureau of Mines, Bulletin 360 (1932).

2. J. Shelton and C. Barczys, "Research Report on Chemical Chimney Cleaners" (Sante Fe: Shelton Energy Research, 1981).

laboratory. Also, new formulations are constantly appearing, and some of them may be substantially different from the products in previous tests. Thus the controversy goes on.

My mind is very much open on the subject. True, the published scientific tests thus far have been negative. But many chimney sweeps, whose observations I respect, feel that the products have some merit, particularly for conditioning the creosote, making it easier to clean. On the other hand many sweeps also observe no effect. Clearly, if there is an effect, it is not universal. It is interesting that virtually no such products exist in most of Europe.

In light of the conflicting feelings on the topic, I cannot take a stand on chemical chimney cleaner products. The jury is still out. If some of the products work some of the time, I hope there will soon be available independent laboratory tests to demonstrate the fact.

Even granting that chemical chimney cleaners may sometimes work, it is clear that they do not always work—both laboratory and real world evidence are unambiguous on this point. Therefore, it is dangerous for those who use chemical chimney cleaners to presume they are working, for if they are not, the complacency may result in less frequent inspection and/or cleaning, and that could lead to a chimney fire. Frequent chimney inspection and cleaning when necessary are essential to heating with solid fuels safely.

Accessory Devices

There are 4 types of accessory devices that have been or are being marketed claiming creosote reducing effects.

- Chimney caps.
- Creosote filtering devices.
- Afterburners (principally catalytic).
- Dilution air devices, such as the barometric draft control.

Some are very effective, and all have other attributes or liabilities.

Chimney Caps. A very few manufacturers of chimney caps have claimed creosote-reducing effects, via improved draft in windy weather. I know of no independent laboratory tests on this issue, but am skeptical of any such claims. The only mechanism I can envision by which a chimney cap could help is by increasing dilution air, which reduces creosote. A good wind cap will increase the draft in the system in windy weather. One effect would be increased air being drawn in through leaks in the chimney, connector, and appliance. However, this effect helps only in windy weather, and only in the case of leaky installations; therefore, I would not expect the average effect to be large.

Creosote Filters. Creosote filtering is a viable concept, but the practical potential is limited. Figure 15-6 illustrates an example of a creosote filter. The device consists of a knitted wire mesh filter attached beneath a cast-iron plate. The assembly can be rotated to lie either at a right angle to or parallel to the smoke flow, similar to a simple stovepipe damper. When the filter is closed (set across the flow), most of the flue gas flows through its passageways.

A catalytic version of this filter may now be available. It has promise for increased performance and decreased maintenance. It can act as a filter and/or afterburner.

The filter traps some of the particles of soot and/or tar in the smoke. (Thus, certain emissions are also reduced.) But the vapors do not tend to condense on the filter because the filter is approximately the same temperature as the flue gases. Thus, a significant portion of the creosote potential in the flue gases is unaffected by the device.

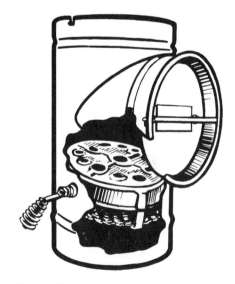

Figure 15-6. The Smoke Consumer, a creosote-reducing accessory. It was first introduced as a filter, but a catalytic version may become available.

In tests at SER, such a device reduced creosote between 23 and 41 percent.[3] Results in your installation may of course be different.

The device requires maintenance. Some of the particle catch is burned off in the recommended daily "continuous burn reactor cycle," but not all. The manufacturer recommends manual cleaning roughly weekly.

The device reduces draft significantly and should not be used where draft is marginal to begin with.

Catalytic Afterburners. Catalytic afterburner accessories generally have the potential to reduce creosote, to improve overall energy efficiency, and to reduce emissions. As explained in chapter 11, in a test at SER, creosote reductions of 41–51 percent were observed for a particular catalytic afterburner under particular operating conditions. Most catalytic afterburners reduce the draft to the appliance somewhat.[4]

Barometric Draft Control. In tests of creosote-reducing accessory devices at SER, the barometric draft control was the most spectacular. Its use resulted in a 65–75 percent drop in creosote buildup, depending on which portions of the venting system were included. In addition, the creosote deposits that did form appeared to be drier—not as shiny—and with a rougher surface, and consequently were easier to brush out of the chimneys. But don't expect to duplicate this level of creosote reduction in your own home for 2 reasons.

First, the venting systems in the SER tests were more airtight than are most in the real world. A barometric draft control works by adding air to the flue. Adding a barometric draft control to a leaky venting system won't make as much difference as will adding one to an airtight stovepipe/chimney setup.

Second, the systems equipped with barometric draft regulators were operated in a manner that would maximize their beneficial effect (which was the overall intent for all the devices tested). Each draft control was set so that the flap was open as far as possible (allowing the greatest amount of dilution air to be admitted), consistent with

achieving the desired burn rate. You may not operate your barometric draft regulators in this optimum (for creosote reduction) manner.

Another important factor to keep in mind about such devices is that they won't work properly unless the chimney is capable of generating *excess* draft. Whenever the flap is open, the draft is reduced. The more excess draft there is available in a given chimney, the more benefit can be realized from fitting it with a regulator (both in reducing creosote and in limiting draft). If you're able to achieve very hot fires easily, and your system doesn't spill smoke into the room, then chances are good that your system has excess draft; many systems do.

A barometric draft control adds dilution air to the flue. Why does this reduce creosote? The air cools the flue, and, other things being equal, this is not beneficial. However, the dilution air also dilutes the smoke, and this tends to reduce creosote. Finally, adding air increases the velocity of the now-diluted flue gases, which is probably beneficial, but no experiments have been done on this.

Before deciding to use a barometric draft control, be sure to read more about the device in chapter 11.

Appliance Design

Three performance characteristics of specific appliance designs can affect creosote accumulation.

- Combustion efficiency.
- Heat transfer efficiency.
- Excess air.

High combustion efficiency minimizes the amount of smoke passing through the chimney, and smoke is the precursor for creosote. Traditional design features—secondary air, liners, and baffles—that are purported to improve combustion efficiency usually do not significantly. In traditionally designed appliances, the operator is much more important than the appliance design for achieving high combustion efficiency.

High heat transfer efficiency increases creosote by resulting in cooler flue gases. Stoves with extra chambers for better heat transfer, appliances equipped with a blower option, and appliances equipped with heat-extracting accessories all are likely to develop more creosote. These features

3. J. Shelton and C. Lewis, "Tests on Creosote Reducing Devices" (Sante Fe: Shelton Energy Research, 1982).

4. Ibid.

may be desirable, but to take full advantage of them it is often important to operate the appliance to achieve relatively good combustion to avoid a heavy creosote problem.

Excess air reduces creosote for 2 reasons. Combustion efficiency tends to be better because the air supply cannot be as restricted, and there is more excess air to dilute the smoke. This is one reason open fireplaces generate so much less creosote than do airtight stoves.

There are 4 appliance design types which break away from tradition and have the potential for substantial creosote reduction.

- Catalytically assisted combustion.
- Improved conventional combustion.
- Heat storage appliances.
- Stokers.

Each was discussed in chapter 7.

CHAPTER 16

ECONOMICS

A common reason people give for their interest in heating with solid fuels is to save money. It can cost hundreds and even thousands of dollars to heat a house for a season using electricity, oil, natural gas, or liquified petroleum (LP) gas, which includes propane and butane. Since a single wood or coal stove can contribute a significant amount of heat, there is the potential for saving at least hundreds of dollars a year.

But saving money by heating with solid fuels is not as easy as you might think. There are many subtle costs which are often forgotten. Some people who *think* they are saving money in fact are not.

In many cases it is difficult to *predict* whether it makes economic sense to heat with solid fuel because many important factors may be unknown—such as the future cost of both the solid fuel and the conventional energy source it displaces, the energy efficiencies of both the solid fuel and the conventional heating system, and how much the solid fuel heater will be used. For these and other reasons, clean, simple yes or no answers to the economic aspects of solid fuel heating are hard to come by in some cases.

For solid fuel heating to be economic, not only must the solid fuel be less expensive *per Btu. of useful heat output*, but the fuel savings must be large enough to pay for the initial cost and continuing maintenance of the system.

COSTS

The 2 primary direct costs associated with solid fuel heating systems are the initial purchase and installation cost, and the continuing fuel costs (Table 16-1). The resulting direct savings are in lower heating bills for natural gas, oil, electricity, or LP gas.

Suppose a $600 stove is purchased along with a $300 prefabricated chimney, and the installation is done by a professional and costs $100. The total installed cost of this system would be $1,000. Assuming the system will last 20 years, how much would have to be saved in fuel costs each year in order to break even? If the conventional system burns oil, the annual fuel savings is the money saved on oil minus the money spent on wood. For simplicity, assume there is no inflation.

(Inflation is likely to result in price increases for both conventional fuels and wood. If both prices rise at the same rate, the relative economics is unchanged. However, the dollar amount of the minimum break-even annual savings will have to increase each year to keep pace with inflation. If the costs of fossil fuels and electricity increase faster than the cost of wood, wood will become more economical.)

Spreading the $1,000 initial cost over the 20

TABLE 16-1
COSTS AND SAVINGS
OF HEATING WITH SOLID FUELS

Costs

Initial Investment	appliance
	chimney
	installation
	smoke detectors
	chain saw
	splitting equipment
	chimney cleaning equipment
	wood storage shed
Continuing Costs	wood fuel
	maintenance:
	chimney cleaning
	stovepipe replacement
	gasket replacement
	stove paint
	replacement glass
	chainsaw and truck fuel
	and maintenance

Savings

Initial Investment Savings	usually none, since a conventional heating system is part of most homes
Continuing Savings	money saved for not having to buy fuel or electricity for conventional heating system

Outline of costs and savings of heating with solid fuels. Note: Not all listed items are applicable in every situation.

years estimated lifetime would suggest $50 per year as the minimum annual savings necessary to break even. But, in fact, about $117 per year would have to be saved. If instead of buying a stove and chimney, the money were deposited in an interest-earning account, the initial $1,000 would be worth much more after 20 years. At an interest rate of 10 percent, you would have $2,349 after 20 years! To be fair, the net annual fuel savings also must be thought of as being deposited in a similar account. The result in this example is that if $117.46 is saved each year, it will amount to $2,349, including interest, after 20 years. Thus, unless the net savings in fuel costs exceeded $117.45 per year, you would have been better off in strict economic terms, to invest the $1,000 in an interest-earning account rather than into the wood heating system.

These numbers have been in reference to a particular example. In general, how much must be saved in fuel costs to break even? For systems with long life expectancies of 20 years or more, the fuel savings each year must be approximately the prevailing interest rate, plus 1 or 2 percentage points, multiplied by the initial cost of the system. If the interest rate is 15 percent and the system costs $800, then the annual savings to break even must be 15 + 1, or 16 percent of $800, which is $128.[1]

The useful life of heating systems is impossible to know in advance, but the better quality appliances and chimneys should last for at least 20 years. Interest rates are less predictable. But the important concept is that money spent on a solid fuel heating system is money not available for earning interest. The savings produced with the solid fuel heating system must exceed more than just the cost of the system in order to break even economically. If one borrows money from a bank for a wood stove purchase, the interest paid on the loan results in the same situation—the fuel savings must equal the amount of the loan *plus interest* to break even.

There are other costs above the initial direct investment in the system. If the chimney is cleaned once a year by a professional sweep, the annual cost of this service will be roughly $30–100. To clean your own chimney, you must invest in equipment (a brush and some rods, for example) that will cost $50–150. Occasional stovepipe replacement is usually necessary. If the pipe is replaced every 4 years at an installed cost of $40, then there will be an average annual cost of $10.

---

1. The required annual savings for each year for the lifetime of the system in order to break even, expressed as a fraction of the initial investment, is

$$(1 + i)^n i \ / \ ((1 + i)^n - 1)$$

where n is the life expectancy of the system in years and i is the prevailing annual interest rate (such as in money market funds). Values for the formula are computed in the table below.

		Years (n in the formula)			
		5	10	20	40
Annual	5	.23	.13	.08	.06
Interest Rate	10	.26	.16	.12	.10
(i in the formula)	15	.30	.20	.16	.15
	20	.33	.24	.21	.21

For example, if the stove and chimney are expected to last 20 years, and the annual fuel savings were invested in an account earning 15 percent interest, then the annual savings must exceed .16 (from the above table) of the initial cost of the system in order to break even economically. If the system initially costs $1,000, one must save $160 a year in fuel costs just to break even compared to putting the same amount of money into an interest-earning account.

These results are expressed in terms of the value of the dollar at the time of the initial purchase. Thus, the break-even annual savings actually must be increased each year to keep pace with inflation.

If you cut your own firewood, the initial cost and maintenance of the chain saw and splitting equipment must be taken into account, and a portion of the cost of the vehicle used for hauling the fuel.

BREAK-EVEN FUEL COSTS

The annual savings in fuel costs is simply the amount you would have spent on fuel energy to run your conventional heating system if you did not have the solid fuel heater, minus the actual expenditures for fuel for both the conventional and solid fuel systems. After a solid fuel heater has been in use for a heating season, there is nothing complicated about determining this savings. But to decide in advance if it will pay you to install a solid fuel heater, the task has some complexities.

For example, suppose a wood stove is to be installed in a house with an oil-fired central heating system. In order to save money on fuel by using the stove, wood must be cheaper than oil *for the same amount of useful heat delivered to the living space of the house.*

No. 2 heating oil has an energy content of about 139,000 Btu. per gallon. If the oil heater has a net overall energy efficiency of 65 percent, then 90,000 Btu. (0.65 × 139,000) of useful heat is obtained for each gallon of oil burned. If oil costs $1.50 per gallon, then $16.67 [($1.50 ÷ 90,000 Btu.) × 1,000,000] is the fuel cost per million Btu. of delivered heat. This figure is easily adjusted for the prevailing price of oil, but the net heating efficiency of a furnace or boiler is not easily determined, and can vary from 30 percent to 85 percent!

The fuel cost per million Btu. of wood stove heat can be estimated also, but with an additional uncertainty – the energy content of the wood. The energy per cord varies over a large range, from less than 15 million Btu. to over 30 million Btu. depending mostly on wood density. Unfortunately, the density of a given type of wood is variable over a range of about 10 percent, and the amount of solid wood per cord has a larger variation. Thus, the actual energy in a cord of a given wood type has an inherent uncertainty of about 20 percent, and the difference between different types of wood can be 100 percent.

As an example, if wood costs $130 per cord and contains 22.5 million Btu. per cord, and is used in a stove with a net efficiency of 50 percent, wood

TABLE 16–2
SOLID FUEL HEATING FACTORS

Solid fuel system net	very high*	(85 percent)	+2
overall energy	high	(65 percent)	+1
efficiency	average	(50 percent)	0
	low	(40 percent)	−1
	very low	(30 percent)	−2
Solid fuel			
Wood energy content†	high	(30 million Btu.)	+1
per cord	average	(22.5 million Btu.)	0
	low	(17 million Btu.)	−1
OR			
Coal energy content†‡	high	(14,500 Btu.)	+1
per pound (moist,	average	(11,000 Btu.)	0
mineral-matter-free	low	(8,500 Btu.)	−1
basis)			
Fossil fuel heating	very low	(40 percent)	+2
system	low	(50 percent)	+1
net overall energy	average	(65 percent	0
efficiency (for	high	(85 percent)	−1
electricity, use 0)			
		TOTAL_____	

---

*At the time of writing (late 1982) there were probably no solid fuel appliances with efficiencies this high. However since such efficiencies are theoretically possible, the entry is included in the table to cover possible technological advancements.
†As is true throughout this book, higher (or gross) heating values are used for fuels.
‡Much of the coal used for home heating fits into the "high energy content" category in this table. Figure 4–1 may be helpful in determining the energy content of your coal.

Factors needed for using figures 16–1 through 16–11.

heat costs $11.56 per million Btu. [130 ÷ (22.5 × 0.5)].

Figures 16–1 through 16–5 summarize the fuel cost comparisons of wood to oil, natural gas, electricity, LP gas, and kerosene. The heavier line (marked "O") in each figure represents the break-even fuel prices in the average case of a wood stove with a conservatively estimated efficiency of 50 percent, a conventional gas or oil furnace or boiler with a generously estimated efficiency of 65 percent (100 percent for electric heat), and wood with 22.5 million Btu. per cord (high heat value). Under these assumptions, wood heat is cheaper (in terms of fuel costs) if the point on the graph corresponding to the actual prevailing fuel prices falls to the upper left of the solid line. For example, if wood costs $150 per cord and oil costs $1.00 per gallon, the corresponding point on the graph is at the intersection of the vertical $150-per-cord line and the horizontal $1.00-per-gallon line. This point is to the right of the solid

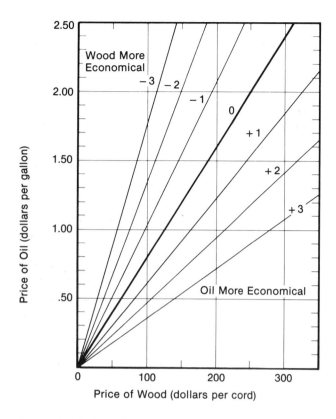

Figure 16-1. Wood versus oil.

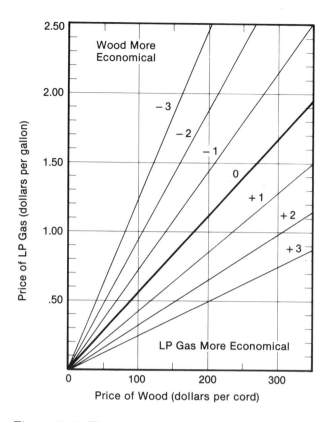

Figure 16-2. Wood versus LP gas.

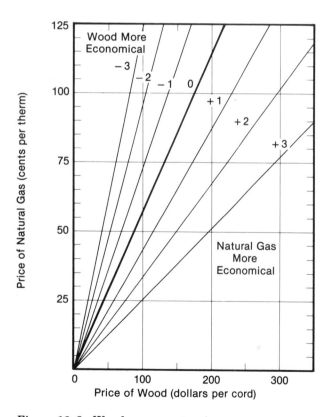

Figure 16-3. Wood versus natural gas.

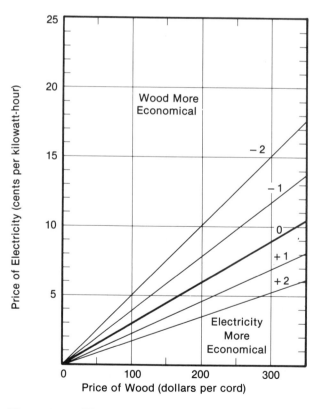

Figure 16-4. Wood versus electricity.

line, indicating oil heat is less expensive than wood heat in this case. If oil were more than $1.20 per gallon, wood would be the better buy, since the intersection of the 2 fuel-cost lines would then lie to the left of the solid line.

The heavy line corresponds to the case of average wood, and average efficiencies for the stove and for the furnace. In practice, most situations are not average in all 3 of these respects. The lighter lines represent the break-even fuel prices for a variety of other conditions. To estimate which line is appropriate in any particular case, assess whether each of the 3 critical characteristics is high, average, or low, assign the appropriate number to each, and total the 3 numbers (see Table 16-2). The appropriate line to use on the graphs is the one labeled with this total. For instance, if the wood stove has a low energy efficiency (for example, a Franklin stove operated mostly with its doors open), the corresponding number for this category is −1. If the available wood has a high energy content (such as hickory or oak), +1 is the score for this category. If the conventional heating system is an old, poorly maintained oil-fired boiler with a low energy efficiency, the number for this category is +1. The sum of these 3 numbers, −1, +1, and +1, is +1;

so in this case the appropriate break-even line on the oil-wood graph is the one labeled "+1." Wood is the more economic fuel in this case if the point on the graph corresponding to the actual prices of oil and wood falls to the left of this "+1" line.

It is apparent from these graphs that, in terms of fuel costs, purchased wood is often competitive with electric heat. For the average case (the heavy line), wood at up to $335 per cord is a better buy than electricity at $.10 per kilowatt-hour. On the other hand, to be competitive with natural gas at $.50 per therm,[2] the average case would require the cost of a cord of wood to be less than $75.

Wood and coal are competitive with each other (approximately) when the price per cord of wood equals the price per ton of coal, *and* when you are comparing comparable quality fuels within each type—for example, comparing high density hardwoods to anthracite or higher rank bituminous coals, or comparing lignite to low density softwoods. For other comparisons, use Figure 16-11.

The comparisons of solid fuels with kerosene (Figures 16-5 and 16-10) apply to unvented kero-

---

2. Natural gas is usually sold by the *therm*, which is 100,000 Btu. of fuel energy, which in turn is approximately 100 cubic feet.

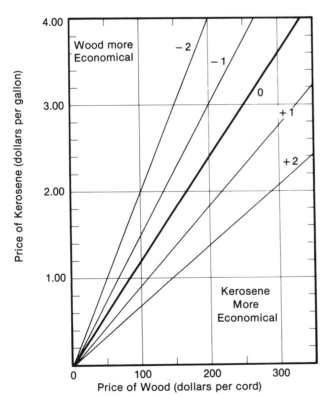

Figure 16-5. *Wood versus kerosene.*

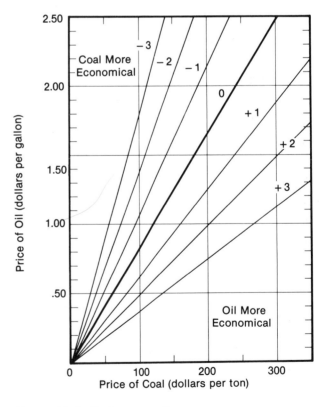

Figure 16-6. *Coal versus oil.*

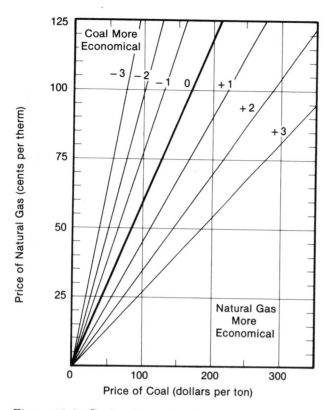

Figure 16-7. Coal versus natural gas.

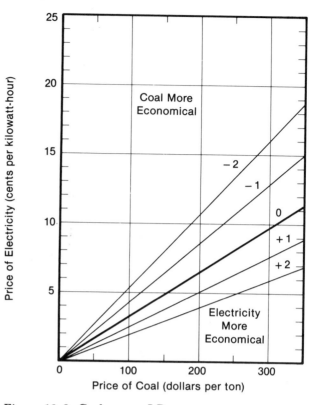

Figure 16-8. Coal versus LP gas.

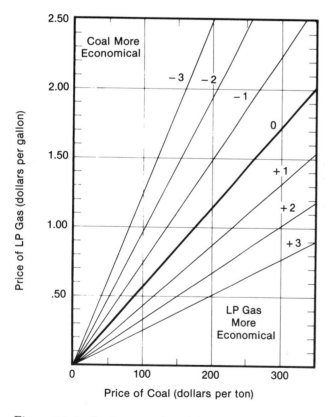

Figure 16-9. Coal versus electricity.

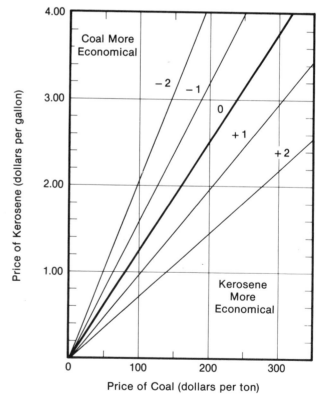

Figure 16-10. Coal versus kerosene

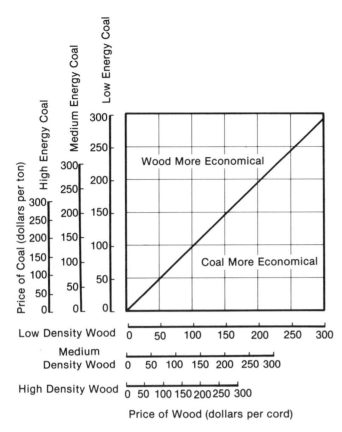

Figure 16-11. Coal versus wood.

sene space heaters, and assume 100 percent energy efficiency for these heaters; since there is no heat loss up the chimney for an unvented heater, heat transfer efficiency is 100 percent. This assumes the humidification effect is desirable so that the latent heat output is counted as being as valuable as the sensible heat output. There is also an assumption here that it is not necessary to open a window for increased ventilation in order to operate the kerosene heater safely. This may not be the case in tight houses especially.

Given these assumptions, and assuming an average stove burning medium density wood (the zero line in Figure 16-5), kerosene at $1.50 per gallon is competitive with wood at $125 per cord.

TOTAL ANNUAL FUEL SAVINGS

If solid fuel is available at a better-than-break-even fuel cost, then how much can one expect to save annually in fuel costs? That depends principally on what fraction of the heating job will be met by the solid fuel appliance.

As an example, suppose a wood stove contributes half of the needed heat to a house. If oil is the other energy source, and if it would have taken $1,000 worth of oil to heat the house without the help of the stove, then $500 is saved by not buying half the oil. The net fuel savings is $500 minus what was spent on wood fuel. If wood was purchased at the break-even price (as determined from the graphs), then the wood would cost $500 and the net fuel savings is zero. If, however, wood was purchased at half the break-even cost, then the wood needed would cost $250, for a net fuel savings of $250 ($500 minus $250). The biggest savings result when the solid fuel has a large cost advantage and a large amount of it is used, displacing more of the conventional energy source.

Many homeowners report much larger savings than would be predicted using the above method. This larger savings is usually possible only with stoves and other room heaters, not with solid fuel central heaters. Room heaters make it possible to heat only a portion of a house. Some seldom-used rooms can be closed off. Outlying rooms far from the heater naturally will be cooler even if not closed off. Not everyone finds these temperature differences to be convenient or comfortable. But when it is possible, a substantial savings results. The total needed heat from whatever source often can be reduced by 50 percent. This is made possible by the haven of warmth in the few rooms heated by the solid fuel appliance. In this case, even if the wood or coal is somewhat more expensive than the oil or gas or electricity it displaces, the net effect can be a substantial savings.

However, *predicting* savings from this effect is difficult. A reasonable procedure is to use published guidelines for savings resulting from thermostat setbacks, using as the setback the *average* decrease in temperature in the whole house. Roughly speaking, each degree (Fahrenheit) of permanent (not just nighttime) setback results in a fuel savings of 2-4 percent.

STEP-BY-STEP: DETERMINE IF IT PAYS TO HEAT WITH SOLID FUELS

1. Initial Costs

Determine or estimate the total initial cost of the system—the appliance itself, the new chim-

Sample Costs of Solid Fuel Heating

Here's an example of costs of solid fuel heating, intended to illustrate the nature of the costs and the computations.

INITIAL COSTS		TOTALS
stove	$ 600	
chimney	300	
installation	100	
	$1,000	$1,000

ANNUALIZED INITIAL COST

Assuming the prime interest rate is 10 percent, add 2 percentage points and multiply times initial costs:

$$0.12 \times 1000 = \$120 \qquad \$ 120$$

ANNUAL UPKEEP COSTS

chimney cleaning	$50	
stovepipe replacement every every few years; average annual cost	10	
chemical chimney cleaner	20	
gasketing, stoveblack, and glass replacement; average annual cost	15	
	$95	$ 95

TOTAL ANNUALIZED INITIAL AND UPKEEP COSTS

$$\$120 + \$95 \qquad \$ 215$$

This is the required annual net savings in fuel costs necessary to break even on the entire enterprise.

NET FUEL SAVINGS

Case A: Suppose the stove has been in use for 1 year. Suppose that records from before the solid fuel heating system was installed indicate about 950 gallons of oil were required to heat the house (you may want to adjust for hot water fuel costs).

Suppose the average oil price over the year just past was $1.75 per gallon.

Then it would cost 950 × 1.75 = $1,662 to heat the house today with oil only (assuming no conservation measures have been added in the meantime, such as storm windows, weatherstripping, or attic insulation).

Now for the year just past with the stove in use, suppose the total oil bill (at $1.75 per gallon) was $925.

Suppose wood fuel was purchased already cut and split, and the total cost for the heating season just past was $450.

Then the total fuel cost was $925 plus $450, or $1,375.

Thus the net fuel savings was $1,662 − $1,375 = $287. Since this exceeds the total annualized initial and upkeep costs of $215, you saved money: $72 ($287 minus $215).

Case B: Suppose you do not yet have a solid fuel heating system and want to predict whether or not it will be cost effective to have a wood stove. We will use the formula for the annual net fuel savings: $fC(1-R)(1-F)$.

- f: If you plan to have the wood stove operating most of the time, then "f" could be 0.5 or larger. Let's assume $f = 0.5$.
- C: Suppose you anticipate it will cost you $1,500 to heat your house with a conventional central, natural gas heating system (based, for instance, on last year's bills).

- R: Suppose natural gas costs $.50 per therm. Suppose the energy efficiency of the wood stove is expected to be high (around 65 percent), that the gas furnace has a typical efficiency of about 65 percent, and that the wood fuel available has a high energy content per cord. Then from Table 16–2, the sum of factors is +2.

Then, using the +2 line on Figure 16–2, the break-even fuel cost for the wood will be about $150 per cord.

Suppose you expect to be able to buy this wood at $125 per *actual* cord (see chapter 3). Then R, the ratio of the actual cost to the break-even cost of the wood, is 125 ÷ 150 = 0.83.

- F: Suppose using the stove will make it possible to let some of the outlying rooms get cool, and that a 25 percent reduction in *total* needed heat will result; then $F = 0.25$.

The annual net fuel savings is then predicted to be:

$$f C (1-R) (1-F) = 0.5 \times \$1500 (1-0.83) (1-0.25) = \$96$$

In this case the net annual fuel savings of $96 is less than the total of the annualized initial costs and the upkeep costs of $215. Thus, heating with wood in this case would result in a net extra cost of $215 − $96 = $119.

ney if applicable, the installation cost, chain saw and splitting equipment if applicable, and so on.

2. Annualized Initial Cost

Multiply the total initial cost by 2 percentage points more than the characteristic prevailing interest rate, such as the prime rate. (Use the decimal form: if the prime rate plus 2 equals 12, use .12.)

3. Annual Upkeep Costs

Add together annual upkeep costs, such as chimney cleaning services, gasketing, stove paint, stovepipe replacement, and so on.

4. Total Annualized Initial and Upkeep Costs

Add together the annualized initial costs and the annual upkeep costs. The total is an estimate of the amount you must save each year in fuel costs in order to break even.

5. Net Fuel Savings

If the system has been installed and in use for a season, use billing records from your pre-solid-fuel years to estimate the cost to heat the house if the solid fuel appliance were not used. Be sure to adjust the cost per gallon or per kilowatt-hour or per therm to values current during the heating season in question. Then subtract the amount spent on solid fuel. The result is annual net fuel savings.

If you are trying to *predict* savings and have not yet purchased a solid fuel heater, estimate with the following formula, which summarizes much of the previous discussion.

The annual net fuel savings is given by:

$$fC (1-R)(1-F)$$

where

C = the annual conventional fuel cost for heating the whole house (with no solid fuel assistance). In new or not yet built houses, a heat loss calculation may be necessary to estimate heat needs and hence costs.

f = the fraction of the total heat supplied to the house which is supplied by the solid fuel appliance. One-fourth to one-half is a reasonable estimate although much more is possible.

R = the ratio of the actual cost of the solid fuel to the break-even cost of the solid fuel. This can be determined using the graphs in this chapter.

F = the fraction of the conventional total house heating which is not now needed because the local or zoned warmth supplied by a wood or coal system makes it possible, without discomfort, to let portions of the house be colder. A reasonable estimate is one-fourth to one-half.

6. The Bottom Line

Take the annual net fuel savings, actual or predicted, as determined by 1 of the 2 methods above. Compare this to annualized initial and upkeep costs computed previously. If the savings exceed the costs, you are in business. If not, you may still be in business; read the next section.

CONCLUSIONS

For many people, heating with wood or coal saves substantial money. But what if you go through all the calculations in this chapter and the conclusion is that you are not saving money by heating with wood or coal, or that you would not save if you bought a solid fuel heating system. Should you give up the idea? No!

Economics is by no means the only factor to be considered. Tending a fire can be a pleasure, not a chore. A room heater (as opposed to a central heater) adds a special kind of comfort to a house. By adjusting either the fire or your distance from the stove, you can always find the desired degree of heating to make your body comfortable. Most stoves and other room heaters can supply heat very quickly, more quickly than can many conventional central heating systems. Most solid fuel heaters also increase your independence—power failures and oil embargoes do not affect the operation of or fuel availability for a wood stove.

So do not be discouraged if heating with solid fuel is not economical for you. It is good to be realistic about the costs, but you may want to take the plunge anyway, because of all the other benefits.

APPENDIXES

APPENDIX 1
Pyrolysis *

A thin slab of wood which is heated slowly and gradually, such as in an oven, so that all of the piece is at the same temperature, will undergo the following changes (it is presumed for the moment that combustion does not occur). For temperatures up to about 212°F. (100°C.), the major effect of heat is to drive out moisture which was present in the wood. This consumes heat energy (such processes are called *endothermic*) since evaporation of water requires energy. Each pound of water driven off requires about 1,150 Btu. of energy — about 1,050 Btu. for the heat of evaporation of water at room temperature, plus about 100 Btu. to break the hygroscopic bond between water and wood.

As the temperature increases from 212°F. to about 540°F. (about 280°C.) a small amount of additional water is driven out, and carbon dioxide, carbon monoxide, formic acid, acetic acid, glyoxal, and probably many other compounds are evolved out of the wood. The processes are still endothermic — heat energy is consumed, not generated. All of the compounds except water and carbon dioxide are combustible, but the actual mixture, including the water vapor and carbon dioxide, may not be flammable (ignitable) due to excessive dilution with the noncombustible gases.

The bulk of pyrolysis occurs between 540°F. (about 280°C.) and about 900°F. (about 500°C.). The reactions evolve heat — they are *exothermic*. Hence, the tempera-

ture of the wood starts to rise spontaneously, even in the absence of oxygen. Large amounts of gases are generated. The most abundant ones are carbon monoxide, methane (the principal ingredient of natural gas), methanol (wood alcohol), formaldehyde, and hydrogen, as well as formic and acetic acids, water vapor, and carbon dioxide. *Many* other compounds are also formed (Table A1–1). The mixture is sufficiently rich in combustibles that it is ignitable and serves as the fuel source for the large wood flames. By the time the temperature of the wood reaches about 900°F., pyrolysis is essentially complete and the final solid product is charcoal. If denied oxygen so that it cannot burn, charcoal is stable to very high temperatures — charcoal does not melt until its temperature reaches about 6,300°F. (3,480°C.) and its boiling point is about 7,600°F. (4,200°C.), temperatures much, much higher than the highest temperature available in a wood fire.

In a normal wood fire, the wood is not heated so slowly and uniformly that it is all the same temperature. The inside of a piece of wood may not even be warm when the outside is already fully charred and burning. Additional "secondary" pyrolysis reactions take place. Gases and tar generated below the surface of the wood must pass through the surface layers on their way out of the wood. Water vapor can react with carbon in charred layers and become carbon monoxide and hydrogen gases. Carbon dioxide can be similarly transformed into carbon monoxide. These reactions consume heat — they are endothermic. The tar droplets from inside the wood also can react with charcoal, and this reaction evolves heat (it is exothermic).

---

* Most of the following information is from F. L. Browne, *Theories of the Combustion of Wood and Its Control*, Report No. 2136 (Madison, WI: Forest Products Laboratory, 1963).

TABLE A1-1
COMPOUNDS IN WOOD SMOKE

Alkenes	Methylfluorene	Methylpyrene	*Furans*
Cycloalkanes	Phenol	Benzopyrenes	Methyl furan
Cycloalkenes	Cresol	Dibenzanthracene	Benzofuran
Aromatic Hydrocarbons	Methoxyphenol	Benn(a)anthracene	Dibensofuran
Benzene	Phenols Dimethylphenol	Trimethylnaphthalene	Furfural
Toluene	Xylenol	Fluoroanthene	Methylfurfural
Xylene	Quaiacol	Benzofluoranthenes	Furanmethanol
N-propylbenzene	Pyrogallol	Acenaphthene	Tetrahydrofuranmethanol
Isopropylbenzene	*Polynuclear Aromatic*	Coronene	*Alcohols*
Mesitylene	*Hydrocarbons*	Benzo(ghi)perylene	Methanol
Methoxybenzene	Naphthalene	Methylcholanthrene	Ethanol
Benzonitrite	Methylnaphthalene		*Aldehydes*
Indane	Phenylnaphthalene	*Carboxylic Acids*	Formaldehyde
Indene	Phenanthrene	Formic acid	Acetaldehyde
Methylindene	Methylphenanthrene	Acetic acid	Benzaldehyde
Indanone	Anthracene	Propionic acid	*Ketones*
Fluorene	Biphenyl	Butyric acid	Acetone
	Pyrene		Diacetyl

These are some of the compounds reported to be in wood smoke.

APPENDIX 2

Human Temperature Sensation

If you grasp a pot or rack in a hot over, it hurts. It is hot. But, there are other hot things which, when touched, do not burn. Aluminum foil or fiberglass in a hot oven can be touched with no accompanying feeling of hotness. Red hot coals that accidentally fall out of a stove can be picked up and juggled back into the fire with your bare hands with no ill effects (but be careful!). The foil is as hot as the pot, and a glowing coal is much hotter — yet they do not *feel* hot, and they do not necessarily burn. Why?

Skin is not a very good thermometer. There is lot of flesh around the few cells in skin which are the body's temperature sensors. Each sensor cell actually does a fairly good job of sensing its own temperature in the skin; but the temperature of the sensor cells is not necessarily the same as that of the surface being touched.

A hot, heavy metal pan is capable of heating your skin to a painful temperature. Because of its heaviness, it has a lot of heat stored in it, and because it is a metal, that heat can conduct quite readily into your skin, making it nearly as hot as the pan.

Hot aluminum foil does not burn when touched. It is so thin that despite the fact that aluminum is an excellent conductor of heat, the total amount of heat in the thin foil is so small that when distributed in your skin, your skin becomes only warm, not hot. A hot glowing coal can be touched momentarily without burning be-

cause its capacity to store heat is not great. It is a poor conductor of heat, and a thin layer of ash and/or cooler charcoal is often between your finger and the coal while you're touching it.

THERMAL COMFORT

Air temperature is not the only factor in your environment that affects your thermal comfort. The other factors that also influence your feeling of adequate warmth are the humidity of the air and its motion, and the amount of radiant energy traveling through surrounding space.

Air *feels* hotter when it is humid, even though its temperature (as measured more objectively with a thermometer) is the same. At room temperature, the effect is less often noticed, but is there. Many experiments have been done using a set of rooms with very carefully controlled temperature and humidity. As human subjects move from room to room, they state whether each new room feels warmer, the same, or cooler than the previous room. From these studies, it was found that the 3 sets of conditions, summarized in the table below, all give the same feeling of warmth — they all feel like the same temperature:

Actual Temperature (°F.)	Relative Humidity (%)	Sensation* of Temperature (°F.)
71.5	0	70
70.0	50	70
68.5	100	70

* This subjective or "effective" temperature scale is made to agree with the actual temperature when the relative humidity is 50%.

In other words, a person who is comfortable at 70°F. and 50 percent relative humidity will be just a as comfortable at a slightly cooler temperature if the humidity is higher. He or she will also be just as comfortable at a higher temperature, if the air is made appropriately drier.

Alternatively, if the actual temperature in the test rooms is the same and the humidity is varied, then a person's subjective temperature sensation varies as follows.

Actual Temperature (°F.)	Relative Humidity (%)	Sensation of Temperature (°F.)
70	0	68.5
70	50	70.0
70	100	71.5

In other words, humid air feels warmer than dry air when the actual temperatures are identical.

The explanation for this effect is related to evaporation of water from the skin. Evaporation is a cooling process—molecules of water need extra energy to break away from their neighbors and become water vapor. This energy comes partly from the air over your skin, but mostly from the skin itself. Having lost energy, it is cooler (unless the energy is replaced from some other source). When the air is dry, evaporation is enhanced and your skin loses more energy; when the air is more humid, evaporation rates becomes less and your skin loses less energy.

The decreased energy loss from your skin in humid air can be compensated for by decreasing the temperature of the air. As indicated previously, in the vicinity of 70°F., an increase in humidity by 50 percentage points can be balanced by a 1.5°F. decrease in temperature.

It does not follow that humidification saves energy.

In a house with a humidifier, the thermostat can be set somewhat lower than in a house without humidification and the feeling of air warmth in both houses will be the same. However, the energy necessary to humidify the air is usually far greater than the decrease in the sensible heat loss from the house.[†] This is true regardless of the type of humidification, including such simple humidifiers as house plants. The net effect is almost always an increase in total energy consumption (usually of the furnace).

Movement of air also influences thermal comfort. Increased air speed enhances conduction of heat out of the skin and evaporation. At typical room temperatures and relative humidities, an air velocity of 3 feet per second makes the air feel about 4°F. cooler than stationary air at the same actual temperature.

The effect of infrared radiation on thermal comfort is especially important, since so much of the heat output of stoves is radiant. Thermal comfort depends on skin temperature. One can be perfectly comfortable in cool, dry air despite the skin's loss of energy by conduction (to the cool air) and evaporation (into the dry air) as long as enough infrared radiation is absorbed by the skin. Very few home heating systems are designed this way, with cooler air temperatures being compensated for by higher infrared radiation intensity. One reason is the difficulty of having the radiation coming from all directions, so that an occupant will be equally warm on all sides. In a house heated only with an open fireplace, one's back, away from the fire, can be much cooler than one's front. This is due both to unequal amounts of radiation coming from the 2 directions, and to drafts. In a typically snug house heated with a closed stove, there is little if any discomfort from such temperature contrasts.

In winter, when exterior walls and windows are cooler, the infrared environment is less intense than in warmer weather. The air is also drier.Thus one's skin loses more energy by both radiation and evaporation. Hence, air temperatures need to be a little higher (particularly in rooms with exterior walls) to produce the same sensation of human comfort. Since thermostats regulate air temperature, they need to be set a little higher in the colder weather to provide the same degree of thermal comfort, other things, such as clothing, being equal.

† J. W. Shelton, "The Energy Cost of Humidification," *ASHRAE Journal*, January (1976), pp. 52–55.

Moisture Content Scales

There are 2 common ways of reporting moisture content in wood. In this book, moisture content is based upon the weight of the moist wood.

Moisture content (moist wood basis) =

$$\frac{\text{weight of moisture removed in oven drying}}{\text{initial weight of wood, including its moisture}}$$

Using this scale, wood that is half water by weight has a moisture content of 50 percent. A second way to report moisture contents is based on the oven-dry weight of the wood:

Moisture content (oven-dry wood basis) =

$$\frac{\text{weight of moisture removed in oven drying}}{\text{weight of oven-dry wood}}$$

Using this scale, wood that is half water by weight has a moisture content of 100 percent.

To facilitate comparisons between writings using the 2 conventions, Table A3–1 gives conversions.

TABLE A3-1
CONVERSIONS BETWEEN MOIST WOOD AND OVEN-DRY WOOD SCALES

Moisture Content on an Oven-Dry Wood Basis in Percents	Moisture Content in Either Scale in Percents	Moisture Content on a Moist-Wood Basis in Percents
0%	0%	0%
5.3	5	4.8
11.1	10	9.1
17.6	15	13.0
25.0	20	16.7
33.3	25	20.0
42.9	30	23.1
53.8	35	25.9
66.7	40	28.6
100.0	50	33.3
150.3	60	37.5
233.0	70	41.2
Infinite	100	50.0
– –	150	60.0
– –	200	66.7
– –	250	71.4

Conversions between moisture contents as expressed in the moist wood and oven-dry wood scales. To use the table for either conversion, find the value to be converted in the center column. Then, to convert from dry to moist basis, read the adjacent number in the right column. To convert from moist to dry, read the adjacent number in the left column. If m and d represent the moisture contents on the moist-wood and dry-wood bases respectively, expressed as fractions, then m = d/(1+d), and d = m/(1 − m).

Efficiency Measuring Methods
For Solid Fuel Heaters *

The objective of a standard test method is to determine a set of test conditions, including fuel species, piece size, moisture content, load size, stacking geometry, appliance control settings, ambient temperature, and measurement methodology

- Which will yield moderately realistic performance results,
- Which will preserve typical real-world rankings of appliances,
- Which gives reproducible results, and
- Which does not make the testing overly expensive.

Determining overall energy efficiency requires knowing the useful heat output and the fuel energy input. Fuel energy inputs are determined by weighing the fuel entered, measuring its moisture content, and measuring its higher heating value.

There are 3 *basic* approaches for determining the useful heat output: direct calorimetry, stack loss determination, and *in situ* co-heating. These approaches are not exclusive and in fact combining elements of 2 of them may yield a better testing approach than any 1 by itself.

ROOM CALORIMETRY

The most direct approach uses a calorimeter room, a device which measures the heat output of whatever is inside it. There are many possible designs, but such rooms are typically very well insulated and very airtight. Either air or water is used to transfer the heat generated inside the room to the outside. The basic measurements needed to determine heat output are the flow and temperature rise of the heat transfer fluid (air or water).

Calorimeter rooms must be carefully designed and operated so that the performance of the appliance is normal. Essentially neutral pressure must be maintained in the room so that the draft is the natural draft due to the chimney. The appliance must not be subjected to any significant forced convection due to the air moving system of the calorimeter room. This can easily be arranged by using plenums, baffles, screens, and so on. The thermal environment – air and radiant temperatures – must be typical of a home environment.

---

* For a more complete discussion, see "Thermal Performance Testing Methods for Residential Solid-Fuel Heaters," an SER publication. See Appendix 7.

Again this is easily done by reasonably straightforward engineering.

A simple calorimeter room test of an appliance yields a continuous record of heat output rate (power output) over time. From this record, one can observe the steadiness of the heat output, the burn duration, the instantaneous effects on power output from adjustments of the controls, and, by integrating the curve, the total heat output over time.

Other Types of Calorimetry

Room calorimetry is 1 way to measure heat output directly. Another method is to measure radiant and convective contributions separately using radiometers and air flow, water flow, and temperature-measuring devices. This method has been used extensively in England for measuring the performance of coal stoves, particularly those installed in masonry surroundings such that only 1 side effectively radiates useful heat. Thus, a hemispherical array of radiometers is adequate. In some instances, sheet metal shrouds have been used to collect the convected heat. Here there is a danger that the heat transfer may be modified by the measuring device.

Another approach for radiant heaters is indirect calorimetry, wherein surface temperatures are measured and resultant radiant and convective heat transfer is calculated. This method is not very accurate usually.

One of the advantages of room calorimetry is that it is indifferent to whether the heat output from the appliance is radiant or convective; it all gets measured regardless of its original form. Stack loss methods are similarly indifferent.

STACK LOSS METHODS

There is a very wide variety of stack-loss methods. They all involve assessing the energy lost up the flue and then using the fact that whatever fuel energy was released inside the appliance, but which did not leave the appliance with the flue gases, must have passed out of the heating system and into the house. Thus, for purposes of measuring heat output, the method is often labeled "indirect."

A common procedure is to measure stack temperature and concentrations of CO_2, CO, and O_2. If the appliance is weighed continuously during the test (to determine the fuel consumption rate), one can then compute the heat output and energy efficiency throughout the burn.

However, this computation involves a large number of approximations and assumptions.

First, the energy content of the fuel and carbon, hydrogen, and oxygen mass fractions must be known. If power outputs (as well as energy efficiencies) are required, the actual release rates of energy, carbon, hydrogen, and oxygen must be known also. In the case of ordinary liquid and gaseous fuels delivered to a combustion chamber at a constant rate, all these quantities are easily knowable.

In the case of a wood stove, continuous weighing of the stove with its fuel can yield the mass release rate of the fuel, but *the inhomogeneous manner in which wood burns makes it impossible to infer the elemental and energy release rates from the mass release rate* (or just to know the energy content and elemental composition of the burning wood), *despite the fact that the composition and energy content of the "raw" wood are known.* In principle, initial warming of the wood can result in only the evaporation of some of its moisture content, resulting in a negative energy release rate and a zero carbon release rate. Towards the end of most burning cycles, the composition of the "wood" remaining approaches that of charcoal, which has nearly twice the energy content of wood; at this stage, the oxygen and hydrogen release rates from the fuel are nearly zero. The composition of the *burning* wood changes.

Second, there must be no unaccounted-for chemical energy in the flue gas. Chemical energy losses from wood stove fires can constitute 30 percent of the wood energy input. A significant portion of this loss is in chemical forms other than CO. Thus it is critical that *when the stack loss method is applied to wood heaters, the chemical energy density of the flue gases must be known. Assessing CO alone does not adequately accomplish this.*

To get around this problem, it is often assumed that the chemical energy in forms other than CO is properly accounted for by assuming it is in the form of CH_4, where the amount of this compound is determined by balancing a chemical equation which itself is an approximation of wood combustion. There are, of course, other compounds than CO and CH_4 in the flue gases which are incompletely burned. The key question is whether or not the actual energy content of the flue gas is adequately obtained via the assumption that CO and CH_4 are the only contributors. The assumption is clearly wrong, but it is possible that the resulting error is negligible or that the error is balanced out by other errors. It may be necessary to measure other flue gas components, such as hydrocarbons, in order to refine the estimate of chemical energy loss.

Third, steady-state thermal conditions must prevail. At times of measurement, there must be no net flow of heat into or out of the material of the stove or stovepipe. Most solid fuel appliances operate most of the time with substantial fluctuations in combustion rate. Particularly in massive appliances, net heat flow into or out of the appliance mass may be substantial. If this is neglected, the computed so-called instantaneous efficiencies will tend to be artificially high while the appliance is heating up, and low while it is cooling down. There is some compensation when the values are averaged, but averaging does not necessarily eliminate all the distortion.

Although each of the above assumptions/approximations may be fairly serious in itself, it is possible that there is compensation in the resulting errors—that the computed average efficiency over 1 or more complete burn cycles comes out to be reasonably accurate, say within 3 or 4 percentage points of the actual value. Then the method would have merit because it could be less expensive than room calorimetry. Laboratory work continues on checking the accuracy of this method and investigating refinements which would improve its accuracy.

IN SITU COHEATING METHOD

The third basic method for measuring thermal performance involves testing *in situ*. The appliance is in a home, not a laboratory, but also not just any home. The heat loss characteristics of the home must be well known, including effects of wind and solar radiation. There must also be some kind of alternative heating system—either conventional or specially designed—whose energy consumption and/or heat output can be accurately monitored. The basic approach is then to monitor the input and/or output of the conventional system over periods both when the solid fuel appliance is in use and when it is not. The method has been used by a number of researchers.

This method seems to go right to the heart of the matter for the homeowner who wants to know how much oil, gas, or electricity can be saved. *Net* overall energy efficiency can be measured, including all losses and gains due to home/appliance interactions—chimney heat contributions, changes in air infiltration, some kinds of standby losses, and altered temperature distributions in the house. However, these effects are different in every house. Thus, the particular results apply only to the test house. The method is also very time consuming. In addition to the time necessary to instrument and calibrate the house, each appliance must generally be tested for weeks to obtain useful data.

The main value of this type of testing is in quantifying the appliance/house interactions, not in quantifying the performance of the appliances themselves. These interactions are very important to understand, for they can be large.

Energy Conservation Myth: Limited Thermostat Setback at Night

There seems to be disagreement over the energy-conserving potential of short-term or temporary (overnight) thermostat setbacks. Many published lists of energy-saving tips seem to suggest that there is some optimum amount of setback which, if exceeded, will result in greater energy consumption. For instance, Western Mass. Electric Company distributed a pamphlet which includes the suggestion, "Set thermostat to provide a comfortable temperature and avoid constant readjustments. Don't lower the thermostat more than 6 to 8 degrees at night."

The suggestion that larger setbacks than the suggested 6–8 degress may result in larger energy consumption seems to be based on the notion that if the structure itself (walls, floors, ceilings) and its contents get too cold during the setback period, more energy may have to be used to reheat all this mass than was saved during the setback period, thus resulting in a net increase in energy consumption.

In fact, this notion is nonsense. The energy necessary for warmup after a setback almost never exceeds the savings during the setback. The reason is based on the fact that temperature losses depend on temperature differences. However, when temperatures are changing rapidly, as they are likely to with frequent thermostat adjustments, the heat capacity of the structure may have to be taken into consideration in calculating conductive losses. Heat conducted out of a building is finally lost only after penetrating the walls or roof, since a sudden drop in temperature inside the structure can permit some of the heat stored in the walls to return to the heated space. Thus, we really need to know the temperature difference (or gradient) at the outermost layer of the building's exterior. Examination of the fundamental heat-conduction equation shows that after a thermostat setback this temperature gradient at the outer part of the walls and roof can never exceed its steady-state value before the setback (assuming constant outdoor conditions). And in fact, it is always less, indicating that energy is always conserved, regardless of the duration or amount of setback. (Not all heat is lost by conduction; in typical homes, half of the total heat loss is due to air exchange. This loss is closely related to the indoor-outdoor air temperature difference and is clearly reduced any time the indoor air temperature is reduced.)

The amount of energy saved by night setback depends on climate, the heat-loss coefficient of the building, the normal set point, the heat capacity of the building, and the heating-plant capacity. A 5–15 percent savings can be expected under typical circumstances for night setbacks of 5–10°F. The savings are not directly proportional to the amount of setback, but can easily exceed 50 percent for large setbacks in mild climates. And in fact, the energy savings are often larger than just the decreased heat loss. The efficiency of most heating systems increases with the intensity of use. During the pickup time after a setback, the heating system is likely to be "on" for a longer time than usual, during which its efficiency is higher. In addition, refrigerators, freezers, and water chillers will consume less energy during setbacks because of their cooler environments.

Larger setbacks *can* result in *larger* energy consumption in systems with proportional control, such as a system in which the high demand during the pickup period might call into service an additional or auxiliary heating system which was less efficient than the basic unit. However, such a circumstance is unnecessary and would most likely be the result of poor design or inadequate maintenance.

The usual practical limit to the amount of high setback is comfort, perhaps at night, but more usually in the morning if one does not have an automatically timed thermostat. The fast response of most hot-air systems may allow larger setbacks. In those circumstances where possible discomfort associated with morning pickup time does *not* provide a limit to the amount of setback, the health of house plants may, or ultimately, the prevention of freezing damage to water pipes.

In engineering new houses, there are some difficult choices. Although temporary thermostat setback always saves energy, the amount of setback tolerable from a comfort standpoint may depend on the pickup time; that is, on the capacity of the heating system. The larger the capacity of the heating system, the shorter the warm-up time and hence the larger possible setback. However, oversized heating systems (except electric resistive heating) have lower efficiencies. The optimum choice thus depends on the style of the inhabitants. Also, massive construction, as for instance in passive south-window solar heating systems, does not permit as much savings from letting the inside temperature fall at night because it will not fall very much. That is a beauty and liability of such systems; they respond slowly to thermal perturbations.

In summary, the *limited* aspect of night setback does not belong in an energy conservation list but perhaps in a discussion of comfort. From the point of view of energy conservation alone, the larger the setback the better, regardless of its duration.

"Available" Energy
In Wood Fuel

Heating values of wood most often are used in 2 ways associated with measuring and calculating the heat outputs of appliances and their overall energy efficiencies. Calculating energy efficiencies always involves comparing the actual heat output to the maximum theoretically possible heat output. The maximum possible heat output is the mass of fuel consumed times its heating value. (For the moment I am being intentionally vague about *which* heating value.) That is, 1 common use of the heating value is in the denominator when computing energy efficiencies.

The other common use of heating values is for computing heat output in cases where it is not measured directly. For instance, if stack losses are measured, then, using energy conservation, the difference between the energy content of the fuel and the stack losses must be the heat output of the appliance.

Both these uses are illustrated in the following equivalent expressions for overall energy efficiency:

$$\frac{\text{heat output}}{\text{wood energy input}} =$$

$$\frac{\text{measured heat output}}{HV \times M} =$$

$$\frac{\text{wood energy input} - \text{stack losses}}{\text{wood energy input}} =$$

$$\frac{HV \times M - L}{HV \times M} = 1 - \frac{L}{HV \times M}$$

where
HV = heating value of fuel
M = mass of fuel
L = measured stack losses

Does green wood have less "available energy" than seasoned wood? In practice (as discussed in detail in chapter 14), using green wood indeed *usually* does result in lower energy efficiencies. But there are exceptions, depending on the type of heater and on how it is operated. And these exceptions indicate that it is *not* true that green wood has less available energy.

In part, to help clarify discussions of the effect of moisture in wood combustion, 2 concepts for the energy content of fuels are used—the higher heating value (also called gross heating value) and the lower heating value (also called net heating value). Only the higher heating value should be used, as explained below. In practice, both are used, and this can lead to con-

siderable confusion. Because of this and other factors, heating values given in various books and articles range from less than 5,000 Btu. to over 9,000 Btu. per pound.

The conceptual definition of "heating value" is the most heat that could possibly be obtained by burning a given amount of fuel. For typical wood at zero moisture content, 8,600 Btu. per pound is typical for the "higher heating value."

The difference between the higher and the lower heating values is in how water vapor is handled in the calculations. When wood is burned, the combustion products contain water vapor from 2 sources. All but oven-dry wood contains moisture. When the wood is burned this moisture is evaporated or boiled off. This process requires energy—about 1,050 Btu. per pound of water. Perfectly dry wood also has water vapor in its smoke; when wood burns, the hydrogen atoms in the wood combine with oxygen to become water molecules. Whatever the source, water vapor contains considerable potential energy, called latent heat, which is released as ordinary sensible heat if the vapor condenses to liquid water. For every pound of water vapor which condenses, about 1,050 Btu. of heat are released.

Condensing the water vapor and using the released heat would increase the energy efficiency of wood heating systems. However, when wood is burned in most wood heaters, the water vapor in the smoke rarely condenses, and when it does, it usually happens in an exterior chimney where the released heat is not used. Thus, since in most heating systems the latent heat part of the energy is not used for heating, its contribution is frequently subtracted out from the higher heating value, yielding the lower heating value.

Despite this rationale for using lower heating values, all test results reported in North America should use *only* the higher-heating-value concept in computing efficiencies and heat outputs for 4 reasons.

• Using lower heating values leads to the possibility of efficiencies exceeding 100 percent. This is intolerable. Even relatively unknowledgeable consumers would laugh at efficiencies over 100 percent. If such appliances do not now exist, they may in the future.

• The most respected organization of heating engineers, the American Society of Heating, Refrigerating and Air Conditioning Engineers, Inc. (ASHRAE), specifies the use of higher heating values in computing efficiencies.

• The gas, oil, and kerosene appliance industries use only higher heating values in computing efficiencies. For the sake of comparability, we should too.

• The Wood Heating Alliance specifies the use of higher heating values in computing efficiencies.

In Europe, lower heating values are used. Thus, one most be careful in interpreting reported efficiencies. Corresponding European (lower heating value) and American (higher heating value) efficiencies are indicated in Table 3–2. Based on the same test data, European-calculated overall efficiencies are about 10 percent (not percentage points) higher than American efficiencies.

Another independent source of confusion is the unit of wood used. In laboratories, higher heating values are most often expressed on a per-pound-of-oven-dry-wood basis. A typical value is 8,600 Btu. per pound. The actual wood may have any moisture content; it need not be at zero moisture content. But the higher heating value is expressed per pound where it is assumed that the pound is the weight of the pure wood fiber in the piece of wood, without its water content. Using this concept for assaying the amount of fuel, the higher heating value of a particular log or of a particular pile of wood is independent of its moisture content. Adding or taking away moisture does not change the amount of fuel, just as adding a little water to a tank of gas does not change the amount of gas. Such added water may make the fuel harder to burn, but if it does burn, the total amount of energy released in the process is the same. (If the resulting water vapor is condensed, then the heat used to evaporate can also be *recovered.*)

The alternative concept for the amount of wood is more common in the field. The unit of wood, such as the pound, for which the higher heating value is given includes the weight of any moisture in the wood. If wood has a moisture content of 50 percent on the moist basis, it is half water by weight. Thus, using this second concept for assaying the wood, the heating value of the wood would be half of 8,600, or 4,300 Btu. per pound—because a pound of wood at 50 percent moisture content contains only half a pound of pure dry wood. This has nothing to do with the energy that may or may not be lost when the water evaporates and is recovered if some of the water vapor condenses; it is just that a portion of the weight of green wood is moisture, not dry wood fiber.

To avoid possible confusion in using heating values, it is wise to use terminology such as the following:

• Higher heating value, dry basis *(HHVDB)*
• Higher heating value, moist basis *(HHVMB)*

Using either of the 2 definitions leads to the same results as long as the appropriate wood weight is always used. The total energy available from W pounds of wood at a moisture content m (moist basis, expressed as a decimal or fraction) is:

$$HHVDB \times ((1-m) \times W) = (HHVDB \times (1-m)) \times W = HHVMB \times W$$

If HHVDB is used, it must be multiplied by the dry wood portion of the total wood weight—that is, by $(1-m) \times W$. But grouping the factor $(1-m)$ with HHVDB yields HHVMB, which is then appropriately multiplied by the full fuel weight, W.

The following numerical examples illustrate the various heating value concepts.

Assume a higher heating value, dry basis of 8,600 Btu./lb. Then for wood at *zero moisture content*, the higher heating value, dry basis, is

$$HHVDB = 8,600 \text{ Btu./lb.}$$

and the lower heating value, dry basis, is

$$LHVDB = HHVDB - 0.54 \times 1050$$
$$= 8,600 - 567 = 8,033 \text{ Btu./lb.}$$

Note: 0.54 is the pounds of water vapor produced in burning 1 pound of wood at zero moisture content, assuming the hydrogen content of the wood is 6 percent.

For wood at zero moisture content, dry basis and moist basis wood weights are the same; thus, for zero moisture content wood only

$$HHVDB = HHVMB = 8,600 \text{ Btu./lb.}$$
$$\text{and } LHVDB = LHVMB = 8,033 \text{ Btu./lb.}$$

For wood at a *moisture content of 20 percent* on the moist basis (which is the same as 25 percent on the dry basis), a 1-lb. log contains 0.8 lb. of dry wood, and 0.2 lb. of water.

Then $HHVDB$ = 8,600 Btu./lb.
$HHVMB = HHVDB\,(1-m)$
$= 8,600\,(.8)$ = 6,880 Btu./lb.
$LHVMB = HHVDB\,(1-m) - 1,050m$
$- 0.54\,(1-m)\,(1,050)$
$= 6,880 - 210 - 454 = 6,216$ Btu./lb.
$LHVDB = LHVMB \div (1-m)$
$= 6,216/.8$ = 7,700 Btu./lb.

Note: The figures on page 21 of *The Woodburners Encyclopedia* are for wood with the same moisture content (there is a typographical error on that page) but on the basis of a piece of wood whose oven-dry weight would be 1 lb. Thus the piece actually weighs 1.25 lb. The calculations above yielded 6,216 as the lower heating value per pound as fired (moist basis). Thus for a piece weighing 1.25 lb., the lower heat content would be $1.25 \times 6,216 = 7,700$ Btu., which is the number given in *The Woodburners Encyclopedia.*

Since fuelwood is usually purchased by the cord, heating values per cord are useful measures. Using a volume basis such as a cord, higher heating values are independent of the moisture content of wood.* The

---

* This neglects the small shrinking/swelling as moisture content changes.

amount of fuel or chemical energy in a cord of wood is not changed by the presence or absence of moisture. Spraying water on a pile of wood does not change the amount of fuel when assessed in terms of its higher heating content.

Figures 3-1 and 3-2 summarize the concepts of both higher and lower heating values, using all 3 common bases—moist wood weight, dry wood weight, and wood volume.

OTHER DEFINITIONS OF HEATING VALUES

Many other definitions of heating values are possible. Some of the possible variables of each definition are

- Higher or lower heating value concept.
- Oven-dry basis or as-fired basis.
- Moist or dry basis for moisture contents.
- Water vapor from evaporation of moisture only or also from combustion of wood.
- Water vapor losses as latent heat only, or latent plus sensible heat.
- Flue-gas temperature if sensible heat in water vapor is included.

Clearly there are a large number of possible definitions of heating values or available energies. Some have stronger scientific rationale than others, but regardless of rationales, only 2 definitions are generally recognized, and of these only 1—the higher heating value—is generally accepted as being appropriate for computing and reporting energy efficiencies in North America.

APPENDIX 7

Shelton Energy Research

Most of the experimental results discussed in this book are from tests conducted by Shelton Energy Research (SER), and many research reports cited in this book are available through SER.

SER evolved from my research on wood stoves while in the physics department of Williams College in Williamstown, Mass. As part of a student project, the first calorimeter room in North America designed for wood stoves was built in the summer of 1974. In January, 1975, I taught a special course at Williams College on testing wood stoves. The traditional stack loss method was used in the laboratory part of the course. The stove rested on a scale so the rate of wood consumption could be measured, while an "Orsat" apparatus was used to measure CO_2, O_2, and CO concentrations in the flue gases.

At this point, my primary interest in testing stoves shifted from the educational value for students to the research itself. Very little scientifically valid research had been done on the performance of residential solid fuel heaters. Thus, there was a real need for the research, and the potential for scientific excitement in discovering facts.

For serious research, the choice was made to depend primarily on room calorimetry as the test method. The stack loss method, as traditionally practiced, involved too many assumptions, the validity of which was unknown (see Appendix 6). Thus, North America's second calorimeter room designed for wood stove testing was designed and built in the spring and summer of 1975 at Williams College. Soon thereafter, selected flue-gas instrumentation was added to increase the amount of data obtained. A system for measuring flue-gas veloc-ity directly and continuously was developed as part of this instrumentation. The combined systems yielded direct and continuous measurement of the heat output rate of the appliance and the sensible heat loss rate up the flue. This, combined with other data, yielded all 3 energy efficiencies—overall, combustion, and heat transfer.

With this equipment, student assistants and I conducted a number of basic studies on wood stoves and combustion. Initially, it was thought that a survey of basic stove design types would be useful, but it soon became apparent that performance was affected substantially by variables other than design.

Thus, approximately 2 years of research were devoted primarily to studying the effects of fuel moisture content, fuel species, fuel load size, and fuel piece size. Some generic design features were also investigated, such as firebrick liners, secondary air, and baffles. This work was supported largely by Williams College, which covered the laboratory overhead, by the National Science Foundation, which funded student summer stipends, and by myself, through the time I contributed.

A proposal (I.D. No. BB-U6-2014) was submitted by Williams College in January, 1976, to the Energy Research and Development Administration. It was not funded for reasons that are interesting in the context of today's knowledge and activity. Two reviewers' comments were: "Stove design and research of stove design is nothing new, and one of the most effective designs still in use was that done by Benjamin Franklin." And, "From the scientific merit aspect, an analytic model for efficiency based on existing theory and data from large units, and related to what would be expected from small units is probably the best approach. Enough existing data from small units is likely available to verify the model."

Since 1979, Shelton Energy Research, an independent laboratory, partially funded by government agencies, manufacturers, and publishers, has been located in Santa Fe, New Mexico. Santa Fe is adjacent to the Santa Fe National Forest, and wood heating is common in both Santa Fe and the entire northern part of the state. Santa Fe is sometimes mistakenly viewed as being similar in climate to Albuquerque, Phoenix, and Tucson. To the contrary, Santa Fe has 4 seasons, with a real winter; snow is common. Summers are mild, not sweltering. Santa Fe's 6,000 degree-day heating season is comparable to Albany, New York, at 6,200. In an average summer, the temperature exceeds 90°F. only 1 percent of the time. The outdoor relative humidity is only about 5–15 percent. Evaporative coolers maintain indoor conditions at SER's laboratory at around 70°F., with 30–50 percent relative humidity.

Most importantly, however, SER, like many other testing laboratories, uses testing methods that are intentionally designed to be insensitive to weather, in order that the results can be reproduced from day to day, season to season, and from laboratory to laboratory. For example, most performance testing at SER involves a silo around the chimney so that the natural draft generated by the chimney is not affected by weather (temperature or wind). Thus, the time of year and the weather need not affect test results.

SER's third generation calorimeter room was constructed in 1980 and is so sensitive that it can detect the heat output from a 15-watt light bulb.

SER has always pioneered in the field, specializing in areas where standards have not yet been established. The laboratory's research and testing experience encompasses a diversity of interests, ranging from basic research on combustion and creosote to testing of products, and development of test methodology and instrumentation. Specifically, SER's experience includes:

- A wide variety of stove designs.
- Fireplaces.
- Inserts.
- Accessory products, including fireplace accessories, creosote-reducing devices, and barometric draft controls.
- Chemical chimney cleaners.
- Prefabricated chimneys.
- Chimney fires.
- Catalytic combustion.
- Variables affecting creosote formation, including fuel species, moisture content, power output, and dilution air.
- Creosote and heat transfer efficiency.
- Operator variables affecting stove performance, including load size, piece size, moisture content, species, and power output.
- Stack loss and calorimeter room testing methods.
- Improved instrumentation for measuring flue-gas velocity.
- Woodstove emissions.

SER reports are available on most of these topics.

Since its inception, one of SER's strong interests has been not only to uncover the facts that consumers and the industry need, but to translate the results of the research into practical, understandable, usable information. SER accomplishes this through frequent magazine articles, training sessions, papers delivered at conferences, and especially, through SER's own publications and slide sets. Publication sales also help support SER's consumer-oriented research projects—projects which are sometimes too controversial for manufacturers or government agencies to fund.

SER constantly updates its publications. If you are interested in obtaining those SER publications cited in this book, or any others, or SER's slide sets, send a stamped, self-addressed envelope to Shelton Energy Research, Dept. B72, P.O. Box 5235, Santa Fe, NM, 87502, for a current brochure.

APPENDIX 8
Periodicals, Equipment Directories, Standards, Building Codes and Trade Organizations

PERIODICALS

Sweeps Line
Canadian Chimney Sweeps Association
RR #5
Rockwood, Ontario
Canada N0B 2K0

Cook Stove News
The Aprovecho Institute
442 Monroe
Eugene, OR 97402

Alternative Energy Retailer
P.O. Box 2180
Waterbury, CT 06722

The National Energy Journal
411 Cedar Road
Chesapeake, VA 23320

Sweeping
National Chimney Sweep Guild
P.O. Box 1078
Merrimack, NH 03054

The Chimney News
41663 Keel Mt.
Lebanon, OR 97335

Wood'n Energy
P.O. Box 2008
Concord, NH 03301

EQUIPMENT DIRECTORIES

Woodstove Directory
P.O. Box 4474
Manchester, NH 03108

International Solid Fuel Buyer's Guide
The National Wood Stove and Fireplace Journal
411 Cedar Road
Chesapeake, VA 23320

Fireplace Insert Directory
Woodfired Energy Systems Directory (furnaces and boilers)
Wood Energy Research Corporation
Box 800
Camdem, ME 04843

Solid Fuel Accessories Directory
Alternative Energy Retailer
P.O. Box 2180
Waterbury, CT 06722

STANDARDS

Underwriters Laboratories, Inc.
333 Pfingsten Road
Northbrook, IL 60062
UL 737: Fireplace Stoves
UL 103: Chimneys: Factory-Built, Residential
         Type and Building Heating Appliance
UL 127: Factory-Built Fireplaces
UL 1482: Solid-Fuel Type Room Heaters
UL 391: Solid-Fuel-Fired Central Furnaces
UL 907: Fireplace Accessories
UL 462: Heat Reclaimers

National Fire Protection Association
Batterymarch Park
Quincy, MA 02269
NFPA 97M: Glossary of Terms Relating to Heat
           Producing Appliances
NFPA 211: Chimney, Fireplaces and Vents
           (Revisions, especially concerning
           stove installations, are currently
           under review by NFPA)
NFPA 90B: Warm-Air Heating and Air
           Conditioning Systems

MODEL BUILDING CODES

Canadian Heating, Ventilating and Air-Conditioning Code, Ottawa, Canada: National Research Council of Canada.

The BOCA Basic Mechanical Code, Chicago: Building Officials and Code Administrators International, Inc.

The National Building Code, New York: American Insurance Association.

The Standard Building Code, Birmingham, Ala.: Southern Building Code Congress International.

Uniform Mechanical Code, Whittier, CA: International Association of Plumbing and Mechanical Officials, and International Conference of Building Officials (ICBO).

TRADE ORGANIZATIONS

Wood Heating Alliance
Suite 700
1101 Connecticut Ave., NW
Washington, DC 20036

Canadian Wood Energy Institute
16 Lesmill Road
Don Mills, Ontario
Canada M3B 2T5

National Chimney Sweep Guild
P.O. Box 1078
Merrimack, NH 03054

Canadian Chimney Sweep Association
223 Scott Street
St. Catherines, Ontario
Canada L2N 1H6

GLOSSARY

Airtight stove. A stove in which a large fire can be suffocated by shutting the air inlets, resulting ultimately in a large mass of unburned fuel remaining in the stove.

Appliance (as used in this book). A solid fuel stove, fireplace, furnace, boiler, water heater, or cookstove.

Aquastat. An automatic device for controlling water temperature.

Baffle plate. A partition inside an appliance to control the flow direction of combustion air, flames, or flue gases.

Barometric draft control or regulator. A device designed to prevent excessive draft in a fuel-burning appliance by automatically admitting the appropriate amount of air to the venting system. Draft regulators are usually installed in chimney connectors.

Boiler, hot water. A hot-water central heating appliance.

Boiler, steam. A steam central heating appliance.

Breaching. Horizontal access into a chimney for a chimney connector. In prefabricated chimneys, a T provides a breaching. In masonry chimneys, breachings are holes that should have sleeves or liners of tile or heavy steel.

Btu. (British thermal unit). A unit for measuring energy, equal to the amount of energy needed to increase the temperature of 1 pound of water by 1 degree Fahrenheit.

Chimney capacity. The maximum safe venting capability of a chimney, most often expressed in terms of the fuel consumption rate of connected appliances (in Btu. per hour), but more fundamentally related to the mass flow (e.g., pounds per minute) of flue gas that will flow up a chimney under given conditions of temperature and barometric pressure.

Chimney chase. An enclosure around a chimney, typically the full height of the chimney and typically used to hide a prefabricated chimney.

Chimney connector. The connector between an appliance and its chimney. Stovepipe is commonly used for chimney connectors. Some appliances have no connector because the chimney connects directly to the appliance (e.g., many prefabricated fireplaces).

Chimney fire. The burning of creosote/soot deposits inside a chimney or connector.

Circulating fireplace. A fireplace with multiple-wall construction around the fire chamber which permits air to circulate between the walls, become heated, and enter the house either directly or via short ducts.

Circulating stove. See stove, circulating.

Collar (or flue collar). The part of a fuel-burning appliance to which the chimney connector or chimney attaches.

Combustible (as applied to walls, floors, and ceilings in the context of solid fuel heater clearances for safety). Constructed of or surfaced with wood, paper, natural or synthetic fibers, plastic, or other material that will ignite and burn, whether flameproofed or not, and whether plastered or unplastered. Combustibility is a relative concept. This definition is adapted from the definition in the NFPA booklet "Glossary of Terms Relating to Heat-Producing Appliances" (Boston: National Fire Protection Association, 1972).

Combustion efficiency. The percentage of the energy content of the fuel consumed, based on the higher heating value of the fuel, that is converted into heat in the combustion process, regardless of whether the heat is used or goes up the chimney.

Combustion products (or products of combustion). The products of the chemical reactions which constitute combustion, typically consisting of carbon dioxide, water vapor, and small amounts of some incompletely burned organic compounds.

Cookstove. A wood-burning or coal-burning appliance with a closed fire chamber, which is intended primarily for cooking and includes an oven.

Cord. A common measure of firewood and pulpwood, equal to the amount of wood in a carefully stacked (parallel) pile of wood that is 4 feet high, 8 feet wide, and 4 feet deep. The amount of solid wood in this 128-cubic-foot pile is usually estimated to be between 80 and 90 cubic feet, but can vary beyond these limits.

Cresote. Chimney and stovepipe deposits originating as condensed components in smoke (including vapors, tar, and soot). Creosote is often initially liquid, but may dry or pyrolyze to a flaky or solid form.

Damper. A valve, usually a movable plate, for controlling the flow of air or smoke.

Draft. Technically, and in this book, the difference in pressure at the same elevation between the inside and the outside of a chimney, chimney connector, or appliance. The term "draft" is also sometimes used to denote the rate of combustion air flow into a fuel-burning appliance, or the rate of flue-gas flow.

Elbow, stovepipe. The stovepipe fittings or sections involving turns or bends. Most common are 90° elbows. Some types of elbows are adjustable from 90° to 0° (no bend).

Energy content. The total chemical energy content of a given amount of fuel, obtained by multiplying the higher heating value of the fuel by the weight of the fuel.

Energy efficiency (or overall energy efficiency). The percentage of the energy content of the fuel consumed, based on the higher heating value of the fuel that becomes useful heat in the house.

Excess air. Air admitted to a burner that is in excess of the amount theoretically needed for complete combustion.

Exterior chimney. A chimney that runs up the exterior of a building. Exterior masonry chimneys typically have 3 of their sides exposed to the weather, the fourth being against the outside of the building. Exterior prefabricated chimneys typically are sup-ported partway up the building, and they typically are entirely exposed to the weather. A prefabricated chimney in a chase may be partly protected from the weather depending on the tightness of the chase.

Factory-built. See Prefabricated.

Fines. The small, less-than-nominal-size pieces of coal generated in handling coal.

Firebrick. Brick capable of withstanding high temperatures, such as in stoves, furnaces, and boilers. Different types of firebrick have different temperature limits.

Fireclay. Clay that will withstand high temperatures without cracking or deforming. NFPA recommends that fireclay chimney flue liners resist corrosion, softening, or cracking from flue gases at temperatures up to 1,800°F.

Fireplace face. The opening of a fireplace to the room.

Fireplace heat exchanger. A device installed inside a fireplace that is intended to extract additional heat from the fire, and that does not cover the face of the fireplace (as glass doors *do*).

Fireplace insert. A device installed in a fireplace that is intended to extract heat from the fireplace and covers the face of the fireplace.

Fireplace stove. A free-standing, solid fuel, room-heating appliance operated either with its fire chamber open or closed to the room. NFPA and most codes use the term "Room Heater—Fireplace Stove Combination" for a fireplace stove.

Fireplace, zero clearance. A factory-built metal fireplace with multi-layer construction, providing enough insulation and/or air cooling so that the base, back, and, in some cases, sides can safely be placed in direct contact ("zero clearance") with combustible floors and walls.

Flue. The passageway for transporting the exhaust (combustion byproducts plus admixed air) from a fuel-burning appliance to the outdoors.

Flue gases. The gases in an operating venting system, consisting of combustion products plus whatever air is mixed with them.

Fly ash. Ash that goes up the chimney, as opposed to ash that remains in the fuel-burning appliance.

Fuel magazine. A fuel supply chamber from which fuel feeds to the combustion chamber. The fuel magazine and the combustion chamber may be the same, but then combustion is normally restricted to a small portion of the chamber.

Furnace. 1. A hot-air central heating appliance (the definition used in this book). **2.** Any central heating appliance. **3.** The combustion chamber of any fuel-burning appliance.

Gasifier. A device or appliance intended primarily to produce gaseous fuel from solid fuel. The gaseous fuel is burned in some other appliance.

Green wood. Freshly cut wood from a live tree, not yet dried.

Hearth. The floor of the combustion chamber in solid fuel appliances.

Hearth extension. The noncombustible co-planar extension of the hearth beyond the opening of a fireplace or fireplace stove. The term is also sometimes used to denote the floor protector under or around any residential solid fuel appliance.

Heat transfer efficiency. The percentage of the total heat generated in the combustion process that is used in (or transferred to) the house for space heating.

Higher heating value. The amount of heat (all forms) released per unit of fuel when the fuel is burned completely. Common units are Btu. per pound.

Ignition temperature. The minimum temperature of a flammable mixture of gases at which it can spontaneously ignite. There is no simple nor universally accepted definition of ignition temperature for solid fuels.

Infrared radiation. The invisible and harmless electromagnetic radiation beyond the red end of the visible spectrum which is emitted by all hot objects. This radiation is converted into heat when it is absorbed.

Insulating brick. Low density (high porosity), low thermal conductivity firebrick intended for use in kilns and furnaces to insulate them, reducing heat losses. Its conductivity and its heat storage capacity are both $\frac{1}{5}$ to $\frac{1}{3}$ that of hard firebrick.

Interior chimney. A chimney that runs up the interior of the building.

Latent heat. The potential energy in water vapor which is converted into (sensible) heat when the vapor condenses. A pound of water vapor at room temperature has about 1,050 Btu. of latent heat.

Liner, chimney (or flue). Usually a high-temperature clay ("fireclay"), round or rectangular sleeve lining the interior of masonry chimneys. Although not recognized by all building codes, other materials such as stainless steel and poured or injected hardening mixes can be used as liners.

Liner, stove. A layer of metal, brick, or other material placed immediately adjacent to a side or bottom of an appliance, intended either to protect the main structure from getting too hot, or to insulate the combustion chamber, making it hotter and, thus, promoting more complete combustion.

Lower heating value. The amount of heat released, excluding the latent heat of water vapor per unit of fuel, when the fuel is burned completely. Common units are Btu. per pound.

Moisture content. The percentage of fuel weight that is moisture.

Net energy efficiency. The *net* heat delivered by the whole heating system, divided by the total energy content of the fuel consumed (based on the higher heating value). This definition yields the efficiency expressed as a fraction; multiplication by 100 results in expression as a percent. Net energy efficiencies may include such factors as heat transfer from interior chimneys, standby losses, and increases in air infiltration due to operation of the appliance.

NFPA. The National Fire Protection Association, an independent not-for-profit organization for fire safety. Many jurisdictions adopt NFPA standards as part of their codes.

Ovendry wood. Wood that has been dried to constant weight at about 200° F. and low humidity. Ovendry wood is defined to have zero moisture content.

Power output. In this book, the rate of heat output from a heater, commonly expressed in Btu. per hour or in watts.

Prefabricated. As applied to fireplaces and chimneys: at least partially manufactured in a factory for final assembly and/or installation at the site. Typically applies to listed zero-clearance fireplaces and to listed double-wall and triple-wall chimneys, all made primarily of metal, in contrast to site-built masonry fireplaces and chimneys. However, distinctions are blurring with increasing interest in both fireplaces and chimneys made of specilized "prefabricated" masonry components.

Primary combustion. The burning of solid wood and some of the combustible gases, which takes place in that portion of the appliance where the wood is. The distinction between primary and secondary combustion is somewhat artificial. Definitions of each are necessarily somewhat arbitrary.

Pyroligneous acid. The acidic brown aqueous liquid obtained by condensing the gaseous products of pyrolysis of wood. Pyroligneous acid is the same as creosote in its wettest form.

Pyrolysis. The chemical alteration of a solid fuel by the action of heat alone, in the absence of oxygen and, hence, without burning. The products of pyrolysis are gases, vapors, and charcoal or coke.

Radiant stove. See Stove, radiant.

Refractory. Any solid ceramic material suitable as a

structural or protective material at high temperatures in a corrosive environment.

Seasoned wood. Wood that has lost a significant amount of its original (green) moisture. The term has no quantitative and universally accepted meaning.

Secondary combustion. The burning of the combustible gases and smoke which are not burned in primary combustion.

Secondary combustion chamber. The place where secondary combustion occurs.

Sensible heat. Energy in the form of random motions of molecules, atoms, and electrons. Sensible heat is the form of energy that is transferred from a warm object to a cooler object when they are in direct physical contact. Sensible heat can be sensed or felt directly by human skin. Infrared radiation (sometimes called radiant heat) is a completely different energy form which is transformed into sensible heat when it is absorbed by a surface such as skin. Latent heat is a form of potential heat which also cannot be sensed by humans, but can be converted into sensible heat.

Smoke. The visible components of flue gas. The term is also used sometimes to refer to all components of flue gas.

Solid fuel. Fuel in solid form, such as wood, coal, paper, and related products.

Soot. Soft, black, velvety, carbon particle deposits inside appliances, chimneys, or connectors. Soot originates in oxygen-poor flames.

Stack effect. The effects resulting from the warm air in buildings on a cold day being relatively buoyant, just as are the flue gases in a chimney or stack. Effects include pressure differences between the inside and outside of the building, airflow into the building in

the lower stories, and airflow out of the building in the upper portions.

Stove. A free-standing, solid-fuel-burning, room-heating appliance intended to be operated with its door(s) closed. NFPA and most codes use the term "solid-fuel room heater" for stoves.

Stove, circulating. A stove with an outer jacket (usually sheet metal) beyond the main structure, with openings at or near the bottom and top so that air can circulate between the stove body and its jacket. For purposes of determining safe clearances, a circulating stove must be fully jacketed on all 4 sides and on the top. The top of the jacket is usually louvered.

Stove, radiant. A stove whose heat output is mostly in the form of radiant energy.

Stove mat (or stoveboard). A prefabricated panel used as a floor or wall protector.

Stovepipe (or smokepipe). Single-walled light-gauge (roughly .019 to .024 inches thick) metal pipe generally intended for use as chimney connectors.

Stovepipe damper. A damper installed in a stovepipe connector to regulate flow and draft.

Thermostat. An automatic device for regulating the temperature in a building by controlling the heating or cooling source or its distribution.

Thimble. A device installed in combustible walls, through which stovepipe passes, intended to help protect the walls from igniting due to stovepipe heat. A thimble by itself is not usually adequate. The simplest thimbles are simply metal or fireclay sleeves or cylinders.

UL. Underwriters Laboratories, Inc.

Water heater. An appliance intended principally for heating domestic (or tap) water.

INDEX

Environmental concerns:
acid rain, 11–12, 40
air pollution, 11 and *illus.*, 12–13, 40, 114
soil erosion, 9 and *illus.*, 10
Equipment:
directories (list), 259
see also Accessories

F

Face cord, 31 and *illus.*
Fans and blowers:
for central heater, 202, 209
for circulating stove, 126
draft inducer, *illus.* 82, 83
and forced convection, 16, 103 and *illus.*, 104 and *illus.*
for heat distribution, 127, 143–44, 145 and *illus.*, 171
for heat exchangers, 168–70 *passim*, 171, 191, 192 and *illus*, 195, 197
and power failure, 223
Fire:
banking of, 214
in chimney, *see* Chimney fire
coal, 213–14
house, 67, 69, 72, 82, 225, 227–28
smoldering, 45, 51, 53–54, 91, 95, 113, 219
-starting, in stove, 123, 211, *illus.* 212
wood, *see* Wood fire
see also Combustion; Flame
Fire alarm, 229–31, *illus.* 231
see also Smoke detector
Fireback, 193, *illus.* 194
Fire extinguisher, 177, 229 and *illus.*
Fireplace, 178–203 and *illus.*, *table*
accessories, 189–91, *illus.* 189–90
banking fire in, 214
and chimney size, *table* 61
control of heat output, 116–17
design principles, 185–89, *illus.* 186–88
doors (glass) for, 83, 147, 189–90 and *illus.*, 191
and dryness of indoor air, 103
energy efficiency, overall, 182–85, *illus.* 182–84, *table* 183
grates, 192–93 and *illus.*
heat exchangers, 191–92 and *illus.*
heat loss from, 178–81, *illus.* 180–81, 186–90
outside air for, 147, 148
parts and construction, *illus.* 179
power output, 178, *illus.* 182, *table* 183, *illus.* 184
shape of, 186, *illus.* 187
size of fire in, 182–83, 218
smoke shelf, *illus.* 179, 188, 189
smoke spillage from, 83, 186, 188–89
with steel shell, 199–200 and *illus.*
vs. stove performance, 183–84
and temperature (outdoor), *illus.* 184

Fireplace, prefabricated, 200–203
Fireplace, Russian, 120 and *illus.*, 125–26, 203
Fireplace furnace, 202–203 and *illus.*
Fireplace insert(s)/stove(s), 194–95
and chimney size, *table* 60, 76
and creosote, 196, 198, 199, 216
and energy efficiency, 146, 195
floor protection for, 151
and flow resistance, 60, 197
installation, 196–99, *illus.* 197–98
outside air for, 148
Firewood, *see* Wood (as fuel)
Flame, 46 and *illus.*, 47–48
color of, 45, 54–55
in combustion process, 46–48, 51
velocity of, *table* 48, 49
see also Combustion; Fire
Floor protection, 151–54, *illus.* 152–54, *table* 152, 162 and *illus.*, 222
Flue, *see* Chimney
Flue damper, 183, 195
Fly ash, 52, 55, 225
Food *vs.* fuel conflicts, 9, 10, 12–13
Franklin stove, 112, *illus.* 113, 117 and *illus.*, *illus.* 123
Fuels:
compared:
coal *vs.* wood, *table* 3, 11–12, 37, 38, 40, 243, *illus.* 245
for cost, 241 and *table*, *illus.* 242, 243 and *illus.*, *illus.* 244, 245 and *illus.*
for energy efficiency, 86, 88
for sulfur and ash content, *table* 11
used in U.S., 1–6 and *illus.*
Fuels, solid:
combustion process, 50–52, *illus.* 52
costs, 241–47 and *illus.*
list of, 222
stoves for, *see* Coal stove(s); Stove(s); Wood stove(s)
unsuitable, 221–22
see also Coal (as fuel); Wood (as fuel)
Fuelwood, *see* Wood (as fuel)
Furnaces, *see* Central heater(s), . . .

G

Gas furnace or boiler, 88, 184–85, 206 and *illus.*, *illus.* 209
Gasification of wood and coal, 5–6, 50, 133; *see also* Pyrolysis
Grate(s):
in coal stove, 136–37 and *illus.*
shaking, 136–37, 213–14
for fireplace, 191–93 and *illus.*
in wood stove, 123 and *illus.* 124
Green wood:
combustion rate, 54
drying, 32–33
ignition of, 54
vs. seasoned wood, 27 and *illus.*
and creosote, 33, 214–16 and

Green wood: (continued)
illus., 219–20 and *tables*
in stove with refractory liner, 128 and *illus.*

H

Hardwoods, 21–22; *see also* Wood (as fuel): species
Health concerns:
cancer, 11, 177, 223
carbon monoxide, 149, 167, 192, 209
coal ash toxicity, 39
from "dry heat" of stove, 103
respiratory ailments, 40
about smoke, 11
see also Safety
Hearth stove, *see* Fireplace insert(s)/stove(s)
Heat:
basic concepts, 14–20 and *illus.*
sensible, 19, 249–50
Heat distribution, 121, 126–27 and *illus.*, 143–45 and *illus.*
Heat exchanger(s), 168–70, *illus.* 169, 191–92 and *illus.*
for water heating, 175 and *illus.*, 176
Heat loss, *illus.* 90
from central heater, 207–208
from chimneys and connectors, *table* 68
from fireplace, 178–81 and *illus.*, 186–89, *illus.* 188
standby loss, 146–47, 178, *illus.* 181, 182, 208
from wood stove, *illus.* 89
Heat shield:
for stove, 157–58 and *illus.*
for stovepipe, 161–62, *illus.* 163
Heat storage, 19, 177–80 and *illus.*, 238
and refueling frequency, 132
and steadiness of heat, 142–43
in various materials, *table* 16
Heat transfer, 88–91 and *illus.*, 112–13
and baffles, 99, 104, 109, and *illus.*, 110
vs. combustion efficiency (design conflict), 128–29 and *illus.*
vs. creosote minimization (design conflict), 91, 129–30, 237–38
efficiency, 101–12 and *illus.*, *table* 101, 237–38
and secondary air, 94, 107–108
and smoke residence time, 104–105, *illus.* 106
and stove construction materials, *table* 6, 102, 126
and stove finish, 101–102 and *illus.*
and stove installation, 110–12
and stove liners, 98
and stovepipe length, 110, *illus.* 111
and surface area of stove, 108–10, *illus.* 109–10
and turbulence, 104, *illus.* 105
Hill, Richard: heating system design, 129 and *illus.*